Laser Additive Manufacturing: Design, Materials, Processes and Applications

Laser Additive Manufacturing: Design, Materials, Processes and Applications

Editors

Jie Yin
Yang Liu
Ping Zhao

MDPI • Basel • Beijing • Wuhan • Barcelona • Belgrade • Manchester • Tokyo • Cluj • Tianjin

Editors

Jie Yin
Gemological Institute
China University of Geosciences
Wuhan
China

Yang Liu
Faculty of Mechanical Engineering & Mechanics
Ningbo University
Ningbo
China

Ping Zhao
Department of Microtechnology and Nanoscience
Chalmers University of Technology
Gothenburg
Sweden

Editorial Office
MDPI
St. Alban-Anlage 66
4052 Basel, Switzerland

This is a reprint of articles from the Special Issue published online in the open access journal *Micromachines* (ISSN 2072-666X) (available at: www.mdpi.com/journal/micromachines/special_issues/laser_additive_manufacturing_LAM).

For citation purposes, cite each article independently as indicated on the article page online and as indicated below:

LastName, A.A.; LastName, B.B.; LastName, C.C. Article Title. *Journal Name* **Year**, *Volume Number*, Page Range.

ISBN 978-3-0365-6070-0 (Hbk)
ISBN 978-3-0365-6069-4 (PDF)

Cover image courtesy of Jie Yin, Yang Liu and Ping Zhao

© 2023 by the authors. Articles in this book are Open Access and distributed under the Creative Commons Attribution (CC BY) license, which allows users to download, copy and build upon published articles, as long as the author and publisher are properly credited, which ensures maximum dissemination and a wider impact of our publications.

The book as a whole is distributed by MDPI under the terms and conditions of the Creative Commons license CC BY-NC-ND.

Contents

About the Editors vii

Jie Yin, Yang Liu and Ping Zhao
Editorial for the Special Issue on Laser Additive Manufacturing: Design, Processes, Materials and Applications
Reprinted from: *Micromachines* 2022, 13, 2057, doi:10.3390/mi13122057 1

Ke Chen, Haoran Wan, Xiang Fang and Hongyu Chen
Laser Additive Manufacturing of Anti-Tetrachiral Endovascular Stents with Negative Poisson's Ratio and Favorable Cytocompatibility
Reprinted from: *Micromachines* 2022, 13, 1135, doi:10.3390/mi13071135 5

Changhui Song, Zehua Hu, Yunmian Xiao, Yang Li and Yongqiang Yang
Study on Interfacial Bonding Properties of NiTi/CuSn10 Dissimilar Materials by Selective Laser Melting
Reprinted from: *Micromachines* 2022, 13, 494, doi:10.3390/mi13040494 21

Youwen Yang, Wei Wang, Mingli Yang, Yingxin Yang, Dongsheng Wang and Zhigang Liu et al.
Laser-Sintered Mg-Zn Supersaturated Solid Solution with High Corrosion Resistance
Reprinted from: *Micromachines* 2021, 12, 1368, doi:10.3390/mi12111368 37

Xiaodong Zheng, Jiahuan Meng and Yang Liu
Strain Rate Dependence of Compressive Mechanical Properties of Polyamide and Its Composite Fabricated Using Selective Laser Sintering under Saturated-Water Conditions
Reprinted from: *Micromachines* 2022, 13, 1041, doi:10.3390/mi13071041 49

Yufeng Tao, Chengchangfeng Lu, Chunsan Deng, Jing Long, Yunpeng Ren and Zijie Dai et al.
Four-Dimensional Stimuli-Responsive Hydrogels Micro- Structured via Femtosecond Laser Additive Manufacturing
Reprinted from: *Micromachines* 2021, 13, 32, doi:10.3390/mi13010032 61

Di Wang, Han Wang, Xiaojun Chen, Yang Liu, Dong Lu and Xinyu Liu et al.
Densification, Tailored Microstructure, and Mechanical Properties of Selective Laser Melted Ti–6Al–4V Alloy via Annealing Heat Treatment
Reprinted from: *Micromachines* 2022, 13, 331, doi:10.3390/mi13020331 73

Wenqi Zhang, Baopeng Zhang, Haifeng Xiao, Huanqing Yang, Yun Wang and Haihong Zhu
A Layer-Dependent Analytical Model for Printability Assessment of Additive Manufacturing Copper/Steel Multi-Material Components by Directed Energy Deposition
Reprinted from: *Micromachines* 2021, 12, 1394, doi:10.3390/mi12111394 89

Minqiang Kang, Yongfa Qiang, Canlin Zhu, Xiangjun Xiang, Dandan Zhou and Zhitao Peng et al.
Hybrid Dissection for Neutron Tube Shell via Continuous-Wave Laser and Ultra-Short Pulse Laser
Reprinted from: *Micromachines* 2022, 13, 352, doi:10.3390/mi13030352 109

Lingjian Meng, Jiazhao Long, Huan Yang, Wenjing Shen, Chunbo Li and Can Yang et al.
Femtosecond Laser Treatment for Improving the Corrosion Resistance of Selective Laser Melted 17-4PH Stainless Steel
Reprinted from: *Micromachines* 2022, 13, 1089, doi:10.3390/mi13071089 123

Xinlin Wang, Jinkun Jiang and Yongchang Tian
A Review on Macroscopic and Microstructural Features of Metallic Coating Created by Pulsed Laser Material Deposition
Reprinted from: *Micromachines* **2022**, *13*, 659, doi:10.3390/mi13050659 139

Yan Zhou, Lifeng Xu, Youwen Yang, Jingwen Wang, Dongsheng Wang and Lida Shen
Microstructure and Corrosion Behavior of Iron Based Biocomposites Prepared by Laser Additive Manufacturing
Reprinted from: *Micromachines* **2022**, *13*, 712, doi:10.3390/mi13050712 159

Yihang Cui, Jiacheng Cai, Zhiguo Li, Zhenyu Jiao, Ling Hu and Jianbo Hu
Effect of Porosity on Dynamic Response of Additive Manufacturing Ti-6Al-4V Alloys
Reprinted from: *Micromachines* **2022**, *13*, 408, doi:10.3390/mi13030408 171

Zhidong Xu, Dengzhi Wang, Wenji Song, Congwen Tang, Pengfei Sun and Jiaxing Yang et al.
Microstructure and Wear of W-Particle-Reinforced Al Alloys Prepared by Laser Melt Injection
Reprinted from: *Micromachines* **2022**, *13*, 699, doi:10.3390/mi13050699 181

Lu Zhang, Yan Li, Simeng Li, Ping Gong, Qiaoyu Chen and Haoze Geng et al.
Fabrication of Titanium and Copper-Coated Diamond/Copper Composites via Selective Laser Melting
Reprinted from: *Micromachines* **2022**, *13*, 724, doi:10.3390/mi13050724 191

Xiao Yang, Shuo Wang, Hengpei Pan, Congyi Zhang, Jieming Chen and Xinyao Zhang et al.
Microstructure Transformation in Laser Additive Manufactured NiTi Alloy with Quasi-In-Situ Compression
Reprinted from: *Micromachines* **2022**, *13*, 1642, doi:10.3390/mi13101642 207

Zheng Li, Hao Li, Jie Yin, Yan Li, Zhenguo Nie and Xiangyou Li et al.
A Review of Spatter in Laser Powder Bed Fusion Additive Manufacturing: In Situ Detection, Generation, Effects, and Countermeasures
Reprinted from: *Micromachines* **2022**, *13*, 1366, doi:10.3390/mi13081366 219

About the Editors

Jie Yin

Prof. Dr. Jie Yin is currently a Professor and Doctoral Supervisor at the Gemological Institute and Advanced Manufacturing Research Center of China University of Geosciences (CUG). He received his Ph.D. degrees from Huazhong University of Science and Technology (HUST) in 2014, then he joined in Wuhan National Laboratory for Optoelectronics (WNLO) as a post-doctoral and associate researcher for six years. His research focuses on laser advanced manufacturing and laser-matter interaction. He is the recipient of the Young Outstanding Talents of CUG Scholars; of the Best Paper Award from the International Conference on Research Advances in Additive Manufacturing (RAAM 2019); and of the Excellent Report Award by the 4D Printing Technology Forum in 2022. He has leaded more than 10 projects including National Natural Science Foundation of China (NSFC), Shanghai Aerospace Science and Technology Innovation Key Fund (SAST), and the Fundamental Research Funds for the Central Universities. He has served as Editorial Board Member of international academic journals such as Micromachines, and reviewers for 20 academic journals. Furthermore, he has published more than 40 academic papers on International Journals (five representative papers with a total IF=55.5 in the past 5 years), with more than 2400 citations in Google Scholar, 3 ESI/F5000 highly cited papers and two cover papers.

Yang Liu

Prof. Dr. Yang Liu is currently an associate professor in the Department of Mechanical Engineering of Ningbo University. He received his PhD degree from South China University of Technology in 2015, and he worked as a visiting scholar in Cardiff University from 2018 to 2019. Prof. Liu is the head of the additive manufacturing research group in NBU, and one of the early scholars to study the properties of additive manufacturing alloys such as Ti, Al, and HEAs under extreme work environments (high pressure, high velocity or high/low temperatures). His work mainly aims to reveal the dynamic deformation behaviors and failure mechanisms of additive manufacturing alloys under dynamic loading, which is helpful to widen the application of additive manufacturing techniques and products. So far, he published more than 25 journal papers as first/corresponding author with more than 1400 citations, and undertaken more than 10 research projects.

Ping Zhao

Dr. Ping Zhao received the B.Sc. and Ph.D. degrees from Huazhong University of Science and Technology in 2009 and 2014, respectively. After doctoral graduation, he joined in Huawei Technologies Co Ltd. as a senior research engineer in advanced optical communication technologies. With 5 years' industrial research experience, he joined the Photonics Lab in Chalmers University of Technology in 2019 as a researcher. He realized the first continuous-wave phase-sensitive amplifier based on CMOS-compatible integrated photonic waveguides with a noise figure well below the conventional 3 dB quantum limit of phase-insensitive amplifiers, which was selected by Optica (Formerly OSA) as "Optics in 2022". His research topics include micro/nanophotonic devices, nonlinear optics, optical communication and advanced laser manufacturing. To date, he authored/coauthored 20 journal and conference papers as well as 10 granted PCT patents. He received H.C. Oersted & Marie Skłodowska-Curie COFUND Fellowship and is in the topical advisory panel of *Micromachines* and a reviewer of various international journals including *Optics Letters*, *Optics Express* and *Journal of the Optical Society of America B*.

Editorial

Editorial for the Special Issue on Laser Additive Manufacturing: Design, Processes, Materials and Applications

Jie Yin [1,2,*], Yang Liu [3] and Ping Zhao [4]

1. Gemological Institute, China University of Geosciences, Wuhan 430074, China
2. Advanced Manufacturing Research Institute, China University of Geosciences, Wuhan 430074, China
3. Faculty of Mechanical Engineering & Mechanics, Ningbo University, Ningbo 315211, China
4. Department of Microtechnology and Nanoscience, Chalmers University of Technology, 41296 Gothenburg, Sweden
* Correspondence: yinjie@cug.edu.cn

Laser-based additive manufacturing (LAM) is a revolutionary advanced digital manufacturing technology developed in recent decades, which is also a key strategic technology for technological innovation and industrial sustainability. This technology unlocks the design and constraints of traditional manufacturing and meets the needs of complex geometry fabrication and high-performance part fabrication. A deeper understanding of the design, materials, processes, structures, properties and applications is desired to produce novel functional devices, as well as defect-free structurally sound and reliable LAM parts.

The topics in this Special Issue include macro- and micro-scale additive manufacturing with lasers, such as structure/material design, fabrication, modeling and simulation, in situ characterization of additive manufacturing processes and ex situ materials characterization and performance, with an overview that covers various applications in aerospace, biomedicine, optics and energy.

In this Special Issue, papers on different subjects were published after the high-quality reviewing process, with a total of 16 contributions (14 original research papers and 2 review papers) and times of viewed over 13K (as of 15 November 2022). Six articles were selected as Editor's Choice and one article was selected as the Issue Cover of Volume 13, Issue 8 (https://www.mdpi.com/2072-666X/13/8 (accessed on 25 August 2022)). Each of them is briefly introduced below according to four aspects (design, processes, materials and applications) of laser additive manufacturing in this Special Issue.

1. Design of Laser Additive Manufacturing in This Special Issue

The design of laser additive manufacturing covered in this Special Issue includes the structural design (e.g., the negative Poisson's ratio endovascular stents [1], the multi-material heterostructures [2]) and the material design (e.g., the in situ synthesis of a novel alloy [3] and composite modification [4]).

Chen et al. [1] reported that anti-tetrachiral auxetic stents with negative Poisson's ratios (NPR) were designed and fabricated via LAM. The cytocompatibility tests indicate the envisaged cell viability and adhesion of the vascular endothelial cell on the LAM-fabricated anti-tetrachiral auxetic stents. This study is a first step toward the structural design and manufacturing of endovascular stents with negative Poisson's ratios, which exhibit predestined biocompatibility.

Song et al. [2] studied the microstructure, element diffusion, microhardness changes and phases at the interface of NiTi/CuSn10 multi-material heterostructures. They found that the formation of the N4Ti3 strengthening phase improves the microhardness of the interface, making the microhardness of the interface significantly higher than that of both sides. Under the optimized process parameters, the maximum strength can reach 309.7 ± 34.4 MPa.

Yang et al. [3] reported that a supersaturated solid solution of a Mg–Zn alloy with good cytocompatibility was achieved using mechanical alloying (MA) combined with

laser sintering. Note that the supersaturated solid solution Mg–Zn powders were firstly prepared using MA, which can break through the limit of phase diagram under the action of forced mechanical impact.

Zheng et al. [4] reported that polyamide 12 (PA12) and carbon-fiber-reinforced polyamide 12 (CF/PA12) composites were fabricated using selective laser sintering (SLS) LAM. They investigated the coupling effects of the strain rate and hygroscopicity on the compressive mechanical properties. Results showed that the CF/PA12 had a shorter saturation time and lower saturated water absorption under the same conditions, indicating that the SLS of CF/PA12 had lower hydrophilia and higher water resistance when compared to the SLS of PA12. This work provided a basic knowledge of the mechanical properties of SLS polyamide under different load and saturated-water conditions and, thus, is helpful to widen the application of SLS PA products in harsh environments.

2. Processes of Laser Additive Manufacturing in This Special Issue

This Special Issue article focuses on continuous-wave (CW) laser additive manufacturing (e.g., laser powder bed fusion (L-PBF), laser direct energy deposition (L-DED)), but also includes pulsed-wave (PW) laser advanced manufacturing (e.g., femtosecond-pulsed laser machining, two-photon polymerization [5]) and heat treatment annealing processes [6]. It is worth mentioning that laser additive composite manufacturing (e.g., CW L-PBF & CW L-DED [7]), laser subtractive composite manufacturing (CW laser and PW femtosecond laser machining [8]) and laser additive & subtractive composite manufacturing (CW L-PBF and PW femtosecond laser surface ablation modification [9]) are also current research frontiers in laser advanced manufacturing.

Zhang et al. [7] established an analytical model to predict the layer-dependent processing parameters for fabricating 07Cr15Ni5 steel via L-DED on a CuCr substrate. Changes in the effective thermal conductivity and specific heat capacity with the layer number, as well as the absorption rate and catchment efficiency with the processing parameters, are considered. The parameter maps predicted by the model are in good agreement with the experimental results.

Kang et al. [8] proposed a hybrid dissection method, where a high-power CW laser was firstly employed to make a tapered groove on the shell's surface and then a femtosecond-pulse laser was used to micromachine the groove in order to obtain a cutting kerf. By using the hybrid method, the cutting efficiency was improved about 49-times compared to the femtosecond laser cutting. This method provides a high-efficiency and non-thermal cutting technique for reclaimed metallic neutron tube shells with millimeter-level-thick walls, which has the advantages of non-contact, minimal thermal diffusion and being free of molten slag.

Meng et al. [9] reported that L-PBFed 17-4PH SS was treated by a femtosecond laser to regulate the surface structures and corrosion-resistance behaviors. The Cr, Cu and other alloying elements precipitated on the laser-ablated surface were beneficial to the formation of a passivation film, leading to an improved corrosion-resistance performance.

Wang et al. [10] reported a review of the macroscopic and microstructural features of metallic coating created by pulse-wave (PW) laser material deposition (LMD). The research status of temperature field simulation, surface quality and microstructural features, including microstructures, microhardness, residual stress and cracking, as well as corrosion behavior of metallic coating created by pulsed-laser material deposition, were reviewed.

3. Materials of Laser Additive Manufacturing in This Special Issue

This Special Issue article covers a wide range of forming materials, including steel [1,9] and other iron-based alloys [11], magnesium alloys [3], titanium alloys [6,12], copper alloys [7], aluminum alloys [13], composites (e.g., CF/PA12 [4], copper/titanium-coated diamond [14]) and, in particular, metallic multi-materials (e.g., NiTi/CuSn10 [2], copper/steel [7]) and 4D-printing materials (e.g., NiTi shape memory alloys [2,15], smart materials for soft interactive hydrogel [5]).

Cui et al. [12] investigated three types of Ti-6Al-4V (TC4) materials with different porosities fabricated via L-PBF using different printing parameters. Experimental results indicate that the porosity significantly affects their dynamic response (the yield strength and spall behaviors).

Xu et al. [13] reported that W-particle-reinforced Al alloys were prepared on a 7075-aluminum alloy surface via laser-melt injection (LMJ) to improve their wear resistance. Results confirmed that a W/Al laser-melting layer of about 1.5 mm thickness contained W particles and Al4W was formed on the surface of the Al alloys.

Zhang et al. [14] used the L-PBF technique to fabricate titanium- and copper-coated diamond/copper composites. The microstructure, roughness, interface bonding, thermal and mechanical performance were studied. The work offered electroless plating and evaporation methods for L-PBF to coat copper and titanium on the diamond particle surface for improving the wettability of the diamond/copper interface, which opened up a new method for laser 3D-printing technology to print a broad range of diamond-particle-reinforced metal matrix composites.

Yang et al. [15] investigated the quasi in situ compression recovery properties of NiTi alloys with the L-DED process. Results showed that the material can be completely recovered under 4% deformation and the B19' martensite phase content and dislocation density are basically unchanged. However, the recovery rate was only 90% and the unrecoverable strain was 0.86% at 8% deformation.

Li et al. [16] reviewed the literature on the in situ detection, generation, effects and countermeasures of spatter in L-PBF, which has an intrinsic correlation with the forming quality. Although many researchers have provided insights into the melt pool, microstructure and mechanical property, reviews of spatter in L-PBF are still limited. Hence, this work is expected to pave the way towards a novel generation of highly efficient and intelligent L-PBF systems.

4. Applications of Laser Additive Manufacturing in This Special Issue

The findings of this Special Issue article are expected to provide guidance and reference for scientists and engineers in the fields of aerospace (e.g., copper/steel multi-material engine thrust chambers [7], titanium alloy lightweight structural components [6,12]), biology (e.g., Mg–Zn [3], Fe/ZnS bone-repair material [11]), energy (e.g., neutron tube shells [8]) and micromachines (e.g., programmable micro-robots [5]).

Tao et al. [5] demonstrated a novel femtosecond LAM process with smart materials for soft interactive hydrogel micro-machines. These programmable stimuli-responsive matrices mechanized hydrogels into robotic applications at the micro/nanoscale (<300 × 300 × 100 μm^3). Benefiting from high-efficiency two-photon polymerization (TPP), nanometer feature size (<200 nm) and flexible digitalized modeling technique, many micro/nanoscale hydrogel robots or machines have become obtainable with respect to future interdisciplinary applications.

Wang et al. [6] investigated the influence of process parameters on the densification, microstructure and mechanical properties of a Ti–6Al–4V alloy printed by L-PBF, followed by annealing treatment. The results show that the microstructure can be tailored by altering the scanning speed and annealing temperature. The maximum elongation of 14% can be achieved at an annealing temperature of 950 °C, which was 79% higher than that of as-printed samples. Meanwhile, ultimate tensile strength larger than 1000 MPa can be maintained, which still meets the application requirements of the forged Ti–6Al–4V alloy.

Zhou et al. [11] introduced zinc sulfide (ZnS) into Fe bone-implants manufactured using the LAM technique, which promoted the collapse of the passive film and accelerated the degradation rate of the Fe matrix. This work indicated that the Fe/ZnS biocomposite is able to act as a promising candidate for bone-repair material.

To conclude, we would like to acknowledge all the authors for their contributions, adding to the success of this Special Issue in *Micromachines* as well as the reviewers whose feedback helped to improve the quality of the published papers.

Acknowledgments: The Guest Editors are very grateful to Min Su from the *Micromachines* publishing office for her great guidance, assistance and support that leads to the success of this Special Issue.

Conflicts of Interest: The authors declare no conflict of interest.

References

1. Chen, K.; Wan, H.; Fang, X.; Chen, H. Laser Additive Manufacturing of Anti-Tetrachiral Endovascular Stents with Negative Poisson's Ratio and Favorable Cytocompatibility. *Micromachines* **2022**, *13*, 1135. [CrossRef] [PubMed]
2. Song, C.; Hu, Z.; Xiao, Y.; Li, Y.; Yang, Y. Study on Interfacial Bonding Properties of NiTi/CuSn10 Dissimilar Materials by Selective Laser Melting. *Micromachines* **2022**, *13*, 494. [CrossRef] [PubMed]
3. Yang, Y.; Wang, W.; Yang, M.; Yang, Y.; Wang, D.; Liu, Z.; Shuai, C. Laser-Sintered Mg-Zn Supersaturated Solid Solution with High Corrosion Resistance. *Micromachines* **2021**, *12*, 1368. [CrossRef] [PubMed]
4. Zheng, X.; Meng, J.; Liu, Y. Strain Rate Dependence of Compressive Mechanical Properties of Polyamide and Its Composite Fabricated Using Selective Laser Sintering under Saturated-Water Conditions. *Micromachines* **2022**, *13*, 1041. [CrossRef] [PubMed]
5. Tao, Y.; Lu, C.; Deng, C.; Long, J.; Ren, Y.; Dai, Z.; Tong, Z.; Wang, X.; Meng, S.; Zhang, W.; et al. Four-Dimensional Stimuli-Responsive Hydrogels Micro-Structured via Femtosecond Laser Additive Manufacturing. *Micromachines* **2022**, *13*, 32. [CrossRef] [PubMed]
6. Wang, D.; Wang, H.; Chen, X.; Liu, Y.; Lu, D.; Liu, X.; Han, C. Densification, Tailored Microstructure, and Mechanical Properties of Selective Laser Melted Ti–6Al–4V Alloy via Annealing Heat Treatment. *Micromachines* **2022**, *13*, 331. [CrossRef] [PubMed]
7. Zhang, W.; Zhang, B.; Xiao, H.; Yang, H.; Wang, Y.; Zhu, H. A Layer-Dependent Analytical Model for Printability Assessment of Additive Manufacturing Copper/Steel Multi-Material Components by Directed Energy Deposition. *Micromachines* **2021**, *12*, 1394. [CrossRef] [PubMed]
8. Kang, M.; Qiang, Y.; Zhu, C.; Xiang, X.; Zhou, D.; Peng, Z.; Xie, X.; Zhu, Q. Hybrid Dissection for Neutron Tube Shell via Continuous-Wave Laser and Ultra-Short Pulse Laser. *Micromachines* **2022**, *13*, 352. [CrossRef] [PubMed]
9. Meng, L.; Long, J.; Yang, H.; Shen, W.; Li, C.; Yang, C.; Wang, M.; Li, J. Femtosecond Laser Treatment for Improving the Corrosion Resistance of Selective Laser Melted 17-4PH Stainless Steel. *Micromachines* **2022**, *13*, 1089. [CrossRef] [PubMed]
10. Wang, X.; Jiang, J.; Tian, Y. A Review on Macroscopic and Microstructural Features of Metallic Coating Created by Pulsed Laser Material Deposition. *Micromachines* **2022**, *13*, 659. [CrossRef] [PubMed]
11. Zhou, Y.; Xu, L.; Yang, Y.; Wang, J.; Wang, D.; Shen, L. Microstructure and Corrosion Behavior of Iron Based Biocomposites Prepared by Laser Additive Manufacturing. *Micromachines* **2022**, *13*, 712. [CrossRef] [PubMed]
12. Cui, Y.; Cai, J.; Li, Z.; Jiao, Z.; Hu, L.; Hu, J. Effect of Porosity on Dynamic Response of Additive Manufacturing Ti-6Al-4V Alloys. *Micromachines* **2022**, *13*, 408. [CrossRef] [PubMed]
13. Xu, Z.; Wang, D.; Song, W.; Tang, C.; Sun, P.; Yang, J.; Hu, Q.; Zeng, X. Microstructure and Wear of W-Particle-Reinforced Al Alloys Prepared by Laser Melt Injection. *Micromachines* **2022**, *13*, 699. [CrossRef] [PubMed]
14. Zhang, L.; Li, Y.; Li, S.; Gong, P.; Chen, Q.; Geng, H.; Sun, M.; Sun, Q.; Hao, L. Fabrication of Titanium and Copper-Coated Diamond/Copper Composites via Selective Laser Melting. *Micromachines* **2022**, *13*, 724. [CrossRef] [PubMed]
15. Yang, X.; Wang, S.; Pan, H.; Zhang, C.; Chen, J.; Zhang, X.; Gao, L. Microstructure Transformation in Laser Additive Manufactured NiTi Alloy with Quasi-In-Situ Compression. *Micromachines* **2022**, *13*, 1642. [CrossRef] [PubMed]
16. Li, Z.; Li, H.; Yin, J.; Li, Y.; Nie, Z.; Li, X.; You, D.; Guan, K.; Duan, W.; Cao, L.; et al. Review of Spatter in Laser Powder Bed Fusion Additive Manufacturing: In Situ Detection, Generation, Effects, and Countermeasures. *Micromachines* **2022**, *13*, 1366. [CrossRef] [PubMed]

Article

Laser Additive Manufacturing of Anti-Tetrachiral Endovascular Stents with Negative Poisson's Ratio and Favorable Cytocompatibility

Ke Chen [1,2], Haoran Wan [1], Xiang Fang [1] and Hongyu Chen [1,*]

[1] Key Laboratory of Impact and Safety Engineering of Ministry of Education of China, Ningbo University, Ningbo 315211, China; billke1995@gmail.com (K.C.); jacksonwan0706@hotmail.com (H.W.); fangxiang@nbu.edu.cn (X.F.)

[2] Department of Vascular Surgery, Second Xiangya Hospital, Central South University, Changsha 410011, China

* Correspondence: chenhongyu@nbu.edu.cn

Abstract: Laser additive manufacturing (LAM) of complex-shaped metallic components offers great potential for fabricating customized endovascular stents. In this study, anti-tetrachiral auxetic stents with negative Poisson ratios (NPR) were designed and fabricated via LAM. Poisson's ratios of models with different diameters of circular node (DCN) were calculated using finite element analysis (FEA). The experimental method was conducted with the LAM-fabricated anti-tetrachiral stents to validate their NPR effect and the simulation results. The results show that, with the increase in DCN from 0.6 to 1.5 mm, the Poisson ratios of anti-tetrachiral stents varied from -1.03 to -1.12, which is in line with the simulation results. The interrelationship between structural parameters of anti-tetrachiral stents, their mechanical properties and biocompatibility was demonstrated. The anti-tetrachiral stents with a DCN of 0.9 mm showed the highest absolute value of negative Poisson's ratio, combined with good cytocompatibility. The cytocompatibility tests indicate the envisaged cell viability and adhesion of the vascular endothelial cell on the LAM-fabricated anti-tetrachiral auxetic stents. The manufactured stents exhibit great superiority in the application of endovascular stent implantation due to their high flexibility for easy maneuverability during deployment and enough strength for arterial support.

Keywords: laser additive manufacturing; Poisson's ratio; anti-tetrachiral stents; biocompatibility

Citation: Chen, K.; Wan, H.; Fang, X.; Chen, H. Laser Additive Manufacturing of Anti-Tetrachiral Endovascular Stents with Negative Poisson's Ratio and Favorable Cytocompatibility. *Micromachines* **2022**, *13*, 1135. https://doi.org/10.3390/mi13071135

Academic Editors: Jie Yin, Yang Liu and Ping Zhao

Received: 27 June 2022
Accepted: 14 July 2022
Published: 18 July 2022

Publisher's Note: MDPI stays neutral with regard to jurisdictional claims in published maps and institutional affiliations.

Copyright: © 2022 by the authors. Licensee MDPI, Basel, Switzerland. This article is an open access article distributed under the terms and conditions of the Creative Commons Attribution (CC BY) license (https://creativecommons.org/licenses/by/4.0/).

1. Introduction

Nowadays, cardiovascular, cerebrovascular, and peripheral vascular diseases have become the causes of death with the highest mortality rates [1]. Atherosclerosis is an important pathophysiological process that causes cardiovascular and cerebrovascular diseases and peripheral vascular diseases [2]. With the generation of atheromatous plaque and accumulation in the arterial intima, the vascular lumen becomes narrow, leading to a reduced blood flow through the blood vessels [2]. This can cause ischemic symptoms in the tissues supplied by the corresponding arteries, with different manifestations according to the branches of the artery and the degree of stenosis, including coronary artery disease, stroke, peripheral arterial disease, and renal artery disease [3]. At present, PTA (percutaneous transluminal angioplasty) has become an important method for treating arteriosclerosis obliterans [4]. The use of a balloon catheter to puncture the diseased artery for treatment is simpler and less risky than traditional medical and surgical treatment, and can partially replace bypass surgery [5]. However, the long-term patency rate of diseased vessels is relatively low due to factors such as elastic recoil, intimal tear, dissection, restenosis, and the progression of arteriosclerosis itself after balloon dilatation alone [6]. Concomitant stent implantation during the procedure can avoid vessel elastic recoil after PTA, residual stenosis, and thromboembolic problems resulting from dissection after application of balloon dilatation [7]. In-stent restenosis is the phenomenon of re-occlusion of the vessel after vascular stent implantation [7,8]. Although vascular stent implantation can support stenotic vessels and prevent elastic recoil of vessels, the vascular wall easily causes inflammatory reaction, intimal hyperplasia, and other problems after stent expansion through rebound, damage, etc. [9], thereby resulting in in-stent restenosis. In-stent restenosis is the main failure form of vascular stents and is a key problem to be solved urgently in vascular stent implantation techniques [9]. Therefore, solving the challenge of in-stent restenosis has become the key scientific issue for vascular stent implant technology [10]. Bare metal stents are prone to causing physical damage to the vessel wall, and non-degradable metal stents may also cause immune rejection in the body [11]. Drug-eluting stents delay in-stent restenosis by adding drugs that inhibit the proliferation of vascular endothelial cells, such as paclitaxel, rapamycin, and their derivatives, on the basis of bare metal stents, but cannot fundamentally solve the problem of restenosis [12,13]. Several studies have shown that the structural design of stents plays an important role in vascular endothelial cell proliferation [14] and long-term stent patency [15,16]. There is also study reporting that stents with negative Poisson's ratio structures provide new clinical application prospects [17].

The negative Poisson's ratio structure has a unique tensile/compressive behavior; that is, both transverse strain and longitudinal strain are positive or negative when subjected to uniaxial tension or compression. Structures with negative Poisson's ratios generally possess high fracture toughness, shear modulus, cracks resistance, and indentation resistance [18]. Modified materials have a surface composition and morphology intended to interact with biological systems and cellular functions. Not only does surface chemistry have an effect on material biological response, surface structures of different morphology can be constructed to guide a desirable biological outcome [10]. In recent years, with intense research on negative Poisson's ratio structures, many typical negative Poisson's ratio structures, such as concave hexagonal structure, chiral structure, arrow structure, and rotating quadrangular structure, have been proposed successively [18]. Materials and structures with negative Poisson's ratios have good application prospects in biomedicine, aerospace, shock absorption and sound insulation, and energy absorption buffering [19]. The application of a negative Poisson's ratio structure on vascular stents can simultaneously reduce the axial radial size when the vascular stents are compressed before implantation, which is of benefit for minimally invasive implantation. When expansion is performed after implantation, the axial radial size is simultaneously increased, which contributes to the high-precision positioning of the vascular stent. Even though structures of endovascular stents with negative Poisson's ratios have been designed [20], it is still difficult to manufacture these stents due to their high complexity in configuration and small size. In other words, it is hard to fabricate the small, complexly shaped endovascular stents with negative Poisson's ratios using traditional manufacturing methods.

In recent years, laser additive manufacturing (LAM) has been developed to manufacture metallic structural parts directly from powder materials [21]. In the work of Roxanne Khalaj et al. [22], coronary stents were manufactured by 3D printing. Weijian Hua et al. [23], showed the 3D printing method to also be a powerful tool for stent manufacturing. Kaitlyn Chua's work also illustrates how the field of 3D printing and biomedicine can create more innovative devices and products [24]. For 3D printing, the manufacturing structure is already very simple. We can design and manufacture a new honeycomb structure with irregular nodes and pillars [25] and can also use alloy to manufacture a bionic crab claw structure [26]. Laser powder bed fusion (LPBF) is a popular LAM technique capable of precisely preparing complex-shaped parts based on a completely melted mechanism [27]. Its typical "layer-by-layer" fabricating process enables the rapid manufacturing of metallic parts with complex geometries whose preparation is either too tedious or impossible by conventional methods such as casting or forging followed by machining [28]. It has been demonstrated that the LPBF is capable of fabricating negative Poisson's ratio structures with high precision successfully [29]. LPBF enables the production of highly complex and tailor-performed endovascular stents for patients. Appropriate deformability and biological performance of vascular stents are important factors for the successful implantation of vascular stents and ensuring the restoration of blood flow patency at the vascular stenosis.

In this study, endovascular stents with negative Poisson's ratios were designed and fabricated via LPBF using 316L stainless steel as the raw material. When the lesion involves a branching vessel, membrane-coated stents could insulate the flow and lead to branch ischemia, whereas a bare metal stent supports the vessels without blocking the flow of branching vessels, though it may cause intimal injury. It depends on the physical situation, and may do more good than harm. Poisson's ratios of models with different diameters of circular node (DCN) were calculated using finite element analysis (FEA). The stress concentration of a model with different DCN during compressive deformation was analyzed. A quantified method of structural optimization of endovascular stents with negative Poisson's ratios was proposed. Then, the cytocompatibility tests were conducted on the LPBF-fabricated endovascular stents. The cell viability and adhesion of the umbilical vein endothelial cells on the LPBF-fabricated anti-tetrachiral auxetic stents were investigated. The stent size of this study was mainly based on the diameters of the femoral artery and iliac artery, whose normal values are 7–12 mm. This study shall be a first step toward the structural design and manufacturing of endovascular stents with a negative Poisson's ratios which fulfill predestined biocompatibility.

2. Materials and Methods

2.1. Simulation

The geometrical configuration of an anisotropic anti-tetrachiral cell with elliptical nodes is shown in Figure 1a. The geometrical parameters of unit cell L_x, L_y, r_x, and r_y represent the ligament length and radius of the elliptical node along x and y-directions; t represents the wall thickness of the ligaments and nodes; and h is the thickness of the unit cell along the direction perpendicular to the x–y plane. Periodically distributing the anisotropic anti-tetrachiral cell into a cylinder so that a model of anti-tetrachiral endovascular stents can be obtained. Finite element analysis using ABAQUS was carried out to predict the mechanical behavior of proposed anti-tetrachiral stent.

The model Figure 1c—whose strut size ratio L_x/L_y was set to 2, and the thickness of the strut was 0.9 mm—was meshed using approximately 200,000 tetrahedral elements. The used mechanical properties of the constituent material of stent for FEA simulation are given in Table 1. Boundary conditions of the unit cell are shown in Figure 1d. When the model was compressed along z-direction, a uniform displacement Uz was imposed on the upper surface of the plate and the displacement of the bottom surface in z-direction was set to zero, and both of the plates could move in the x–y plane; other surfaces were unconstrained. Variables used to calculate the Poisson ratio are shown in Figure 1b, where Δy is the displacement of the trace surface in the y axis and Δz is the displacement in the z axis. From the total displacement contour plot, it could be found that all the lateral overhanging struts had the same deforming tendency; moreover, the newly modified re-entrant structure was axisymmetric against the vertical center line, and hence, one end of the strut was chosen to trace the deformation in y and z axes, and the deformation in the y–z plane can be used to calculate the Poisson ratio of the structure because of the feasibility of geometric symmetry. The initial lengths of the model in y and z axes were measured by computer aided design (CAD) software and denoted as l_y and l_z, respectively. Then, the strains in two directions can be calculated using Equation (1) and (2) below.

$$\varepsilon_{transverse} = \Delta y / l_y \tag{1}$$

$$\varepsilon_{vertical} = \Delta z / l_z \tag{2}$$

Table 1. Mechanical properties of the constituent material of the stent for FEA simulation.

Material	Young's Modulus(E)	Poisson Ratio	Yield Stress	Limit Stress	Limit Nominal Strain	Density
316L stainless steel	196,000 MPa	0.3	205 MPa	5151 MPa	60%	7.89×10^{-9} ton/mm^3

Then the Poisson ratio can be obtained using Equation (3) below:

$$v = \varepsilon_{\text{transverse}}/\varepsilon_{vertical} \quad (3)$$

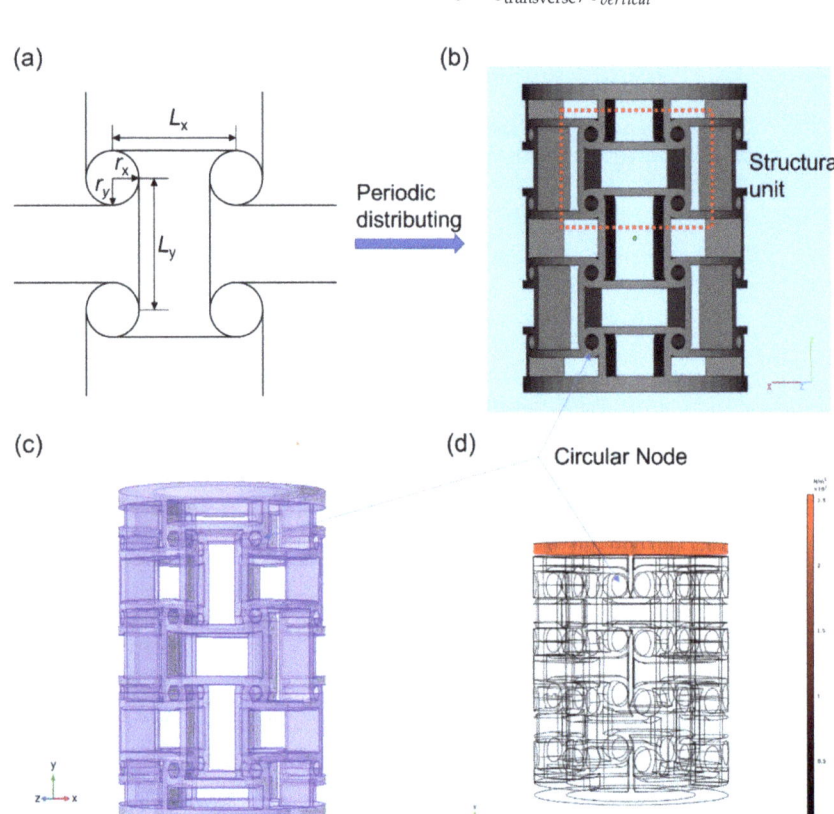

Figure 1. (**a**) Effect of node radius on Poisson's ratio at the same displacement distance; (**b**) three-dimensional model of anisotropic anti-tetrachiral; (**c**) three-dimensional model in FEM software; (**d**) schematic diagram of boundary conditions.

2.2. LPBF Process

The raw material 316L with a spherical shape and a particle range of 15–53 μm was applied in the LPBF process. The 316L powder used was from CarTech. Its UNS Number is S316003. The main specification parameters of 316L: carbon (0.03%), manganese (2.00%), phosphorus (0.045%), sulfur (0.030%), silicon (1.00%), chromium (16.00 to 18.00%), nickel (10.00 to 14.00%). The LPBF processing system consisted of an IPG YLR-500-SM ytterbium fiber laser with a power of ~500 W and a spot size of 70 μm, a SCANLAB hurry scan 30 scanner, an automatic powder spreading apparatus, and a computer control system for LPBF process. The entire process was conducted in an argon atmosphere, and sixteen specimens with dimensions of ϕ 9 mm × 12 mm were fabricated layer by layer using laser powder of 160 W, laser scan speed of 800 mm/s, layer thickness of 30 µm, and hatching space of 50 µm. Anti-tetrachiral stents with different radii of node (0.6, 0.9, 1.2, 1.5 mm) were fabricated using the same laser processing parameters. In order to improve the processing quality of the fabricated anti-tetrachiral stents, the model was titled 45 degrees with the substrate and supports were used to ensure the successful fabrication of the stents, as shown in Figure 2a. The as-fabricated anti-tetrachiral stents are shown in Figure 2b.

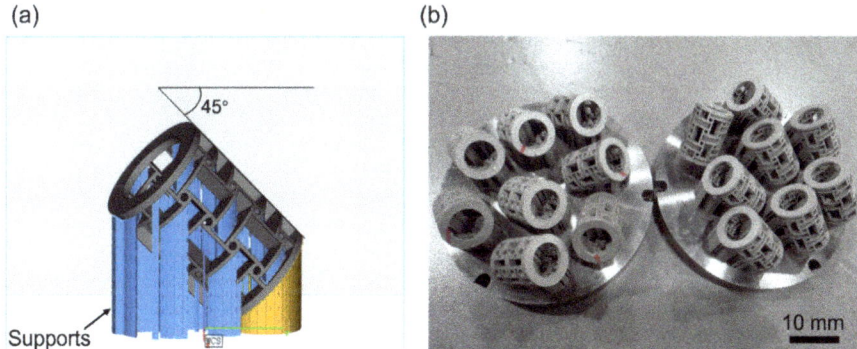

Figure 2. (**a**) Schematic diagram of the support model; (**b**) as-fabricated anti-tetrachiral stents.

2.3. Experimentation

In order to confirm that the anisotropic anti-tetrachiral cells with elliptical nodes conformed to the simulation results, we prepared samples of each model to be tested experimentally through the instrument. We used four groups of experiments for control comparison: (1) 500 mm/s laser scanning speed and 130 W laser power; (2) 500 mm/s laser scanning speed and 170 W laser power; (3) 1000 mm/s laser scanning speed and 130 W laser power; (4) 1000 mm/s laser scanning speed and 170w laser power. Finally, the best process parameters were selected after four groups of specimens were compared and analyzed: 1000 mm/s laser scanning speed and 170 W laser power. The tests of the Poisson ratio were performed on a Hysitron Tl Premier nanoindenter. By means of matrix punches, the same constraints as in the simulation were applied to the upper surfaces of the specimens in sample navigation, and a pressure of 1000 N was applied to the upper surface, and they were compressed downward by 6 mm at the same time, and the displacement in the y-direction and the displacement in the x-direction of the middle point of the model were calculated by the instrument, and the Poisson ratio of the model could be obtained by dividing the y-direction displacement by the x-direction displacement, and comparing that with the simulation results.

2.4. Biocompatibility Test

To investigate the cell viability and adhesion on LPBF-fabricated endovascular stents with negative Poisson's ratios, polyethylene as a negative control and polyurethane (Sigma-Aldrich, Burlington, MA, USA, 81367, 5 g) as a positive control were obtained according to ISO 10993-12. Human umbilical vein endothelial cells (HUVEC) were cultured in the stents, and the Cell Counting Kit-8 (CCK-8, Dojindo Molecular Technologies, Rockville, MD, USA) assay was used to quantitatively investigate the cytotoxicity of stents [30]. The procedure was as follows: Place the stent in a freeze-dryer at −80 °C; dry for 24 h; then transfer it to a sterile 96-well plate. Use a sterile PBS solution and absolute ethanol to soak the stent for 30 min each. Then, suck away the liquid, place the stent under ultraviolet lamp for irradiation for 24 h for sterilization, and then obtain a sterile printed stent for future use. HUVECs were quantitatively seeded on stents at 1×10^5 cells/mL and supplemented with an appropriate amount of complete medium, and 96-well plates were transferred to a CO_2 incubator for constant temperature culture, and the medium was changed every day. According to GB/T16886.5-ISO16886-5 Biological evaluation criteria for medical devices [31], polyurethane was selected as a positive control, and polyethylene was selected as a negative control [32]. Pure CCK reagent and endothelial cell medium (ECM) medium were made into a CCK assay reagent at a ratio of 1:10 for future use. We used a tweezer to transfer the cell scaffold construct to be tested inoculated with endothelial cells into a 96-well plate of blank control group previously cultured with the same number of cells; immediately added 110 µL of prepared sterile CCK detection reagent into the experimental group, negative control group, and positive control group at 37 °C; and then transferred the well plate into a CO_2 incubator for incubation for 3 h. Then, we used a pipette to transfer the liquid in the well plate into another clean well plate, covered the well plate, and transferred it into a microplate reader. We selected the incident wave length as 490 nm, and recorded the measured OD value. Cell survival rate = (OD value of experimental group/OD value of control group) × 100%.

LPBF-fabricated endovascular stents with HUVECs segments were bisected longitudinally to expose the lumen surface and photographed. Both halves of the stents were rinsed in sodium phosphate buffer (pH 7.4) and were then dehydrated in a graded series of ethanol–water. After critical point drying, the tissue samples were mounted and sputter-coated with gold. The samples were visualized using a scanning electron microscopy (SEM, TESCAN, Mira4, Brno, Czech Republic) at 10 keV.

3. Results and Discussion

Based on the research on the in-plane deformation behaviors of the hexachiral honeycomb structures demonstrated in literatures, the deformation behavior of anti-tetrachiral structures with elliptical or round nodes under in-plane loading was assumed as ligament-bending deformation mode, which can be analyzed using Euler–Bernoulli beam bending theory [20]. When the node rotates by an angle φ, the strains ε_x and ε_y along the x and y-directions can be expressed as:

$$\varepsilon_x = \frac{2(r_y - \frac{t}{2})\varphi}{L_x} \quad (4)$$

and

$$\varepsilon_y = \frac{2(r_x - \frac{t}{2})\varphi}{L_y} \quad (5)$$

Accordingly, the in-plane Poisson ratio ν_{xy} can be calculated as:

$$\nu_{xy} = -\frac{\varepsilon_y}{\varepsilon_x} = -\frac{2(r_x - \frac{t}{2})L_x}{2(r_y - \frac{t}{2})L_y} \quad (6)$$

when round nodes are used in the anti-tetrachiral stents, i.e., $r_x = r_y$, the ν_{xy} can be calculated as $-\frac{L_x}{L_y} = -\gamma$, which is a constant. Thus, theoretically, the variation of the size of the nodes has no influence on the Poisson ratio of the single in-plane anti-tetrachiral structure. However, after periodically distributing the in-plane anti-tetrachiral structure into three-dimensional anti-tetrachiral stent, its Poisson's ratio is not a constant and depends on the size of the nodes, as shown in the following.

Figure 3a shows the effect of the node radius on the Poisson ratio of the anti-tetrachiral stent under same displacement distance and the comparison between experimentally obtained and simulated results. It is obvious that all the anti-tetrachiral stents exhibited a negative Poisson ratio; the absolute value of the Poisson ratio increased from 1.077 to the highest value of 1.115 with an increase in node radius from 0.6 to 0.9 mm; then the value decreased to 1.044 as the node radius further increased to 1.2 mm, and finally, to and 1.035 for 1.5 mm. The experimentally obtained Poisson ratio of the anti-tetrachiral stent shows high agreement with the simulated results, as shown in Figure 3b. In order to reveal the influencing mechanism of node radius on the absolute value of Poisson's ratio of designed anti-tetrachiral stent, the Mises stress field and displacement field of the stent along x and y-directions after compressive deformation were analyzed, as shown in Figure 4. One can see that, after periodically distributing the in-plane anti-tetrachiral structure into the three-dimensional anti-tetrachiral stent, each node of stents is not only subjected to the vertical pressure, but also subjected to the pressure given radially during compression, which causes the inward collapse of the single anti-tetrachiral structure [20]. Table 2 shows the horizontal and vertical displacements of the intermediate nodes of the model for different node radii. The Poisson ratio of the model is calculated by dividing the horizontal displacement by the value of the vertical displacement during the simulation analysis. One can see that the inward collapse displacement (horizontal displacement) of the model with a node radius of 0.9 mm is the largest among the four models, and the evolution of inward collapse displacement agrees with variation in Poisson's ratio for the four given models. This is to say, the Poisson ratio depends on the inward collapse displacement of models during compressive deformation. When a different node radius is designed for the single anti-tetrachiral structure, the Mises stress and resulting horizontal and vertical displacements are different, so that the responses of different regions and nodes to the pressure and stress can be different (Figure 4). In the case of choosing the same node, the maximum stress in the 0.9 mm radius model was 5.747×10^{11}, and the stress in the rest of the models decreased with the increase in the node radius. It can be seen in the data that the 0.9 mm model was subjected to the highest stress, and the variations in stress values for the four models are also in accordance with the law of Poisson's ratio variation. Therefore, the Poisson ratio in different anti-tetrachiral stents with different node radii cannot be a constant. When conducting compression via step-by-step mode, obviously, there is a small fluctuation on Poisson's ratio for all the anti-tetrachiral stent models. One can speculate that the rings at the top and bottom inhibit the overall shrinkage and have a certain impact on the Poisson ratio. This gives rise to a speculation that the varied negative Poisson's ratio in anti-tetrachiral stents can be influenced by the suppression effect of the top and bottom rings. The following study shows the simulated results of the model whose original top and bottom rings were removed, so that the suppression effect of rings on the Poisson ratio of anti-tetrachiral stents can be revealed.

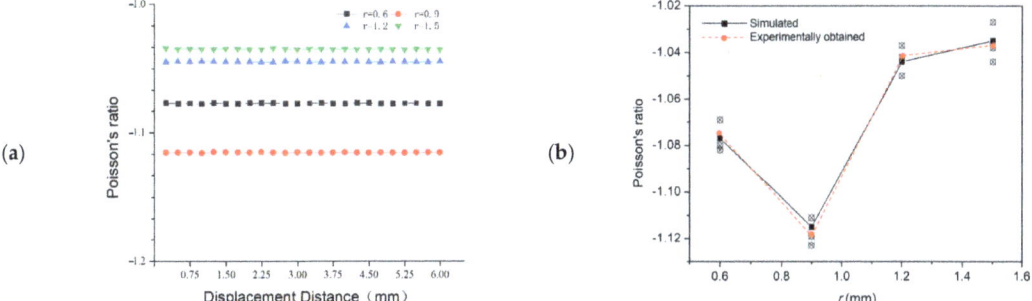

Figure 3. (a) Effect of node radius on Poisson's ratio at the same displacement distance. (b) Comparison of experimentally obtained Poisson's ratio with simulation results.

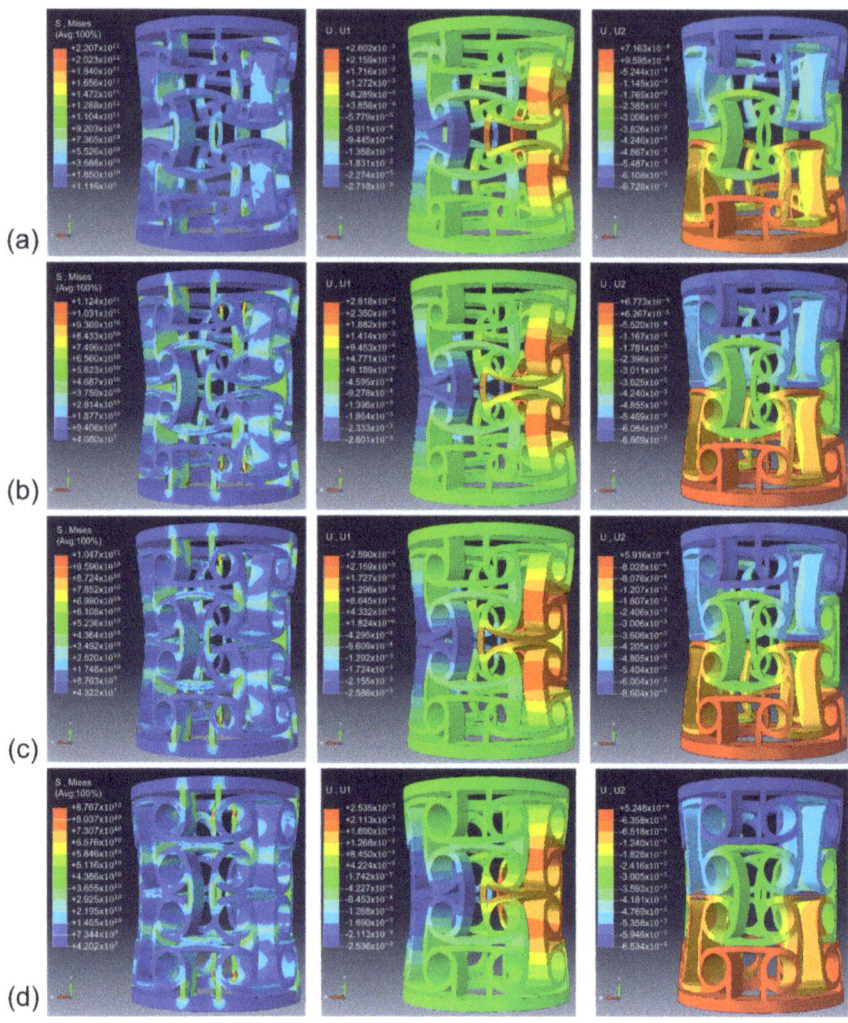

Figure 4. Physical field at each node radius, displacement field in x-direction and displacement field in y-direction: (**a**) r = 0.6 mm; (**b**) r = 0.9 mm; (**c**) r = 1.2 mm; (**d**) r = 1.5 mm.

Table 2. Horizontal and vertical displacements for different node radii.

Node Radius (mm)	Horizontal Displacement (cm)	Vertical Displacement (cm)
0.6	−0.272	0.673
0.9	−0.280	0.670
1.2	−0.259	0.660
1.5	−0.254	0.653

During the study of deformation behavior of anti-tetrachiral structures with round nodes, we proposed the hypotheses: (1) nodes (or circles) are considered rigid; (2) internal forces oriented in a direction perpendicular to the externally applied stress vanish; (3) internal forces are dictated by the observed kinematic behavior; (4) axial and shear deformations of the ligaments are neglected; (5) all deflections are small. This was similar to the work performed by Prall and Lakes [2]. The deformations of the anti-tetrachiral structures with round nodes under uniaxial tensile loading conditions were assumed as ligament bending dominated, and the shearing and tension deformation of the ligaments were not included. We speculate that the upper and lower rings limit the variation of the overall Poisson ratio. Figure 5 shows the analysis of the Poisson ratios of models, which had fixed constraints imposed on their lower bottom surfaces, before and after the ring was removed. The vertical downward displacement of model was set to 6 mm. Based on the simulation results and Table 3, the ring structure inhibits the shrinkage of the whole model. With an increase in the node radius, the inhibition effect became more significant; the absolute values of Poisson's ratios in the ring-free anti-tetrachiral stents with note radii of 0.6, 0.9, 1.2, and 1.5 mm increased by 3%, 8%, 12%, and 15%, compared to the ring-attached stents. Figure 6 shows the stress and displacement nephogram of the ring-free model with a node radius of 1.5 mm. Combining the above data, it is easy to find that the upper and lower rings have a suppressive effect on the overall deformation of the model. No suppression of the upper and lower rings will make the absolute value of Poisson's ratio larger for each model, which will make the stresses on the circular nodes more concentrated and more forces in the middle part, leading to more inward collapsing displacements, thereby obtaining larger values of negative Poisson's ratio.

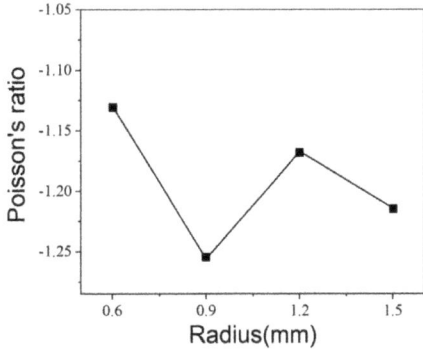

Figure 5. Poisson's ratio for the four models without rings.

Table 3. Comparison of Poisson's ratios for models with and without rings.

Radius (mm)	Horizontal Deformation (cm)	Vertical Deformation (cm)	Change Rate (%)
0.6	−0.272(with) −0.283(without)	0.673(with) 0.719(without)	3
0.9	−0.280(with) −0.301(without)	0.670(with) 0.717(without)	8
1.2	−0.259(with) −0.297(without)	0.660(with) 0.679(without)	12
1.5	−0.254(with) −0.289(without)	0.653(with) 0.650(without)	15

As shown in Figures 3 and 4, the varying of Poisson's ratio in different anti-tetrachiral stents can be caused by different Mises stress and deformation behavior at the joint fillet position. Thus, the convergence and variation of von Mises stress of anti-tetrachiral models with different node radii were analyzed, as shown in Figure 7. Apparently, the stress variation in anti-tetrachiral stents with a node radius of 0.6 mm during deformation was not that significant compared with that of other stents with larger node radii. As the node radius increases, the stresses acting on each node between each other also increase, and the stress increase was most obvious for the model with r = 0.9 mm. It was also found that the stress concentration most easily occurs at the joint positions between the circular node and other cells. Thus, the circular node in the anti-tetrachiral stent mode plays an important role in the resulting value of negative Poisson's ratio during the deformation process. The corresponding results show the anti-tetrachiral stent model with an r of 0.9 mm possessed the most uniform stress distribution and largest area of force acting on the joint fillet. This conclusion highly agrees with the results, as shown in Figure 3.

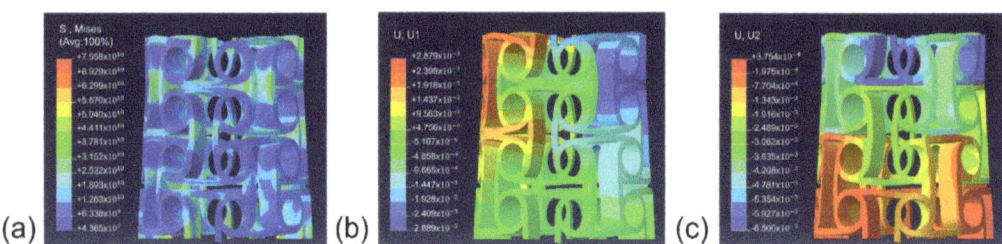

Figure 6. Physical fields of the 1.5 mm model without rings: (**a**) Mises stress field; (**b**) displacement field along the x-direction; (**c**) displacement field along the y-direction.

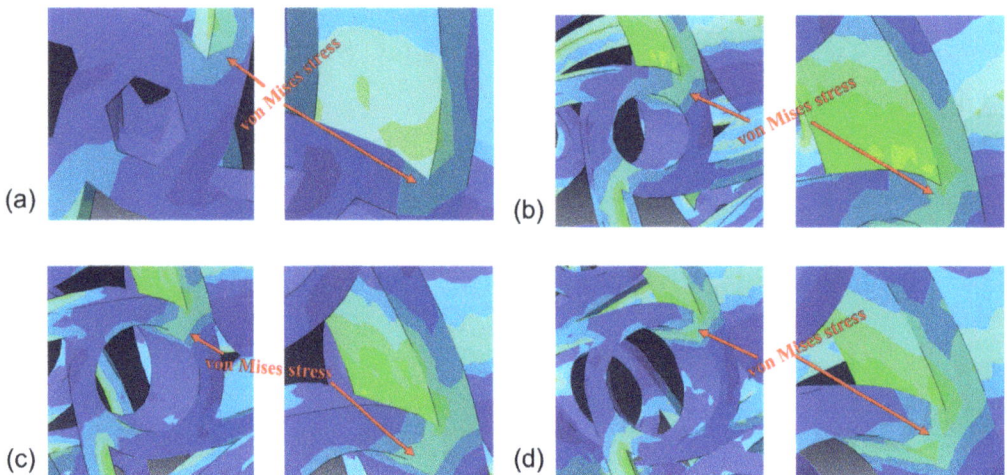

Figure 7. Effect of different radii on stress distribution: (**a**) r = 0.6 mm; (**b**) r = 0.9 mm; (**c**) r = 1.2 mm; (**d**) r = 1.5 mm.

Figure 8 shows the human umbilical vein endothelial cells' (HUVECs) morphology in the LPBF-manufactured scaffolds, and polyethylene was normal and grew well during hours 48 to 72, though there was massive death in the HUVECs co-culture with liquid extracts of polyurethane during hours 48 to 72. According to GB/T 16886.5-2017 qualitative morphological grading of cytotoxicity of extract (Table 4), the LPBF-manufactured scaffolds and polyethylene showed no cell lysis and no reduction in cell proliferation during hours 24 to 72. By comparison, the polyurethane showed obvious lysis of HUVICs. Nearly complete lysis of cell layers took place during hours 24 to 72. According to the toxicity grading method of United States pharmacopeia [33], the relative growth rates (RGR) value evaluation criteria for the cytotoxicity grade of the system, the polyethylene, and LPBF-fabricated scaffolds, the RGR was >95% during the process of cell co-culture, the cytotoxicity was grade 0, and there was no significant difference in the HUVEC counts of the groups ($p > 0.05$). The RGR of polyurethane was 48% at 24 h, and almost zero from 48 h to 72 h, so the cytotoxicity grade could be determined at cytotoxicity grade III, and there was significant difference in the HUVEC counting between polyurethane and LPBF-manufactured scaffolds ($p < 0.05$). The results showed that the cultured cells showed a good growth state on the LPBF-fabricated scaffolds, and the tissue-engineered scaffold constructed after LPBF fabrication of the material had good biocompatibility, a small proportion of apoptosis, and less cytotoxicity, about the same cytotoxicity compared with MIM 316L stents [34].

Figure 8. Microscopic view of in vitro cytotoxicity. The morphology of the HUVECs in the stent and polyethylene indicates that the cells were normal. HUVECs in the polyurethane group started to be lysed from 24 h.

Table 4. Qualitative morphological grading of cytotoxicity of extract [35].

Level	Extent of Reaction	Morphology of Cultured Cells
0	None	Discrete intracytoplasmic volume, no cell lysis, no reduction of cell proliferation
1	Minor	Not more than 20% of the cells are round, loosely attached, and without cytoplasmic or morphological changes; occasional lysed cells are present; only slight inhibition of cell growth is observed
2	Mild	Not more than 50% of the cells are round, devoid of intracytoplasmic granules, no extensive cell lysis, not more than 50% growth inhibition observable
3	Moderate	Not more than 70% of the cell layers contain rounded cells or are lysed; cell layers not completely destroyed, but more than 50% cell growth inhibition may be observed
4	Severe	Nearly complete or complete destruction of cell layers

Figure 9 shows the cell viability and adhesion of the HUVECs on the LPBF-manufactured scaffolds with node radius of 0.9 mm as compared to the control. The cell viability of the LPBF-manufactured scaffold was not significantly different from that of the negative control (polyethylene), and significantly different from the positive control (polyurethane), indicating the excellent biocompatibility of the LPBF-manufactured scaffold. The LPBF-manufactured scaffold co-cultured with HUVICs is shown via light microscopy in Figure 10. The morphologies of the cell adhesion on to the surfaces of the LPBF-manufactured stents after cell culturing for 3d and 5d are shown in Figure 11. The HUVECs adhered to the surfaces of the stents and exhibited a polygonal and full spreading shape; HUVECs are scattered sporadically after culturing in 3d in Figure 11a, indicating that the cells were compatible with the surfaces and grew in a healthy way. By increasing cell culture time to 5d, the cells spread across the surfaces of the stents and covered all of the surface area, as shown in Figure 11b. Overall, the HUVECs on the surfaces of the LPBF-manufactured stents exhibited healthy morphology, attachment, and spreading, as indicated by the flourishing growths of filopodia with widely extended microspikes solidly attached to the surfaces of the scaffolds; see the highly magnified image in Figure 11b. It should be noted that the cell adhesion density in relation to the anti-tetrachiral stents was lower than that of the control. Based on Schulz's research, HUVECs were reported to favor surfaces with roughness within a few micrometers. The surface roughness of the SLM-manufactured was usually much greater than a few micrometers [36], which would have significantly affected the cells' seeding on these surfaces, making it more difficult for the cells to settle on the surfaces (Figure 11b). This raises a challenge for the LPBF process: the surface roughness of an LPBF-manufactured product should be effectively reduced, and any partially melted metal particles deposited on the surfaces of the resultant stents should be removed as well as possible. In this way, the cells can be favored to adhere on the surfaces of the LPBF anti-tetrachiral scaffolds and the vascular lumen will become enlarged, leading increased blood flow through the blood vessels to cure vascular occlusive disease, cardiovascular and cerebrovascular diseases, and peripheral vascular diseases. It should be noted that there is still a long way until use in actual patients for the LPBF-fabricated anti-tetrachiral endovascular stents. This study took the first step to prove cytotoxicity, and in the next step, we will undertake animal experiments—for example, implanting stents into the aortas of the dogs; endothelial cell staining; testing the expression levels of CD31, CD34, CD133, and eNOS; and investigating mechanical fatigue reliability. The ultimate aim is developing a more customizable and advanced way to manufacture vascular stents.

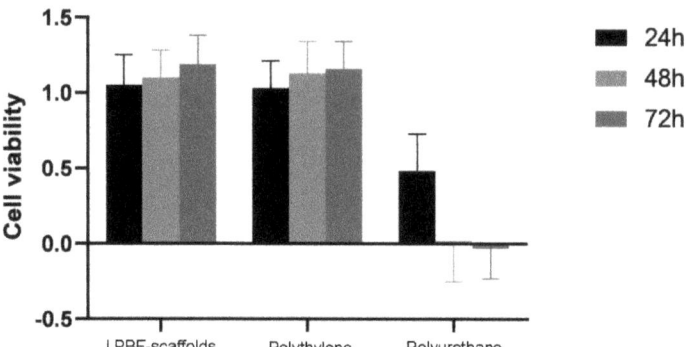

Figure 9. Cytotoxic experiments indicating cellular viability of HUVECs treated with liquid extracts of the LPBF-manufactured scaffolds, polyethylene, and polyurethane at 24, 48, and 72 h. The viabilities of the LPBF-manufactured scaffolds and polyethylene group were higher than 95%, and that of polyurethane was lower than 49%.

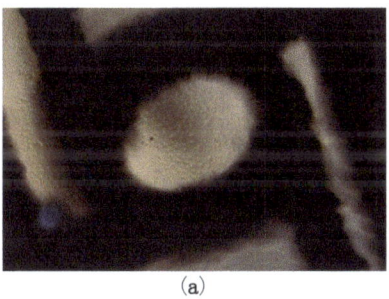
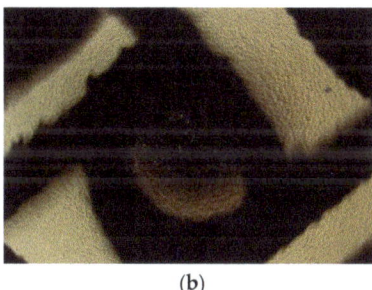

Figure 10. LPBF-manufactured scaffold co-cultured with HUVICs under light microscope: (**a**) co-cultured for 3 days; (**b**) co-cultured for 5 days.

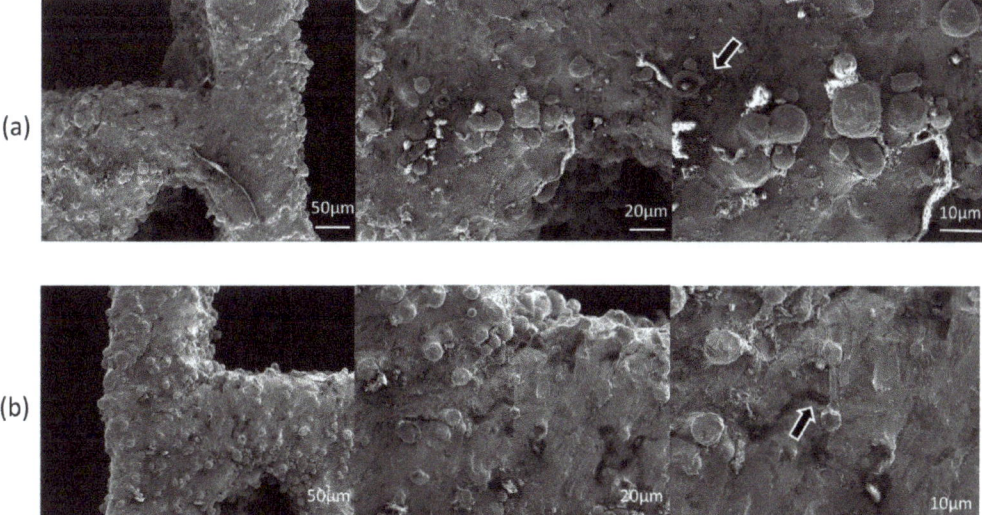

Figure 11. SEM of LPBF-manufactured scaffold loaded with randomly seeded HUVECs. The HUVECs (black arrow) appear to have a spread morphology with cord-like muti-cellular formations. (**a**) Co-cultured for 3 days; (**b**) co-cultured for 5 days.

4. Conclusions

(1) All given anti-tetrachiral stent models possess a negative Poisson's ratio when subjected to compression force. The stent with r of 0.9 mm has the largest absolute value of negative Poisson's ratio. The upper and lower rings of the model have a suppressive effect on the overall shrinkage of the model, and the absolute value of the negative Poisson ratio increases when the rings are removed. However, the existence of the ring structure is not the reason for the Poisson ratio not being a constant.

(2) As the node radius increases, the stress concentration phenomenon appears. Stress concentration causes the absolute value of Poisson's ratio of the model to become smaller. The stress concentration phenomenon appeared least pronounced on the model with a 0.9 mm radius, followed by those with 0.6, 1.2, and 1.5 mm radii. The node radius of 0.9 mm had a larger absolute value of Poisson's ratio because of the reduced stress concentration at the nodes and other unit connections.

(3) The cultured cells showed good growth state on the LPBF-fabricated scaffolds. They demonstrated favorable structural, physical, and chemical stability and biocompatibility, indicating their promising potential for application in animal and clinical practice.

Author Contributions: K.C.: methodology, data curation, writing—original draft preparation, writing—reviewing and editing. H.W.: methodology, data curation, writing—original draft preparation, validation. X.F.: funding acquisition, validation, supervision. H.C.: conceptualization, supervision, data curation, writing—reviewing and editing, validation. All authors have read and agreed to the published version of the manuscript.

Funding: National Science Foundation of China (NSFC) (Grant No. 12102212, 12102208), Natural Science Foundation of Zhejiang Province (LQ22A020010).

Institutional Review Board Statement: Not applicable.

Informed Consent Statement: Not applicable.

Data Availability Statement: All reasonable requests for materials and data will be fulfilled by the corresponding author of this publication.

Acknowledgments: The authors are grateful for financial support from the National Science Foundation of China (NSFC) (Grant No. 12102212, 12102208), Natural Science Foundation of Zhejiang Province (LQ22A020010).

Conflicts of Interest: The authors declare no conflict of interest.

References

1. Roth, G.A.; Dwyer-Lindgren, L.; Bertozzi-Villa, A.; Stubbs, R.W.; Morozoff, C.; Naghavi, M.; Mokdad, A.H.; Murray, C.J.L. Trends and Patterns of Geographic Variation in Cardiovascular Mortality Among US Counties, 1980–2014. *JAMA* **2017**, *317*, 1976–1992. [CrossRef] [PubMed]
2. Lee, J.M.; Choi, K.H.; Koo, B.-K.; Park, J.; Kim, J.; Hwang, D.; Rhee, T.-M.; Kim, H.Y.; Jung, H.W.; Kim, K.-J.; et al. Prognostic Implications of Plaque Characteristics and Stenosis Severity in Patients With Coronary Artery Disease. *J. Am. Coll. Cardiol.* **2019**, *73*, 2413–2424. [CrossRef] [PubMed]
3. Libby, P.; Buring, J.E.; Badimon, L.; Hansson, G.K.; Deanfield, J.; Bittencourt, M.S.; Tokgözoğlu, L.; Lewis, E.F. Atherosclerosis. *Nat. Rev. Dis. Primers* **2019**, *5*, 56. [CrossRef] [PubMed]
4. Schneider, P.A.; Laird, J.R.; Doros, G.; Gao, Q.; Ansel, G.; Brodmann, M.; Micari, A.; Shishehbor, M.H.; Tepe, G.; Zeller, T. Mortality Not Correlated With Paclitaxel Exposure: An Independent Patient-Level Meta-Analysis of a Drug-Coated Balloon. *J. Am. Coll. Cardiol.* **2019**, *73*, 2550–2563. [CrossRef]
5. Lookstein, R.A.; Haruguchi, H.; Ouriel, K.; Weinberg, I.; Lei, L.; Cihlar, S.; Holden, A. Drug-Coated Balloons for Dysfunctional Dialysis Arteriovenous Fistulas. *N. Engl. J. Med.* **2020**, *383*, 733–742. [CrossRef] [PubMed]
6. Tepe, G.; Laird, J.; Schneider, P.; Brodmann, M.; Krishnan, P.; Micari, A.; Metzger, C.; Scheinert, D.; Zeller, T.; Cohen, D.J.; et al. Drug-coated balloon versus standard percutaneous transluminal angioplasty for the treatment of superficial femoral and popliteal peripheral artery disease: 12-month results from the IN.PACT SFA randomized trial. *Circulation* **2015**, *131*, 495–502. [CrossRef]
7. Bausback, Y.; Wittig, T.; Schmidt, A.; Zeller, T.; Bosiers, M.; Peeters, P.; Brucks, S.; Lottes, A.E.; Scheinert, D.; Steiner, S. Drug-Eluting Stent Versus Drug-Coated Balloon Revascularization in Patients With Femoropopliteal Arterial Disease. *J. Am. Coll. Cardiol.* **2019**, *73*, 667–679. [CrossRef]
8. Yang, X.; Yang, Y.; Guo, J.; Meng, Y.; Li, M.; Yang, P.; Liu, X.; Aung, L.H.H.; Yu, T.; Li, Y. Targeting the epigenome in in-stent restenosis: From mechanisms to therapy. *Mol. Ther. Nucleic Acids* **2021**, *23*, 1136–1160. [CrossRef]
9. Byrne, R.A.; Joner, M.; Kastrati, A. Stent thrombosis and restenosis: What have we learned and where are we going? The Andreas Grüntzig Lecture ESC 2014. *Eur. Heart J.* **2015**, *36*, 3320–3331. [CrossRef]
10. Slepicka, P.; Kasalkova, N.S.; Siegel, J.; Kolska, Z.; Bacakova, L.; Svorcik, V. Nano-structured and functionalized surfaces for cytocompatibility improvement and bactericidal action. *Biotechnol. Adv.* **2015**, *33*, 1120–1129. [CrossRef]
11. Wu, X.; Yin, T.; Tian, J.; Tang, C.; Huang, J.; Zhao, Y.; Zhang, X.; Deng, X.; Fan, Y.; Yu, D.; et al. Distinctive effects of CD34- and CD133-specific antibody-coated stents on re-endothelialization and in-stent restenosis at the early phase of vascular injury. *Regen. Biomater.* **2015**, *2*, 87–96. [CrossRef] [PubMed]
12. Yin, R.-X.; Yang, D.-Z.; Wu, J.-Z. Nanoparticle drug- and gene-eluting stents for the prevention and treatment of coronary restenosis. *Theranostics* **2014**, *4*, 175–200. [CrossRef] [PubMed]
13. Torrado, J.; Buckley, L.; Durán, A.; Trujillo, P.; Toldo, S.; Valle Raleigh, J.; Abbate, A.; Biondi-Zoccai, G.; Guzmán, L.A. Restenosis, Stent Thrombosis, and Bleeding Complications: Navigating Between Scylla and Charybdis. *J. Am. Coll. Cardiol.* **2018**, *71*, 1676–1695. [CrossRef] [PubMed]

14. Holy, E.W.; Jakob, P.; Eickner, T.; Camici, G.G.; Beer, J.H.; Akhmedov, A.; Sternberg, K.; Schmitz, K.-P.; Lüscher, T.F.; Tanner, F.C. PI3K/p110α inhibition selectively interferes with arterial thrombosis and neointima formation, but not re-endothelialization: Potential implications for drug-eluting stent design. *Eur. Heart J.* **2014**, *35*, 808–820. [CrossRef] [PubMed]
15. Pant, S.; Limbert, G.; Curzen, N.P.; Bressloff, N.W. Multiobjective design optimisation of coronary stents. *Biomaterials* **2011**, *32*, 7755–7773. [CrossRef] [PubMed]
16. MacTaggart, J.; Poulson, W.; Seas, A.; Deegan, P.; Lomneth, C.; Desyatova, A.; Maleckis, K.; Kamenskiy, A. Stent Design Affects Femoropopliteal Artery Deformation. *Ann. Surg.* **2019**, *270*, 180–187. [CrossRef]
17. Stoeckel, D.; Bonsignore, C.; Duda, S. A survey of stent designs. *Minim. Invasive Ther. Allied Technol.* **2002**, *11*, 137–147. [CrossRef]
18. Stergaard, M.B.; Hansen, S.R.; Januchta, K.; To, T.; Smedskjaer, M.M. Revisiting the Dependence of Poisson's Ratio on Liquid Fragility and Atomic Packing Density in Oxide Glasses. *Materials* **2019**, *12*, 2439. [CrossRef]
19. Scarpa, F.; Smith, C.W.; Ruzzene, M.; Wadee, M.K. Mechanical properties of auxetic tubular truss-like structures. *Phys. Status Solidi (B)* **2008**, *245*, 584–590. [CrossRef]
20. Wu, W.; Song, X.; Liang, J.; Xia, R.; Qian, G.; Fang, D. Mechanical properties of anti-tetrachiral auxetic stents. *Compos. Struct.* **2018**, *185*, 381–392. [CrossRef]
21. Zhang, W.; Zhang, B.; Xiao, H.; Yang, H.; Wang, Y.; Zhu, H. A Layer-Dependent Analytical Model for Printability Assessment of Additive Manufacturing Copper/Steel Multi-Material Components by Directed Energy Deposition. *Micromachines* **2021**, *12*, 1394. [CrossRef] [PubMed]
22. Khalaj, R.; Tabriz, A.G.; Okereke, M.I.; Douroumis, D. 3D printing advances in the development of stents. *Int. J. Pharm.* **2021**, *609*, 121153. [CrossRef] [PubMed]
23. Hua, W.; Mitchell, K.; Raymond, L.; Godina, B.; Zhao, D.; Zhou, W.; Jin, Y. Fluid Bath-Assisted 3D Printing for Biomedical Applications: From Pre-to Postprinting Stages. *ACS Biomater. Sci. Eng.* **2021**, *7*, 4736–4756. [CrossRef] [PubMed]
24. Chua, K.; Khan, I.; Malhotra, R.; Zhu, D. Additive manufacturing and 3D printing of metallic biomaterials. *Eng. Regen.* **2021**, *2*, 288–299. [CrossRef]
25. Yang, J.; Gu, D.; Lin, K.; Wu, L.; Zhang, H.; Guo, M.; Yuan, L. Laser additive manufacturing of cellular structure with enhanced compressive performance inspired by Al–Si crystalline microstructure. *CIRP J. Manuf. Sci. Technol.* **2021**, *32*, 26–36. [CrossRef]
26. Ma, C.; Gu, D.; Lin, K.; Dai, D.; Xia, M.; Yang, J.; Wang, H. Selective laser melting additive manufacturing of cancer pagurus's claw inspired bionic structures with high strength and toughness. *Appl. Surf. Sci.* **2019**, *469*, 647–656. [CrossRef]
27. Chen, H.; Gu, D.; Xiong, J.; Xia, M. Improving additive manufacturing processability of hard-to-process overhanging structure by selective laser melting. *J. Mater. Process. Technol.* **2017**, *250*, 99–108. [CrossRef]
28. Chen, H.; Gu, D.; Deng, L.; Lu, T.; Kühn, U.; Kosiba, K. Laser additive manufactured high-performance Fe-based composites with unique strengthening structure. *J. Mater. Sci. Technol.* **2021**, *89*, 242–252. [CrossRef]
29. Xiong, J.; Gu, D.; Chen, H.; Dai, D.; Shi, Q. Structural optimization of re-entrant negative Poisson's ratio structure fabricated by selective laser melting. *Mater. Des.* **2017**, *120*, 307–316. [CrossRef]
30. Plaza, A.; Merino, B.; Del Olmo, N.; Ruiz-Gayo, M. The cholecystokinin receptor agonist, CCK-8, induces adiponectin production in rat white adipose tissue. *Br. J. Pharmacol.* **2019**, *176*, 2678–2690. [CrossRef]
31. Li, C.R.; Li, L.; Cao, H.J.; Qin, L.; Long, J.; Lai, Y.X.; Wang, X.L.; Li, Y. In vitro biosafety assessment of PLGA/TCP/Mg porous scaffold for bone regeneration. *Session Biomater. Implants.* **2015**, *78*, 81.
32. Bäcker, H.C.; Wu, C.H.; Krüger, D.; Gwinner, C.; Perka, C.; Hardt, S. Metal on Metal Bearing in Total Hip Arthroplasty and Its Impact on Synovial Cell Count. *J. Clin. Med.* **2020**, *9*, 3349. [CrossRef] [PubMed]
33. United States Pharmacopeial Convention. *Committee of Revision. USP XXII, or, NF XVII, or The United States Pharmacopeia, or, The National Formulary*; US Pharmacopeial Convention: Rockville, MD, USA, 1990.
34. Shu, C.; He, H.; Fan, B.-W.; Li, J.-H.; Wang, T.; Li, D.-Y.; Li, Y.-M.; He, H. Canine implant biocompatibility of vascular stents manufactured using metal injection molding. *Trans. Nonferrous Met. Soc. China* **2022**, *32*, 569–580. [CrossRef]
35. Ablikim, Z.; Ablikim, G.; Zhu, Q.F.; Chen, T.; Gang, W.U.; Hua, Y.Y.; Wang, S.H.; Hospital, C.; Pharmacy, S.O. Preparation of five kinds of acellular fish skin matrices and in vitro toxicity analysis. *Acad. J. Second Mil. Med. Univ.* **2019**, *12*, 162–168.
36. Fu, X.; Liu, P.; Zhao, D.; Yuan, B.; Zhang, X. Effects of Nanotopography Regulation and Silicon Doping on Angiogenic and Osteogenic Activities of Hydroxyapatite Coating on Titanium Implant. *Int. J. Nanomed.* **2020**, *15*, 4171–4189. [CrossRef] [PubMed]

Article

Study on Interfacial Bonding Properties of NiTi/CuSn10 Dissimilar Materials by Selective Laser Melting

Changhui Song *, Zehua Hu, Yunmian Xiao, Yang Li and Yongqiang Yang

School of Mechanical and Automotive Engineering, South China University of Technology, Guangzhou 510640, China; 201920100681@mail.scut.edu.cn (Z.H.); 201910100307@mail.scut.edu.cn (Y.X.); meyangli@scut.edu.cn (Y.L.); meyqyang@scut.edu.cn (Y.Y.)
* Correspondence: chsong@scut.edu.cn

Abstract: The dissimilar materials bonding of NiTi alloy with shape memory effect (SME) and CuSn10 alloy with good ductility, electrical conductivity, and thermal conductivity can be used in aerospace, circuits, etc. In order to integrate NiTi and CuSn10 with greatly different physical and chemical properties by selective laser melting (SLM), the effects of forming interlayers with different SLM process parameters were explored in this study. The defects, microstructure, and component diffusion at the interface were also analyzed. Columnar grain was found along the molten pool boundary of the interfacial region, and grains in the interfacial region were refined. Elements in the interfacial region had a good diffusion. Phase identifying of the interface showed that Ni_4Ti_3 was generated. The analysis showed that the columnar grain, refined grains in the interfacial region, and a certain amount of Ni_4Ti_3 could strengthen the interfacial bonding. This study provides a theoretical basis for forming NiTi/CuSn10 dissimilar materials structural members.

Keywords: NiTi alloy; CuSn10 alloy; selective laser melting; dissimilar materials; interfacial bonding properties

1. Introduction

SLM is the most common metal additive manufacturing technology. In this technology, a high energy density laser is utilized as a heat source. The laser spot is concentrated in a very small range to selectively melt the spherical metal powder on the powder bed, by which parts with quite complex shapes can be obtained [1]. It is widely used in aerospace and medical fields because of high forming accuracy [2–4]. However, SLM is mainly used to manufacture single metal parts.

SLM of single material has been unable to meet the performance requirements of people with the continuous development of its commercial application. Therefore, the research of SLM of two or even multiple materials has appeared in recent years. Chen et al. [5] studied the forming of CuSn10/316 L dissimilar materials and found that the generation of interfacial defects was relevant to process parameters of the interface. The defects near the interface were the main factors affecting the bonding strength. Finally, dissimilar materials' tensile parts with tensile strength close to that of CuSn10 single alloy tensile part were formed successfully. Liu et al. [6] formed 18Ni300 maraging steel on traditional machined Cr8Mo2SiV cold work die steel by SLM. They found that a toothed mosaic structure could be formed at the interfacial area, which could promote interfacial bonding. Tan et al. [7] studied the manufacturing of maraging steel/copper dissimilar materials by the combination of selective laser melting and subtractive process. They found that the improvement of interfacial bonding performance was related to the intense Marangoni flows at the interface. This kind of manufacturing can combine the advantages of multiple metal materials, and therefore is very promising.

Shape memory alloy (SMA) refers to the alloy with SME due to the reversibility of phase transition [8], among which NiTi alloy is the most widely used. It has been applied

in sensors, actuators, medical devices, and many other fields [9,10]. The integrated forming of NiTi alloy and other materials can combine the functionality of NiTi with the excellent properties of other materials, thus having broad application prospects. Some studies on the combination of NiTi alloy with other alloys have been conducted in recent years. Li et al. [11] studied the laser welding process of NiTi SMA wire and stainless-steel wire using nickel as the interlayer. It found that appropriate Ni interlayer thickness could reduce the brittle intermetallic compounds and pores, thus improving the bonding properties. Gao et al. [12] proposed a laser welding method using Nb/Cu multi-interlayer. Through a single pass welding, a hybrid joint composed of three metallurgical bonding zones was obtained according to different welding mechanisms. This completely prevented the formation of Ti-Fe and Ti_xNi_y intermetallic compounds, and thus significantly improved the mechanical properties of the joint. Panton et al. [13] studied the laser welding of Nitinol wire and MP35N wire and found that the position of the laser beam controlled the composition of the molten pool, while the intensity of the laser beam affected the size and mixing of the molten pool. These variables greatly influenced the hardness and cracking sensitivity of the interfacial region, and thus seriously affected the bonding strength. At present, one of the research hotspots of NiTi alloy-related dissimilar materials is the combination of Ti6Al4V and NiTi alloy. Because it can be used in biomedical and aerospace fields, thus it is quite promising. The main problem of NiTi bonding with Ti6Al4V is that brittle intermetallic compounds can be easily formed during the process, which affects the bonding properties. Oliveira et al. [14] used niobium as an interlayer to prevent the generation of brittle phase during laser welding of NiTi and Ti6Al4V. The melting temperature of the niobium interlayer is much higher than that of the matrix material. As a result the bulk niobium did not melt during the joining process, which played a role in the diffusion barrier between these two materials. It ensured that no cracks in the welds and brittle intermetallic compounds were discovered. Zoeram et al. [15] utilized thin copper as an interlayer to conduct laser welding of NiTi and Ti6Al4V. It found that selecting an appropriate interlayer thickness could inhibit the formation of Ti-Cu intermetallic compounds and shrinkage cavities, thus improving the interfacial bonding properties. Xie et al. [16] studied the ultrasonic spot welding of Ti6Al4V/NiTi dissimilar joints with Al coating as an interlayer and found that the firm bonding of Ti6Al4V/NiTi dissimilar materials can be realized by selecting appropriate welding pressure. Bartolomeu et al. [17] fabricated NiTi-Ti6Al4V multi-material cellular structures with a Ti6Al4V inner region and a NiTi outer region by SLM. The bonding between NiTi and Ti6Al4V in the transition region was declared to be successful by morphology analysis. There were also some studies on other Ni-based alloys. Tabaie et al. [18–20] studied joining SLM Inconel 718 with forged AD730 TM Nickel-based superalloy using linear friction welding (LFW). Joints free of micro-porosity, micro-cracks, and oxides were obtained. It was found that the strength of AD730 TM alloy depended significantly on the grain size while that of SLM Inconel 718 was dominated by shape (or size) and the presence of secondary phases (γ'/γ'' and Laves). However, both the size and the volume fraction of all phases affected the hardness and mechanical properties of the joint. After welding, a new post-weld heat treatment cycle was designed, resulting in uniform hardness across the dissimilar joint. It can be found from the above research that welding is a quite commonly used technology for the combination of NiTi alloy and other Ni-based alloys related dissimilar materials, in which an interlayer is often applied. SLM technology is seldom used for the combination. This is partly due to the forming of the NiTi alloy by SLM is still immature [21]. Conversely, the studies about the combination of two kinds of materials using SLM are still in their nascence; there are still many mechanisms that need to be researched.

CuSn10 is a type of tin bronze with good ductility, electrical conductivity, and thermal conductivity. NiTi/CuSn10 dissimilar materials parts can combine the excellent shape memory performance of NiTi alloy with the advantages of CuSn10, which has a wide application prospect. The technical experts of NASA are developing a sort of smart radiator that controls heat-loss rate flexibly by deforming. The thermosensitive materials can be utilized to realize the change of shape. When the thermosensitive material experiences the temperature change, the radiator can automatically change its shape to dissipate heat or store heat. This expectation can be achieved by using NiTi alloy to achieve the shape change and using CuSn10 alloy to dissipate heat. As for electrical conductivity, an intelligent fuse can be realized based on the mobile deformation of NiTi alloy and the high electrical conductivity of CuSn10 alloy, which can control the circuit on and off in real-time, according to the temperature to achieve protection. There are also many other possibilities. However, there are few studies on the bonding between NiTi alloy and CuSn10 alloy. This is because the interface is a part of the dissimilar materials parts at which defects are most likely to occur and a significant factor determining the reliability of the whole part. The bonding properties of the interface are mainly affected by the differences between the physical and chemical properties of the two materials, which were found to be large [6,22,23]. In this study, the effects of adding CuSn10 interlayers with different process parameters between NiTi alloy matrix and CuSn10 alloy by SLM on the defects, microstructure, phases, and properties at the interface were investigated, providing a theoretical basis for forming NiTi/CuSn10 dissimilar materials structural parts. The results showed that appropriate Ni_4Ti_3 content and columnar grain at the interface can be realized by selecting appropriate parameters, the combination of which can improve the interfacial bonding properties. Refined grains in the interfacial region can also play a strengthening role.

2. Materials and Methods

2.1. Experimental Equipment and Materials

The SLM equipment used in this study is the Dimetal-100H device jointly developed by the additive manufacturing team of South China University of Technology and Laseradd Technology (Guangzhou, China) Co Ltd. The main equipment parameters were shown in Table 1. The materials used in this study were equiatomic NiTi50 powders and CuSn10 powders. Equiatomic NiTi50 powders were manufactured by Electrode Induction Melting Gas Atomization (EIGA) technology (Beijing Avimetal Powder Metallurgy Technology Co., Ltd., Beijing, China) and the particle size ranges from 15 to 53 µm (degree of sphericity ≈ 90%). CuSn10 powders were also manufactured by EIGA technology (Dongguan Hyper Tech Co Ltd., Dongguan, China) and the particle size ranges from 13 to 53 µm (degree of sphericity > 90%). Inductive Coupled Plasma Emission Spectrometer (ICP) was used to measure the chemical compositions which were listed in Tables 2 and 3. The scanning electron microscope (SEM) images (1000×) and particle size distributions of the powders were shown in Figure 1.

Table 1. Main equipment parameters of SLM Dimetal-100H device.

Item	Parameter
Building size/mm	$100 \times 100 \times 100$
Laser	500 W infrared ytterbium doped fiber laser
Wavelength/nm	1075
Beam Quality Factor	$M^2 \leq 1.1$
Focus beam size/µm	40–60
Layer thickness/µm	20–100
Processing speed/(cm^3/h)	4–20

Table 2. Chemical composition of equiatomic NiTi50 powder.

Element	Ni	Fe	C	O	N	Ti
Proportion (wt.%)	54.5	0.0074	0.006	0.0607	0.0018	Bal.

Table 3. Chemical composition of CuSn10 powder.

Element	Cu	Sn	Others
Proportion (wt.%)	Bal.	10.11	<0.06

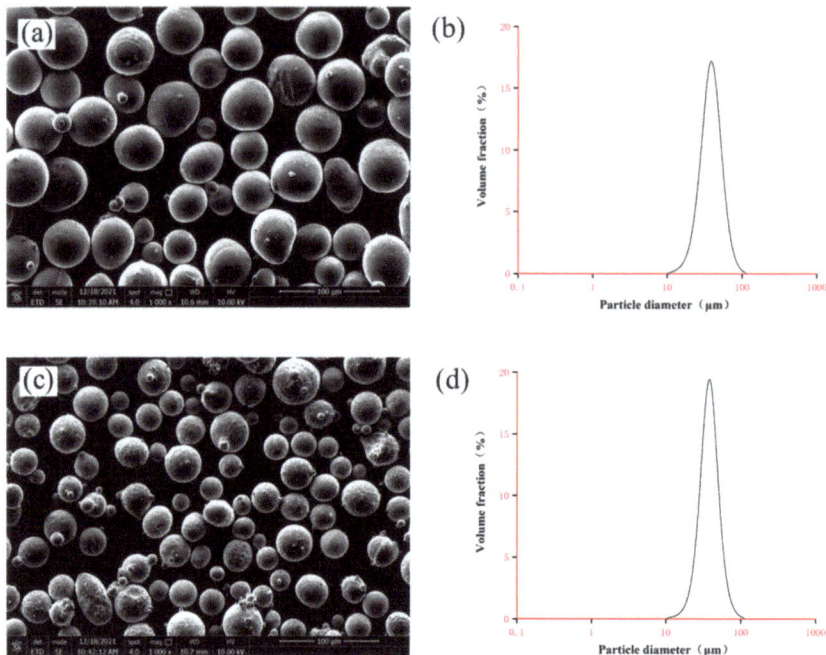

Figure 1. (**a**) SEM image of equiatomic NiTi50 powder (1000×); (**b**) particle size distribution for equiatomic NiTi50 powder; (**c**) SEM image of CuSn10 powder (1000×); (**d**) particle size distribution for CuSn10 powder.

2.2. Experimental Process Design and Related Parameters

The optimum SLM process parameters of individual materials have been studied in previous experiments, and the optimized results were shown in Table 4.

Table 4. Optimum SLM process parameters of NiTi alloy and CuSn10 alloy.

Item	NiTi	CuSn10
Laser power (W)	140	160
Scanning speed (mm/s)	500	300
Scanning space (mm)	0.08	0.08
Layer thickness (mm)	0.03	0.03
Relative density (%)	98.83	99.95

At present, connection strategies of dissimilar materials forming include direct joining, gradient path method, and intermediate section method. Direct joining refers to the method of directly forming another material on one material. The gradient path method refers to that although the transition part is added between the two materials, a third material is not introduced in the transition part. The connection is improved by changing process parameters and other factors. The intermediate section method is to add a third material between the two to improve the connection. Direct joining is usually used where the difference between the physical and chemical properties of the two materials is relatively low while the intermediate section method is opposite. The gradient path method is usually used when the situation is in between. Considering that the physical and chemical properties of NiTi and CuSn10 are quite different, direct joining is useless. The intermediate section method requires complex process exploration, so the gradient path method was selected here for forming. That is, the gradient path method was used to combine the two materials into a part. The strategy of forming CuSn10 alloy after NiTi alloy was adopted in this study. This is mainly because when CuSn10 is formed first, the heat of the laser beam will be rapidly dissipated by CuSn10 with good thermal conductivity when NiTi is formed so that enough heat cannot be gathered to form NiTi with high quality. At the same time, due to that NiTi is very sensitive to heat, the small part of NiTi formed first is easy to become separate from CuSn10 matrix due to thermal stress. The substrate was composed of NiTi alloy so the first layer could be easily connected to the substrate. Direct forming was carried out after remelting once in this experiment. An orthogonal scanning strategy was adopted to form all samples. That is, the laser scanning directions of adjacent layers were perpendicular to each other. The reason why this scanning method was adopted was that different heat flow directions would refine the grain so that the strength could increase [24]. In this study, small cuboids of dissimilar materials were formed, and the analysis of microstructure, element diffusion, microhardness, and phases was conducted on them. The process parameters of the first formed 20 layers of CuSn10 were taken as variables, the influence of which on the interfacial properties was studied. Considering that there was no relevant experiment on forming NiTi/CuSn10 dissimilar materials by SLM, the number of variables was set to only two. Laser power and scanning speed with small fluctuation were adopted to obtain different energy density, trying to preliminarily conclude how the forming of NiTi/CuSn10 dissimilar materials is affected by energy density. The calculation formula of energy density was shown in (1), in which E is energy density, P is laser power, v is scanning speed, h is scanning space, and t is layer thickness. The selected parameters were shown in Table 5. The scanning space is $h = 0.08$ mm and the layer thickness is $t = 0.03$ mm. Except for these layers, other parts were formed with the parameters in Table 4.

$$E = \frac{P}{vht} \tag{1}$$

Table 5. Process parameters of interlayers.

Experiment No.	Laser Power/(W)	Scanning Speed/(mm/s)	Energy Density/(J/mm^3)
1	160	350	190.48
2	120	250	200.00
3	160	300	222.22
4	140	250	233.33
5	180	300	250.00
6	160	250	266.67

Small cuboids were 10 mm × 4 mm × 10 mm in size. A total of 5-mm high NiTi part was formed first, then the NiTi powders were cleared quickly, and CuSn10 powders were put in the powder chamber. Next 20 layers of CuSn10 (0.6 mm) was formed on the formed NiTi playing the role of transient inter-alloys. Finally, a 4.4-mm high CuSn10 was formed on the formed CuSn10. Tensile samples for the bonding strength test were formed in the same way as these small cuboids. For comparison, NiTi and CuSn10 single alloy tensile samples of the same size were formed with the parameters shown in Table 4. Figure 2 shows the diagram of forming the small cuboids and the tensile samples. Each parameter was used to fabricate three tensile samples to avoid occasionality. Figure 3 shows the sizes of the small cuboids and the tensile samples.

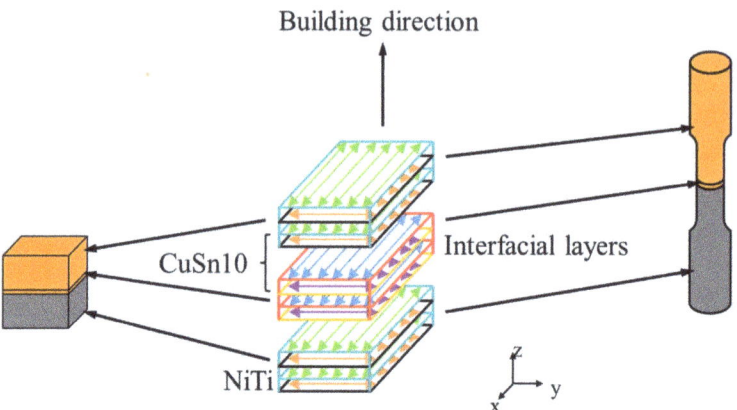

Figure 2. Diagram of forming the small cuboids and the tensile samples.

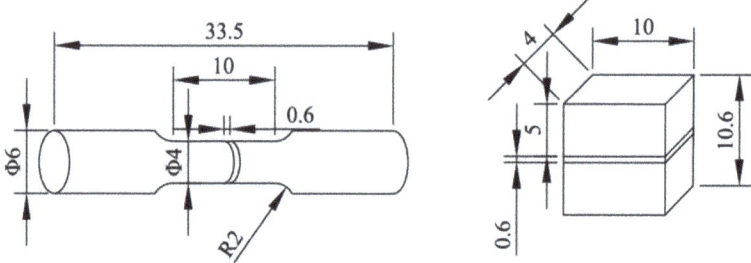

Figure 3. Sizes of the small cuboids and the tensile samples.

2.3. Microscopic Feature

The cuboids were removed from the substrate using Wire Electrical Discharge Machining. The sides of the cuboids were sanded and polished. After polishing, the metallurgical defects at the interface of the dissimilar materials were observed by optical microscope (OM) (DMI3000M, Leica, Weztlar, Germany). After observation, samples were etched for 2 s with a mixed solution (6 g $FeCl_3$ + 10 mL HCl + 150 mL H_2O) and 10 s with another mixed solution (1 mL HF + 2 mL HNO_3 + 10 mL H_2O) to further observe the microscopic features. The microstructure at the interface was observed with SEM (QUANTA250, FEI, Hillsboro, OR, USA). Energy Dispersive Spectroscopy (EDS) (Noran System 7, Thermo Fisher Scientific, Waltham, MA, USA) was used for point scanning along the direction perpendicular to the interface to analyze the element distribution at the interfacial region. X-ray diffraction (XRD) tests were conducted in a step-by-step manner to analyse phases at the interface: step scan 2θ = 20–90°, scanning speed = 0.6°/min, step width = 0.01°. The results were analyzed with jade 6.0.

2.4. Mechanical Properties

Vickers hardness tester (Wilson VH1202, Buehler, Lake Bluff, IL, USA) was used to measure the microhardness from CuSn10 alloy to NiTi alloy at the interfaces of the cuboids at 50 μm intervals. Tensile tests were performed on a universal testing machine (CMT5105, SUST) at a speed of 0.1 mm per min to measure the interfacial bonding strengths and the tensile strength of CuSn10 and NiTi single alloy.

3. Results and Discussion

3.1. Microstructure of NiTi/CuSn10

The microstructure of these cuboids observed with the OM was shown in Figure 4. When E = 190.48 J/mm^3, due to the fast-scanning speed, the cooling speed was also fast. So, a large temperature gradient was formed along the scanning route, thus a large thermal stress was formed. Cracks caused by thermal stress did not originate in CuSn10 alloy due to its good ductility. On the contrary, vertical cracks originated in the process of thermal stress release in NiTi alloy the ductility of which is relatively worse and extended to CuSn10 alloy, as shown in Figure 4a [25,26]. When E = 200.00 J/mm^3, the scanning speed and energy density were moderate, so the interface quality was good and no obvious defects were observed, as shown in Figure 4b. When E = 222.22 J/mm^3, vertical microcracks also occurred due to high scanning speed, as shown in Figure 4c. When E = 233.33 J/mm^3 and 250.00 J/mm^3, it can be seen that with the increase of energy density, movement of the molten pool became more intense and Marangoni convection appeared. The two materials penetrated each other, and the generation of an island can be observed, as shown in Figure 4d,e [7]. When E = 266.67 J/mm^3, movement of the molten pool was further intensified and the gas in the building chamber was involved in the part, forming large pores, as shown in Figure 4f. Two types of defects, pores, and vertical cracks appeared in the process of gradually increasing energy density. While the cracks in the direction perpendicular to the interface generally have little effect on the bonding strength of the joint, samples No.1–No.5 were selected for the follow-up tests [5].

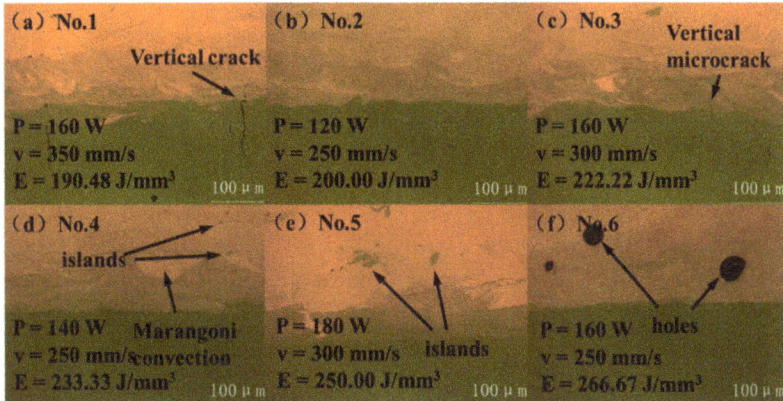

Figure 4. Optical micrographs of interfaces: Sample (**a**) No.1; (**b**) No.2; (**c**) No.3; (**d**) No.4; (**e**) No.5; (**f**) No.6.

The microstructure of samples No.1–No.5 was observed with SEM after etching. Sample No. 1 was selected as a typical sample for microstructure analysis. In Figure 5a, the NiTi/CuSn10 interface was divided into four zones: CuSn10 region, upper area of interfacial region, bottom area of the interfacial region, and NiTi region, as shown in Figure 5b–e. CuSn10 region was composed of equiaxed grains. Upper area of the interfacial region was also composed of equiaxed grains, and both NiTi and CuSn10 could be seen in this zone. Bottom area of the interfacial region was composed of columnar grains, which formed along the molten pool boundary. Columnar grains formed due to the orientation of the thermal gradient. They were formed from CuSn10 to NiTi (along the direction of heat dissipation), playing the role of a stiffener. That is, the columnar grains were embedded into NiTi alloy, making the bonding firmer. So, it can be concluded that the interfacial bonding properties may be improved. NiTi region was also composed of columnar grains. The average grain size and standard deviation for each microstructure zone were listed in Figure 5. The width of columnar grain was regarded as its grain size. It can be seen that the average grain sizes of upper area of the interfacial region and bottom area of the interfacial region were smaller than those of CuSn10 region and NiTi region. A possible explanation is that the mixing of multiple elements in the interfacial region may promote a large increase in the number of crystal nuclei and significantly refine the grains [5]. This may also improve the interfacial bonding properties.

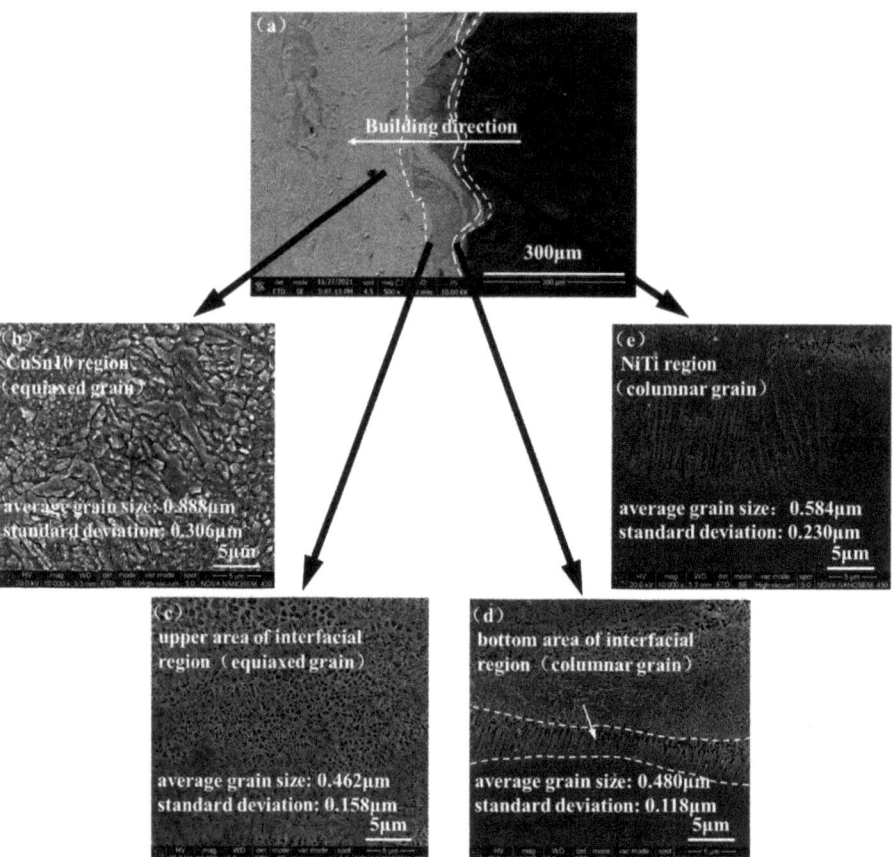

Figure 5. The microstructure of (**a**) NiTi/CuSn10 interface (500×); (**b**) CuSn10 region (10,000×); (**c**) upper area of interfacial region (10,000×); (**d**) bottom area of interfacial region (10,000×); (**e**) NiTi region (10,000×).

3.2. The Element Diffusion at the NiTi/CuSn10 Interface

The samples observed with SEM were shown in Figure 6 (500×). The widths of interfacial regions of five samples were measured and the results were indicated in Figure 6. The maximum was 255.32 ± 27.32 µm while the minimum was 120.13 ± 18.64 µm.

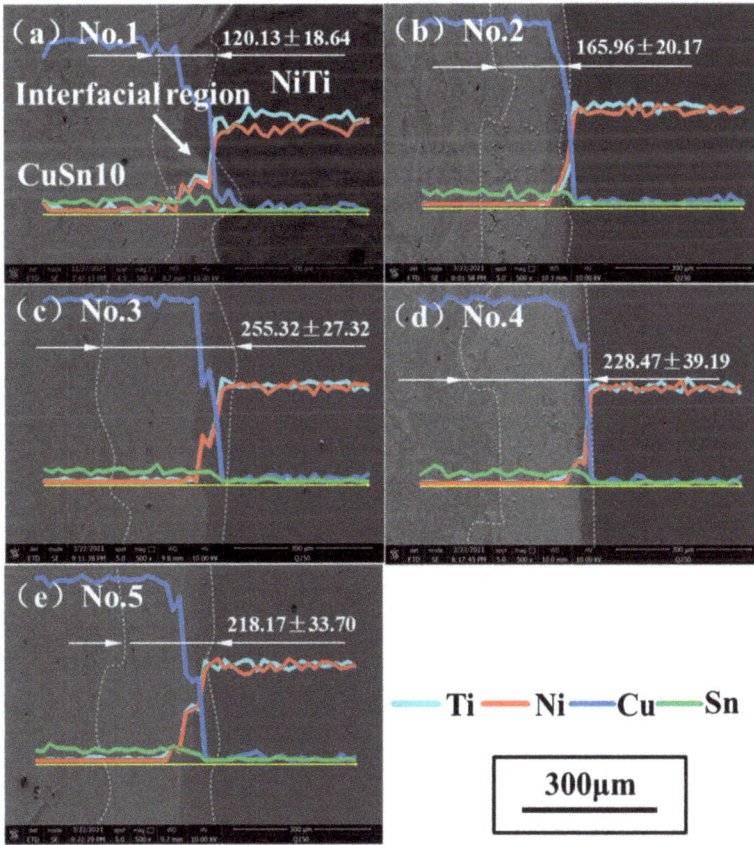

Figure 6. Interfacial SEM images of sample (**a**) No.1; (**b**) No.2; (**c**) No.3; (**d**) No.4; (**e**) No.5.

It can be seen that the width of the interfacial region tended to increase with the increase of energy density. This was mainly because as the energy density increased, the molten pool movement became more violent and thus promoting the mixing of these two alloys. However, this promoting effect had a certain range. When the energy density was higher than a certain value, it can be seen that interfacial region thickness did not increase and tended to be stable. When it comes to a single parameter, it can be seen when comparing samples No.1 and No.3 that when the laser power was constant, the higher the scanning speed, the lower interfacial region thickness was. This was because the diffusion time was shorter when the scanning speed was higher [5]. It can also be seen when comparing samples No.2 and No.4 that when the scanning speed was constant, the larger the laser power and the higher the interfacial region thickness was. This was because the larger the laser power was, the larger the energy input was; thus the molten pool was deeper. Furthermore, the element distribution at the interface was analyzed. The main elements Ni, Ti, Cu, and Sn of the two alloys were taken as the analysis objects. The chemical composition was measured at a point every 2.88 μm along the building direction and 50 points were selected for each sample. The distribution results of interface elements were obtained as shown in Figure 6. Due to the small content of Sn, the change trend was not obvious, but it can still be seen that the content decreased when crossing the interface. The content of Ni and Ti decreased continuously while the content of Cu increased continuously when crossing the interface, indicating that the element mixing at the interfacial region was relatively sufficient.

3.3. The Microhardness at the NiTi/CuSn10 Interface

The microhardness of each sample from CuSn10 alloy to NiTi alloy was measured at 50 μm intervals, and the results were shown in Figure 7. It can be seen that the microhardness was almost constant at first. Then it suddenly became larger at the interface, next it suddenly became smaller, and finally it became constant again. According to the change trend of microhardness, the area near the interface can be divided into CuSn10 zone, interfacial region, and NiTi zone. It is worth noting that the microhardness of the interfacial region was much higher than that of both sides. Sing et al. [27] found intermetallic compounds at the interface, which can make microhardness become higher. Zhou et al. [22] found that less brittle and hard Cu and Ni rich intermetallics (i.e., Cu_3Ti, Ni_3Ti) and more brittle Ti-rich intermetallics (i.e., Ti_2Cu, Ti_2Ni) could be formed in the process of dissimilar laser welding of NiTi and copper. This indicated that intermetallic compounds were possible to form between Cu alloy and NiTi alloy. Therefore, it can be considered that some kind of intermetallic compound was formed at the interface. In order to test the existence of interfacial intermetallic compound, XRD analysis was conducted on the interfaces of the five samples, and the experimental results were shown in Figure 8.

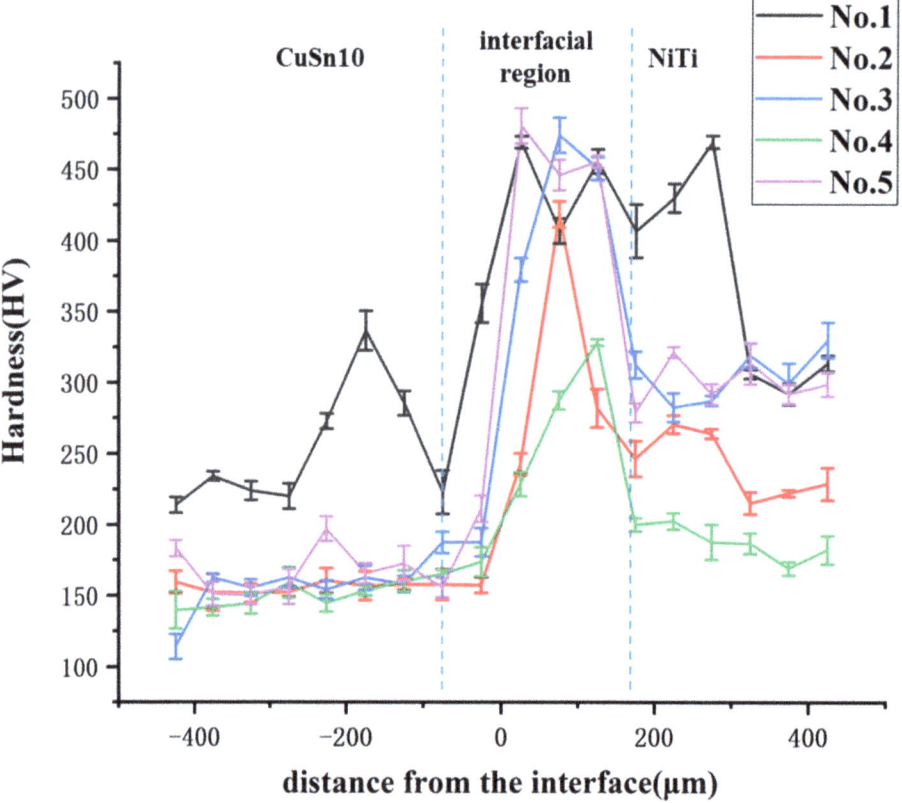

Figure 7. Microhardness change at the interface.

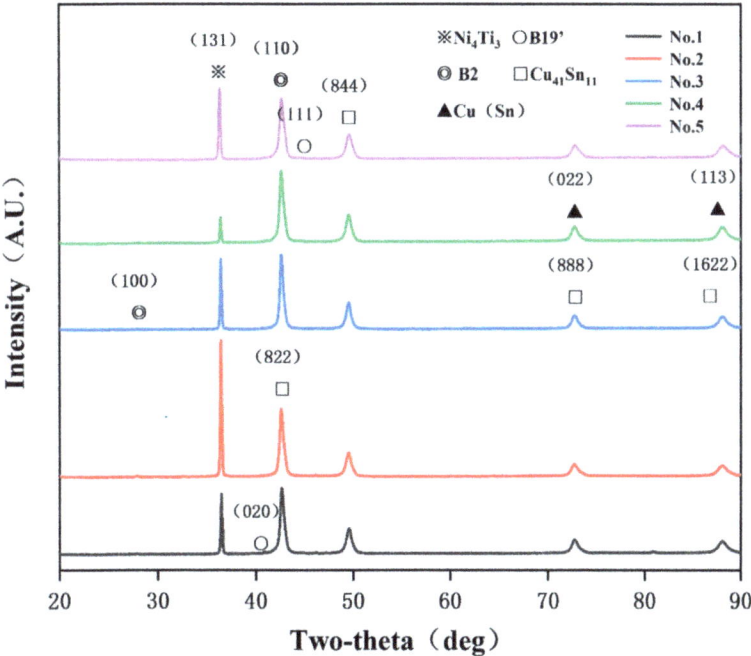

Figure 8. XRD pattern of the NiTi/CuSn10 interfaces.

First, the related phases of NiTi alloy and CuSn10 alloy were determined by phase identifying. Cu(Sn) and Cu41Sn11 were phases that often appeared in SLM CuSn10 alloy while B2 and B19′ were two phases of NiTi alloy. After this, it was found that a new phase appeared, which was found to be Ni_4Ti_3 after searching and comparing. This was because the scanning speed and laser power used at the interface were not high so the heating and cooling rate of the molten pool was low, which was conducive to the formation of Ni_4Ti_3 [28]. Ni_4Ti_3 is a very fine phase, which can produce dispersion strengthening and increase the strength of the sample interface [29]. Therefore, the formation of Ni_4Ti_3 intermetallic compound played a key role in the increase of microhardness at the interface. At the same time, it can be inferred that the interfacial bonding performance had been improved.

3.4. Bonding Strength Test of NiTi/CuSn10 Interface

The interfacial bonding strength between NiTi and CuSn10 has a great influence on the properties of the whole part. In order to study the bonding strength of NiTi/CuSn10, tensile tests of dissimilar materials samples were conducted, and the results were shown in Figure 9a.

The strengths of NiTi and CuSn10 single alloy formed vertically by SLM had been optimized and were used as the control groups with values of 508.6 ± 42.2 MPa and 422.9 ± 10.8 MPa, respectively. All the tensile parts broke at the interface of the two materials. It can be seen that sample No.1 had the worst interfacial bonding strength (156.0 ± 7.0 MPa), equivalent to 37% of that of CuSn10 single alloy. Sample No.3 had the best interfacial bonding strength (309.7 ± 34.4 MPa), equivalent to 75% of that of CuSn10 single alloy. Based on previous results, although there were vertical microcracks at the interface of process parameters corresponding to sample No.3, the interfacial bonding properties of sample No.3 was still better than that of other samples with no observed defects. This indicated that the vertical microcracks did not have a significant impact on interfacial bonding properties. The relationship between interfacial bonding strength and interfacial region thickness was shown in Figure 9b. It can be obtained from Figure 9b that the interfacial bonding strength was positively correlated with interfacial region thickness. This was because the higher the interfacial region thickness was, the more solid metallurgical bonding could be formed at the interface so that the interfacial bonding strength would be higher. Columnar grains at the interface and grain refinement in the interfacial region played a strengthening role. Based on the phase analysis, the improvement of interfacial bonding strength was also related to the formation of fine Ni_4Ti_3 strengthening phase.

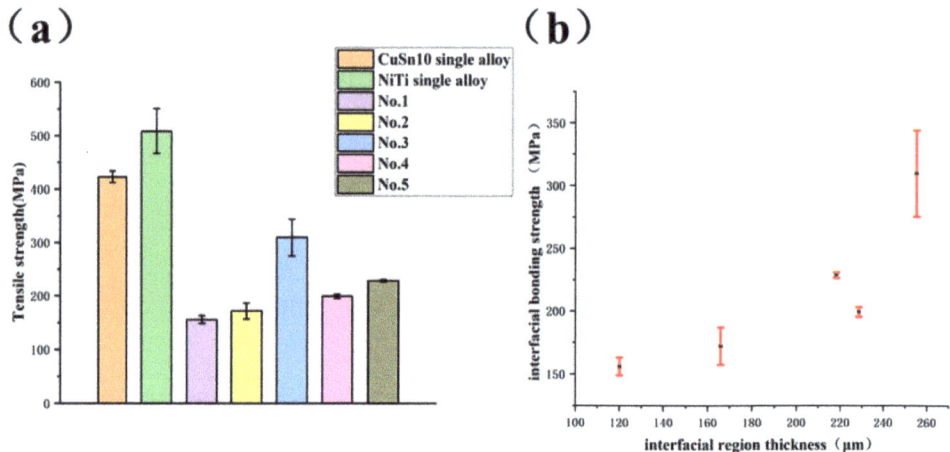

Figure 9. (a) The interfacial bonding strengths of different samples compared with the tensile strength of CuSn10 and NiTi single alloy (b) relationship between interfacial bonding strength and interfacial region thickness.

Low magnification and high magnification SEM images were taken for each fracture, as shown in Figure 10.

At low magnification (200×), it can be seen that the fracture was very flat, and some unmelted metal powder can be seen occasionally, which indicated a brittle fracture as shown in Figure 10a. The whole fracture was divided into many small planes. At high magnification (2000×), some holes due to the gas not escaping and gullies between planes due to the existence of vertical cracks in these samples can be seen. On account of brittle fracture, these samples cannot achieve strength comparable to that of a single metal material. However, it is worth noting that a few dimples can be found in the fracture of sample No.3 at high magnification (5000×); that is, the sample had certain ductility. This indicated that a sample could have good interfacial bonding properties under the strengthening of columnar grain at the interface, grain refinement in the interfacial region and Ni_4Ti_3 with appropriate content.

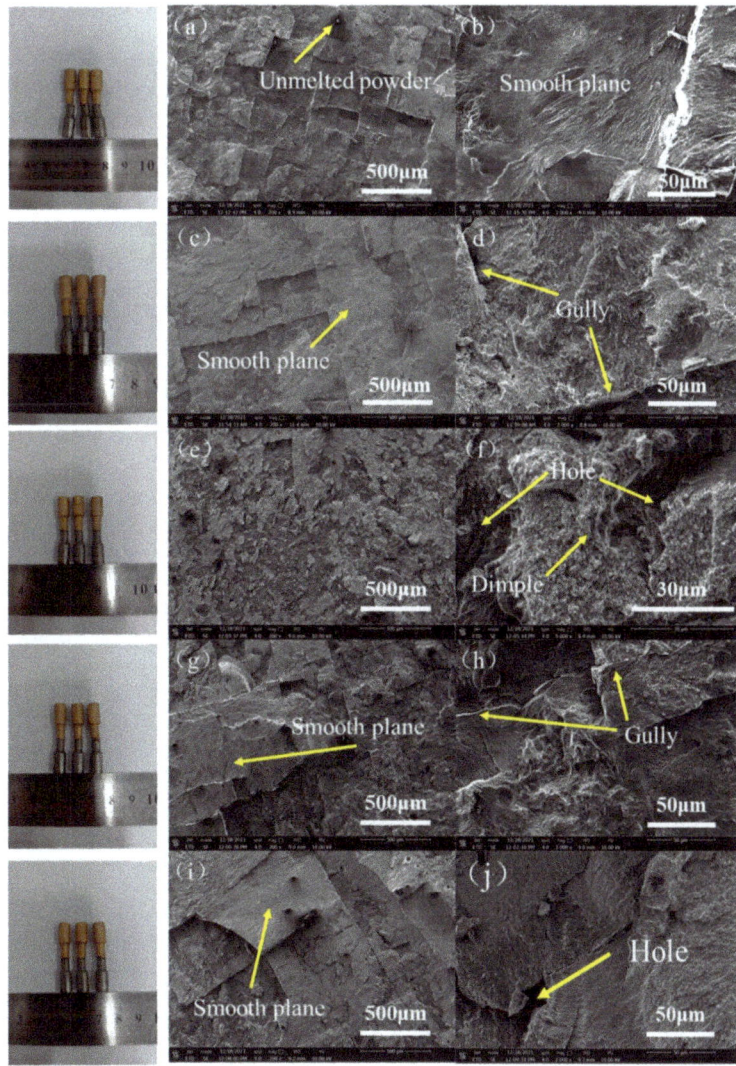

Figure 10. SEM images of fracture of sample (**a**) No.1 (200×); (**b**) No.1 (2000×); (**c**) No.2 (200×); (**d**) No.2 (2000×); (**e**) No.3 (200×); (**f**) No.3 (5000×); (**g**) No.4 (200×); (**h**) No.4 (2000×); (**i**) No.5 (200×); (**j**) No.5 (2000×).

4. Conclusions

The combination of different materials by SLM makes it possible to combine the excellent properties of different materials. In this paper, the selective laser melting technology was used to form CuSn10 alloy on the formed NiTi alloy. The microstructure, element diffusion, microhardness change, and phases at the interface were studied. Tensile parts were formed to test the interfacial bonding properties, and the conclusions were drawn as follows:

(1) Different interfacial structures can be obtained by adjusting the forming energy density input of the interface by adjusting process parameters. Columnar grains can form along the boundary of molten pool in the interfacial region and can effectively improve

the interfacial bonding. Grain refinement in the interfacial region can also play a strengthening role.
(2) In a certain range, the width of the interfacial region increases with the increase of energy density. Beyond a certain range, the width of the interfacial region remains basically unchanged. The wider the interfacial region is, the stronger metallurgical bonding can be formed at the interface.
(3) The formation of N_4Ti_3 strengthening phase improves the microhardness of the interface, making the microhardness of the interface significantly higher than that of both sides. The interfacial bonding strength is different under different process parameters. Under the optimized process parameters, the maximum strength can reach 309.7 ± 34.4 MPa.

In general, the interfacial bonding is still weak, but on the premise of the large gap between the physical and chemical properties of CuSn10 and NiTi alloy, it can be said that the interfacial bonding strength has reached a good level with a value of 309.7 ± 34.4 MPa. This direction still has good application prospects. In addition, how different process parameters affect the interfacial Ni_4Ti_3 strengthening phase, columnar grain, and grain refinement, thus affecting the interfacial bonding strength still needs further quantitative research.

Author Contributions: Conceptualization, C.S.; Investigation, C.S. and Y.X.; Methodology, Z.H.; Resources, Y.L. and Y.Y.; Validation, Z.H. and Y.X.; Writing—original draft, Z.H. and Y.X.; Writing—review & editing, C.S. and Y.L. All authors have read and agreed to the published version of the manuscript.

Funding: This research was financially supported by Key-Area Research and Development Program of Guangdong Province (2020B090923001), National Natural Science Foundation of China (U2001218), Guangdong Province Special Support Plan for High-Level Talents (2019TQ05Z110), Capital's Funds for Health Improvement and Research (CFH 2020-2-5131).

Acknowledgments: The authors gratefully acknowledge China-Ukraine Institute of Welding.

Conflicts of Interest: The authors declare no conflict of interest.

References

1. Yang, Y.; Chen, J.; Song, C.; Wang, D.; Bai, Y. Current Status and Progress on Technology of Selective Laser Melting of Metal Parts. *Laser Optoelectron. Prog.* **2018**, *55*, 011401. [CrossRef]
2. Yap, C.Y.; Chua, C.K.; Dong, Z.; Liu, Z.; Zhang, D.; Loh, L.E.; Sing, S.L. Review of selective laser melting: Materials and applications. *Appl. Phys. Rev.* **2015**, *2*, 041101. [CrossRef]
3. Pinkerton, A.J. Lasers in additive manufacturing. *Opt. Laser Technol.* **2016**, *78*, 25–32. [CrossRef]
4. Gradl, P.R.; Protz, C.; Greene, S.E.; Brandsmeier, W.; O'Neil, D. *Additive Manufacturing Overview: Propulsion Applications, Design for and Lessons Learned*; National Aeronautics and Space Administration: Washington, DC, USA, 2017.
5. Chen, J.; Yang, Y.; Song, C.; Wang, D.; Wu, S.; Zhang, M. Influence mechanism of process parameters on the interfacial characterization of selective laser melting 316L/CuSn10. *Mater. Sci. Eng. A* **2020**, *792*, 139316. [CrossRef]
6. Liu, L.; Song, C.; Yang, Y.; Weng, C. Study on Mechanism of Strengthening Interface Structure of Dissimilar Materials by Selective Laser Melting. *J. Mech. Eng.* **2020**, *56*, 189–196.
7. Tan, C.; Zhou, K.; Ma, W.; Min, L. Interfacial characteristic and mechanical performance of maraging steel-copper functional bimetal produced by selective laser melting based hybrid manufacture. *Mater. Des.* **2018**, *155*, 77–85. [CrossRef]
8. Chen, H.; Luo, B.; Zhu, Z.; Li, B. 4D Printing: Progress in Additive Manufacturing Technology of Smart Materials and Structure. *J. Xi'an Jiaotong Univ.* **2018**, *52*, 1–12.
9. Sun, L.; Huang, W.; Ding, Z.; Zhao, Y.; Wang, C.; Purnawali, H.; Tang, C. Stimulus-responsive shape memory materials: A review. *Mater. Des.* **2012**, *33*, 577–640. [CrossRef]
10. Saedi, S.; Saghaian, S.E.; Jahadakbar, A.; Moghaddam, N.S.; Andani, M.T.; Saghaian, S.M.; Lu, Y.C.; Elahinia, M.; Karaca, H.E. Shape memory response of porous NiTi shape memory alloys fabricated by selective laser melting. *J. Mater. Sci. Mater. Med.* **2018**, *29*, 40. [CrossRef]
11. Li, H.; Sun, D.; Cai, X.; Dong, P.; Wang, W. Laser welding of TiNi shape memory alloy and stainless steel using Ni interlayer. *Mater. Des.* **2012**, *39*, 285–293. [CrossRef]
12. Gao, X.; Wang, X.; Liu, J.; Li, L. A novel laser welding method for the reliable joining of NiTi/301SS. *Mater. Lett.* **2020**, *268*, 127573. [CrossRef]
13. Panton, B.; Pequegnat, A.; Zhou, Y.N. Dissimilar Laser Joining of NiTi SMA and MP35N Wires. *Metall. Mater. Trans. A Phys. Metall. Mater. Sci.* **2014**, *45*, 3533–3544. [CrossRef]

14. Oliveira, J.P.; Panton, B.; Zeng, Z.; Andrei, C.M.; Zhou, Y.; Miranda, R.M.; Brazfernandes, F.M. Laser joining of NiTi to Ti6Al4V using a Niobium interlayer. *Acta Mater.* **2016**, *105*, 9–15. [CrossRef]
15. Zoeram, A.S.; Mousavi, S. Effect of interlayer thickness on microstructure and mechanical properties of as welded Ti6Al4V/Cu/NiTi joints. *Mater. Lett.* **2014**, *133*, 5–8. [CrossRef]
16. Xie, J.; Chen, Y.; Yin, L.; Zhang, T.; Wang, S.; Wang, L. Microstructure and mechanical properties of ultrasonic spot welding TiNi/Ti6Al4V dissimilar materials using pure Al coating. *J. Manuf. Process.* **2021**, *64*, 473–480. [CrossRef]
17. Bartolomeu, F.; Costa, M.M.; Alves, N.; Miranda, G.; Silva, F.S. Additive manufacturing of NiTi-Ti6Al4V multi-material cellular structures targeting orthopedic implants. *Opt. Lasers Eng.* **2020**, *134*, 106208. [CrossRef]
18. Tabaie, S.; Rezai-Aria, F.; Flipo, B.C.D.; Cormier, J.; Jahazi, M. Post-Weld Heat Treatment of Additively Manufactured Inconel 718 Welded to Forged Ni-Based Superalloy AD730 by Linear Friction Welding. *Metall. Mater. Trans. A Phys. Metall. Mater. Sci.* **2021**, *52*, 3475–3488. [CrossRef]
19. Tabaie, S.; Rezai-Aria, F.; Flipo, B.C.D.; Jahazi, M. Grain size and misorientation evolution in linear friction welding of additively manufactured IN718 to forged superalloy AD730 (TM). *Mater. Charact.* **2021**, *171*, 110766. [CrossRef]
20. Tabaie, S.; Rezai-Aria, F.; Flipo, B.C.D.; Jahazi, M. Dissimilar linear friction welding of selective laser melted Inconel 718 to forged Ni-based superalloy AD730 (TM): Evolution of strengthening phases. *J. Mater. Sci. Technol.* **2022**, *96*, 248–261. [CrossRef]
21. Hu, Z.; Song, C.; Liu, L.; Yang, Y.; Hu, P. Progress in the technology of nitinol formed by selective laser melting. *Chin. J. Lasers* **2020**, *47*, 104–115.
22. Zhou, Y.N.; Panton, B.; Han, A.; Zeng, Z.; Oliveira, J.P. Dissimilar laser welding of NiTi shape memory alloy and copper. *Opt. Lasers Eng.* **2015**, *24*, 125036.
23. Zeng, Z.; Oliveira, J.P.; Yang, M.; Song, D.; Peng, B. Functional fatigue behavior of NiTi-Cu dissimilar laser welds. *Mater. Des.* **2017**, *114*, 282–287. [CrossRef]
24. Thijs, L.; Sistiaga, M.L.M.; Wauthle, R.; Xie, Q.; Kruth, J.-P.; Van Humbeeck, J. Strong morphological and crystallographic texture and resulting yield strength anisotropy in selective laser melted tantalum. *Acta Mater.* **2013**, *61*, 4657–4668. [CrossRef]
25. Kempen, K.; Thijs, L.; Vrancken, B.; Buls, S.; Kruth, J.P. Producing crack-free, high density M2 HSS parts by selective laser melting: Pre-heating the baseplate. In *2013 International Solid Freeform Fabrication Symposium*; University of Texas at Austin: Austin, TX, USA, 2013.
26. Yadroitsev, I.; Yadroitsava, I. Evaluation of residual stress in stainless steel 316L and Ti6Al4V samples produced by selective laser melting. *Virtual Phys. Prototyp.* **2015**, *10*, 67–76. [CrossRef]
27. Sing, S.L.; Lam, L.P.; Zhang, D.; Liu, Z.; Chua, C.K. Interfacial characterization of SLM parts in multi-material processing: Intermetallic phase formation between AlSi10Mg and C18400 copper alloy. *Mater. Charact.* **2015**, *107*, 220–227. [CrossRef]
28. Dadbakhsh, S.; Speirs, M.; Kruth, J.P.; Humbeeck, J.V. Influence of SLM on shape memory and compression behavior of NiTi scaffolds. *Cirp Ann. Manuf. Technol.* **2015**, *64*, 209–212. [CrossRef]
29. Saedi, S.; Turabi, A.S.; Andani, M.T.; Haberland, C.; Karaca, H.; Elahinia, M. The influence of heat treatment on the thermomechanical response of Ni-rich NiTi alloys manufactured by selective laser melting. *J. Alloys Compd.* **2016**, *677*, 204–210. [CrossRef]

Article

Laser-Sintered Mg-Zn Supersaturated Solid Solution with High Corrosion Resistance

Youwen Yang [1], Wei Wang [1], Mingli Yang [1], Yingxin Yang [1], Dongsheng Wang [2], Zhigang Liu [3,*] and Cijun Shuai [1,4,5,*]

[1] Institute of Bioadditive Manufacturing, Jiangxi University of Science and Technology, Nanchang 330013, China; yangyouwen@jxust.edu.cn (Y.Y.); wangwei@mail.jxust.edu.cn (W.W.); yangmingli@mail.jxust.edu.cn (M.Y.); 9519930003@jxust.edu.cn (Y.Y.)
[2] Key Laboratory of Construction Hydraulic Robots of Anhui Higher Education Institutes, Tongling Univesity, Tongling 244061, China; wangdongsheng@tlu.edu.cn
[3] School of Electrical Engineering and Automation, Jiangxi University of Science and Technology, Ganzhou 341000, China
[4] State Key Laboratory of High Performance Complex Manufacturing, Central South University, Changsha 410083, China
[5] Double Medical Technology Inc., Xiamen 361026, China
* Correspondence: 9120030053@jxust.edu.cn (Z.L.); shuai@csu.edu.cn (C.S.)

Abstract: Solid solutions of Zn as an alloy element in Mg matrixes are expected to show improved corrosion resistance due to the electrode potential being positively shifted. In this study, a supersaturated solid solution of Mg-Zn alloy was achieved using mechanical alloying (MA) combined with laser sintering. In detail, supersaturated solid solution Mg-Zn powders were firstly prepared using MA, as it was able to break through the limit of phase diagram under the action of forced mechanical impact. Then, the alloyed Mg-Zn powders were shaped into parts using laser sintering, during which the limited liquid phase and short cooling time maintained the supersaturated solid solution. The Mg-Zn alloy derived from the as-milled powders for 30 h presented enhanced corrosion potential and consequently a reduced corrosion rate of 0.54 mm/year. Cell toxicity tests confirmed that the Mg-Zn solid solution possessed good cytocompatibility for potential clinical applications. This study offers a new strategy for fabricating Mg-Zn solid solutions using laser sintering with MA.

Keywords: Mg-Zn solid solution; laser sintering; mechanical alloying; corrosion resistance

1. Introduction

Mg-based alloys are recognized as revolutionary biomaterials because of their natural degradability, high specific strength, suitable Young's modulus, and favorable biocompatibility [1–3]. In order to increase their clinical applicability, one issue deserving greater attention is their rapid degradation in vivo [4,5]. In fact, the rapid degradation of Mg is mainly due to its low electrode potential [6,7]. According to electrochemical kinetics, a solid solution of high electrode potential substances (such as Zn, Fe) in a Mg matrix is able to improve the overall electrode potential, thereby enhancing the corrosion resistance and reducing the degradation [8,9]. Zn is a biometal that fulfills a series of vital biofunctions in vivo [10,11]. It has been reported that Zn is able to stimulate new bone growth [12]. Nevertheless, the solubility of alloy elements, including Zn in Mg, is very limited (less than 1 wt.% at room temperature) [13]. As such, direct alloying with Zn will cause precipitation of the intermetallic phase and adverse formation of galvanic corrosion, thereby accelerating the degradation.

In order to enhance the corrosion potential of Mg, it is necessary to increase the solid solubility of Zn in the Mg matrix. Mechanical alloying (MA) is a non-equilibrium powder solid-state alloying method, which can improve the solid solubility of the second component in the parent phase to form a supersaturated solid solution [14]. During

MA, the powders form composite particles with a layered structure through continuous deformation, fragmentation, and welding [15,16]. In this condition, the lamellar spacing is decreased, reducing the diffusion distance of solid atoms, thereby speeding up the alloying process [17]. At the same time, the mechanical force causes massive strains and defects in the particles and increases the chemical activity of the powders. As a consequence, the energy barrier of atomic diffusion is reduced. In a previous study, a channel was further provided for the rapid diffusion of alloying elements [18].

In fact, MA-prepared powders should be consolidated into samples with desirable shapes to achieve specific functions. Laser sintering, as a laser-additive manufacturing process, is able to fabricate parts with complicated structures [19–21]. It utilizes laser energy to partially melt the powder particles to obtain a limited liquid molten pool, which subsequently undergoes fast cooling. It is expected that the partial melting mechanism and rapid solidification are able to maintain the original supersaturated solid solution, thereby obtaining Mg-Zn parts with high corrosion resistance [22].

According to the above considerations, in this study, Mg-Zn powders were firstly prepared using MA, which were subsequently shaped into samples using laser sintering. The microstructures of the MA-processed Mg-Zn supersaturated solid solution powders were investigated. The corrosion behavior of the laser-sintered Mg-Zn parts was investigated using electrochemical tests and immersion experiments. Additionally, in vitro cell tests were carried out to assess the biocompatibility of the parts for potential bone implant applications.

2. Materials and Methods

2.1. MA Processing of Mg-Zn Powder

In this study, spherical Mg powders (15–53 µm, 99%) and Zn powders (~10 µm, 99%) supplied from Tangshan Wei Hao Co., Ltd., in China were used as the original materials. The Mg-Zn mixture powders with 14 wt.% Zn were synthesized using a planetary ball mill from Retsch GmbH (Dusseldorf, Germany). The ratio of the ball to powder weight was 10:1 and the vial speed was 350 rpm. During MA, 3 wt.% of alcohol was added to the vial to avoid the occurrence of cold welding. Additionally, the milling process was interrupted for 5 min out of every 20 min to avoid overheating.

The surfaces of as-milled Mg-Zn powders after 10, 20, 30 and 40 h were observed via scanning electron microscope (SEM, EVO 18) equipped with an energy-dispersive spectrometer (EDS, FEI Nova Nano SEM450) at 20 kV. The cross-section of the as-prepared powders was studied using SEM. In detail, the Mg-Zn powders were inlaid using a hot inlay machine and polished to 2000 grit using metallographic sandpaper. Furthermore, the microstructure of Mg-Zn powders was analyzed via transmission electron microscope (TEM, JEM2100, Tokyo, Japan) at 200 kV. The phase composition was researched utilizing a D8 Advance X-ray diffractometer (XRD) with Cu-Kα radiation. The scanning speed and scanning range were set to 5°/min and 20~80°, respectively.

2.2. Laser Processing of Mg-Zn Samples

The as-prepared powders after milling for 10, 20, and 30 h were adopted for laser experiments, as schematically shown in Figure 1. The system was composed of a fiber laser, a closed glove box, and a computer control system. The laser adopted in this work was a YLR-500-WC (IPG, Burbach, Germany), which is a continuous wave laser. The spot size was about 60 µm. After a series of pilot experiments, the cubic parts were fabricated at a laser power of 62 W, a scanning rate of 110 mm/s, a hatching space of 50 µm, and a layer thickness of 50 µm. During laser processing, high purity argon was offered at a flow rate of 0.1 L/min to mitigate the oxidation. To facilitate the description, the as-built Mg-Zn parts that developed from as-milled powders for 0, 10, 20, and 30 h were designated as M0, M10, M20, and M30, respectively.

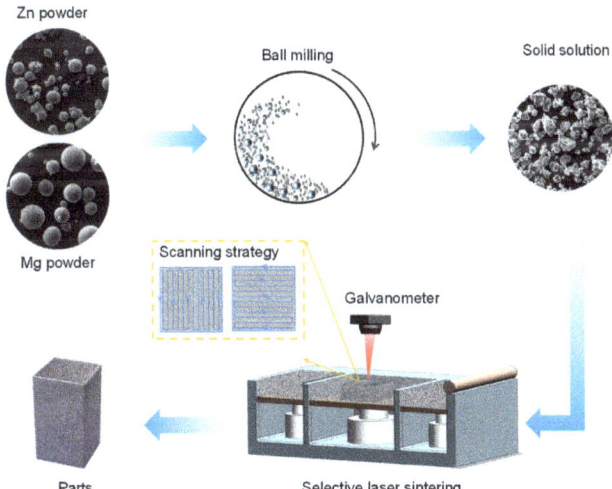

Figure 1. Schematic map showing the preparation process for laser-sintered Mg-Zn parts.

2.3. Electrochemical Tests

The laser-sintered samples were polished to 2000 grit with metallographic sandpaper and then cleaned with ultrasonic waves before drying. The electrochemical testing was carried out in simulated body fluid (SBF) using a three-electrode electrochemical workstation (CHI660C, Chenhua Instruments Inc., Shanghai, China). The testing samples, platinum sheet, and Ag/AgCl acted as the working electrode, counter electrode, and reference electrode, respectively. The samples were treated at open circuit potential for 5 min to obtain stable open potential. Then, the Tafel polarization curve was obtained by polarization test at a rate of 1 mV/s. The corrosion current density and potential were analyzed by fitting the Tafel polarization curve using NOVA software. Electrochemical impedance spectroscopy (EIS) was measured in the range of 100 kHz to 0.01 Hz, and the corresponding Nyquist and Bode diagrams were obtained.

2.4. Immersion Experiments

The immersion experiment was carried out in SBF at 37 °C. The ratio of the solution volume to the exposed area was 20 mL/cm^2. During immersion, the pH of the immersion solution was measured using a pH meter every 24 h. The concentrations of Mg^{2+} and Zn^{2+} in solution were detected using an inductively coupled plasma optical emission spectrometer (ICP-OES, Cambridge, UK). After immersion for 7 d, the corrosion surface was observed using SEM and the composition of the corrosion product was analyzed using EDS. Then, the corrosion products were removed using chromic acid consisting of 200 g/L of CrO_3 and 10 g/L of $AgNO_3$. The corrosion rate (R_{corr}) was calculated via the weight loss method [23]:

$$R_{corr} = \frac{8.76 \times 10^4 \, w}{s \rho t}, \quad (1)$$

where w, t, and s are weight loss, the immersion time, and the area exposed to the solution, respectively. Here, ρ represents the density of the specimens.

2.5. Toxicity Tests

MG-63 cells were used to evaluate the cytocompatibility. The samples were immersed into DMEM medium containing 10% fetal bovine serum, 1% penicillin, and streptomycin. The ratio of the exposure area to solution was 1.25 cm^2/mL and the extract was obtained after immersion for 3 days. Then MG-63 cells were cultured in the extracts at 37 °C and 5%

CO_2 for 1, 3, and 5 d, respectively. Afterwards, the cells were stained with Calcein-AM/PI for 30 min and then observed using a fluorescence microscope (Olympus Co., Ltd., Tokyo, Japan). In addition, the cell viability was evaluated using a cell counting kit-8 (CCK-8) assay. Cells were cultured in 24-well culture plates containing DMEM for 1 day and replaced with extracts to further culture them for 1, 3, and 5 d, respectively. After this, the CCK-8 solution was dropped into each well for 3 h and the absorbance at 450 nm was measured using a microplate reader (Beckman, Brea, CA, USA).

2.6. Statistical Analysis

All tests were performed 3 times to achieve the averages and the obtained data were displayed as means ± standard deviation. Statistical significance was estimated using SPSS 20.0 soft, where $p < 0.05$ was considered to be of statistical difference.

3. Results

3.1. Microstructure of Mg-Zn Mixed Powders

The morphologies of Mg-Zn mixed powders before and after milling are shown in Figure 2. The initial mixed powders exhibited a regular spherical shape, as shown in Figure 2a. After milling for 10 h, the powders obviously changed into a flat block shape owing to the impact of the powerful mechanical forces, as presented in Figure 2b. As the milling time increased to 20 and 30 h, the powders fractured because of the continuous collision and deformation, thereby reducing the particle size and generating numerous new particle surfaces, as shown in Figure 2c,d. In particular, some particles were welded together, considerably reducing the inter-layer space, which was favorable for the solution atom diffusion. However, as the milling time further increased to 40 h, the as-milled powders experienced severe cold welding, as confirmed by the enlarged particle size shown in Figure 2e. Usually, the fracturing and welding process would achieve a balance over excessive milling periods, resulting in work hardening, which should be avoided during MA.

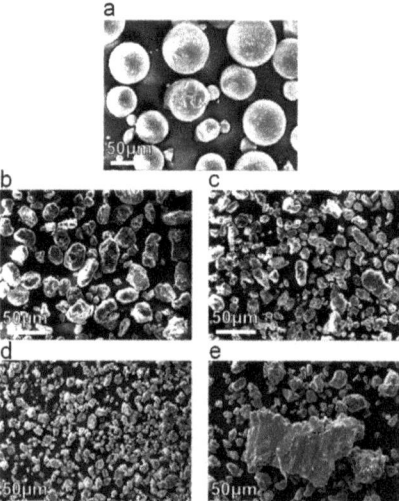

Figure 2. SEM morphologies of (**a**) initial mixed powders and (**b**–**e**) mixed powders after milling for 10, 20, 30, and 40 h, respectively.

The cross-section morphologies of Mg-Zn powders were investigated via SEM in back scattered mode. Herein, the particles were distinguished by dark contrast areas for Mg particles and bright contrast areas for Zn particles. Clearly, the majority of particles remained unalloyed after a short milling time of 10 h, as shown in Figure 3a, since the mechanical collisions and kinetic energy input were inadequate. After milling for 20 h, some Zn particles were trapped in Mg particles, as shown in Figure 3b, whereby the Zn particles showed a fine and lamellar-like structure, indicating severe deformation under the sustained collision of the grinding balls. As the milling time further increased to 30 h, the Zn particles almost disappeared. It was indicated that Zn particles were completely dissolved in the Mg matrix, forming a homogeneous distribution.

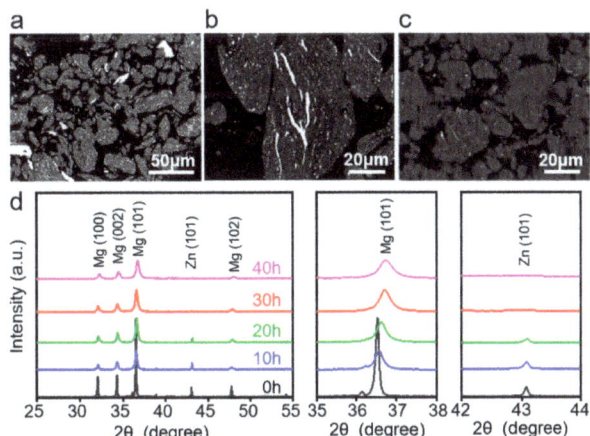

Figure 3. SEM image showing the cross-sections of Mg-Zn powders after milling for (a) 10 h, (b) 20 h, and (c) 30 h. (d) XRD patterns of as-milled powders.

The phase compositions of the as-milled Mg-Zn powders were investigated, with the collected XRD spectrum exhibited in Figure 3d. Strong Mg and Zn diffraction peaks were detected after milling for 10 h. With the increase in milling time, the Zn diffraction peaks clearly became weaker, since partial Zn particles were gradually dissolved into the Mg matrix with the aid of continuous mechanical energy input [24]. A careful examination of the XRD spectrum showed that the primary diffraction peak of Mg shifted towards a higher 2θ position during MA. It was believed that the Zn atoms were dissolved in the Mg matrix, which caused lattice distortion [25]. Notably, the Zn diffraction peaks disappeared after milling for 30 h, which further confirmed the formation of a homogeneous Mg-Zn solid solution.

Based on previous XRD data, the lattice strain, crystallite size, and lattice parameters for as-milled Mg-Zn powders were calculated utilizing the Scherer formula and Williamson–Hall formula, as follows [26]:

$$T = \frac{0.89 \cdot \lambda}{b \cdot \cos \delta} \quad (2)$$

$$b \cdot \cos \theta = \frac{K \cdot \lambda}{T} + 4 \cdot \varepsilon \cdot \sin \delta \quad (3)$$

where T is the crystallites size, b is half maximum width, δ is the peak angle, ε is the lattice strain, and K and λ are constants. As listed in Table 1, as the milling time gradually increased, the lattice strain increased from 7.43×10^{-4} to 23.15×10^{-4}. Meanwhile, the crystallite size of the Mg was gradually refined to 24.2 nm at 30 h, although no significant variation occurred at 40 h, owing to the predominance of cold welding at this stage. Additionally, the lattice parameters (a, c) of Mg declined at the initial 30 h point and similarly remained unchanged at 40 h. These results prove that MA caused numerous lattice distortion or defects, which was beneficial in promoting the diffusion of Zn atoms into the Mg matrix, thereby forming the Mg-Zn supersaturated solid solution.

Table 1. Crystallite size, lattice strain, and constant values of the Mg matrix.

Milling Time	Lattice Strain	Crystallite Size (nm)	Lattice Parameters (Å)	
			a	c
0 h	7.43×10^{-4}	68.2	3.2049	5.2105
10 h	16.60×10^{-4}	37.9	3.2032	5.2001
20 h	20.56×10^{-4}	26.8	3.2001	5.1952
30 h	23.31×10^{-4}	24.2	3.1973	5.1898
40 h	23.15×10^{-4}	24.6	3.1971	5.1892

The microstructure of the Mg-Zn powder after milling for 30 h was studied utilizing TEM, as shown in Figure 4. The powder exhibited a nano-scaled grain size, as shown in Figure 4a. The Debye–Scherrer rings are shown in Figure 4b, proving the existence of $(101)_{Mg}$ and $(002)_{Mg}$. The HRTEM image with the fast Fourier transformation (FFT) inset is presented in Figure 4c. Obviously, some lattice distortions were observed in the red region, while the interplanar spacing was measured for 0.238 nm. This value was smaller than for pure Mg (0.2452 nm) [27]. Additionally, the FFT image of the yellow area shows a hexagonal close-packed structure, suggesting good structural integrity of the magnesium matrix. The majority of the lattice dislocations were caused by the solid solution of Zn. These results directly confirm the formation of the Mg-Zn supersaturated solid solution after MA.

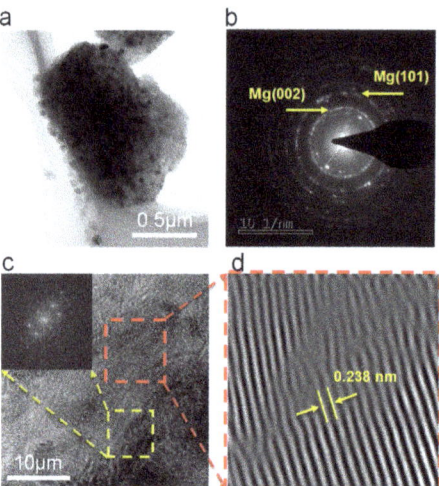

Figure 4. TEM analysis of Mg-Zn powder after milling for 30 h: (a) bright-field TEM image; (b) SAED pattern; (c) high-resolution image with FFT inset; (d) corresponding inverse FFT image.

3.2. Microstructure of As-Build Samples

Laser-sintered samples were fabricated using as-milled powders and the corresponding surfaces after polishing are presented in Figure 5a. It can be observed that there were some microspores in the parts, since the particles were partially melted. Nevertheless, a relatively more homogeneous microstructure was observed for M30, since the as-milled powder at 30 h was refined and was beneficial for the laser forming. The phase composition was investigated using XRD, as shown in Figure 5b. It was found that the XRD patterns of the sintered parts were basically similar to those of the as-milled powders. Particularly, no diffraction peak corresponding to the Zn phase was detected in the M30 part, which indicated that the Mg-Zn solid solution still existed after laser shaping. The laser sintering involved a partial melting mechanism, in which the as-milled powders were partially melted. Subsequently, the molten pool with the limited liquid phase experienced a rapid solidification, which was able to limit the precipitation of the solid solution [28]. As such, the Mg-Zn solid solution was maintained after laser processing.

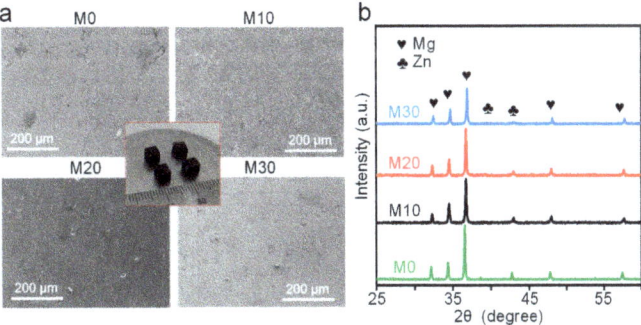

Figure 5. (**a**) Laser-sintered parts and corresponding surfaces after polishing. (**b**) XRD patterns of laser-sintered samples.

3.3. Electrochemical Behavior

The corrosion behavior of the sintered parts was investigated via electrochemical experiments in SBF solution. The polarization curves for all parts are displayed in Figure 6a, and the corresponding corrosion potential (E_{corr}) and corrosion current density (I_{corr}) were calculated via Tafel region extrapolation, as listed in Table 2. The M0 specimen possessed a relatively low E_{corr} of -1.50 ± 0.06 V. Nevertheless, the E_{corr} corresponding to M10, M20, and M30 shifted positively and reached a maximum of -1.34 ± 0.02 V for M30, which revealed a relatively low corrosion tendency. The Mg matrix with the maximum atomic density originated from the solid solution of Zn atoms, which led to the crystallographic texture change, thereby reducing its surface energy [29]. In this case, the low surface energy required massive dissolution activation energy to cause matrix degradation, which had an effective protective effect [30]. In contrast, the I_{corr} evidently decreased to 16.4 ± 1.3 μA/cm^2 as compared with M0 (61.7 ± 2.8 μA/cm^2), indicating superior corrosion resistance.

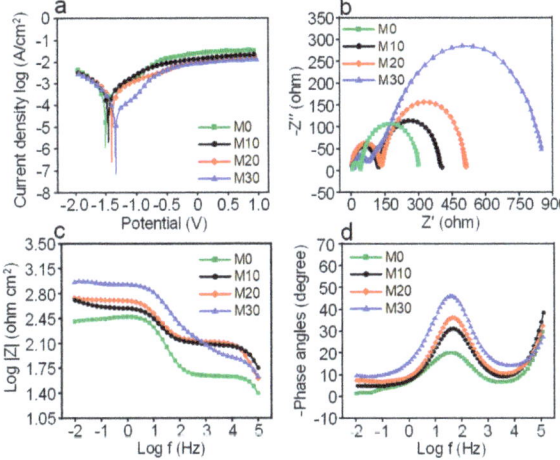

Figure 6. (**a**) Potentiodynamic polarization curves. (**b**) Nyquist diagrams. (**c**) Bode impedance plots. (**d**) Bode phase angle curves.

Table 2. Tafel fitting results derived from the polarization curves of laser-sintered specimens.

Samples	E_{corr} (V)	I_{corr} (μA/cm^2)
M0	−1.50 ± 0.06	61.7 ± 2.8
M10	−1.46 ± 0.03	−1.46 ± 0.03
M20	−1.40 ± 0.04	31.4 ± 2.5
M20	−1.34 ± 0.02	16.4 ± 1.3

An electrochemical impedance spectroscopy (EIS) test was performed, with the obtained results presented in Figure 6b–d. All Nyquist curves presented two typical impedance loops across the whole frequency range, revealing the similar degradation mechanisms. No obvious differences were exhibited over the high-frequency region for the as-built samples, indicating the same dissolution process, as shown in Figure 6b. However, the M0 sample possessed a relatively small capacitive arc in the low-frequency region, which indicated a low charge transfer resistance [31]. In contrast, the diameter of the capacitive arcs corresponding to M10, M20, and M30 gradually increased to the maximum point. As evident in Figure 6c, a relatively large impedance moduli appeared in the low-frequency region. Furthermore, the phase angle values in the low-frequency region also reached the maximum point, as shown in Figure 6d, which revealed an improvement of the charge transfer resistance for M30 [31]. Notably, the phase angle values in the high-frequency region were similar and far below 60° for all samples, indicating the formation of porous oxide film during the corrosion process [32].

3.4. Degradation Performance

To evaluate the degradation behavior, the immersion tests were employed in SBF solution. The corrosion rates were calculated, with the results displayed in Figure 7a. Clearly, the corrosion rates corresponding to M10, M20, and M30 gradually decreased to 0.54 mm/year as compared with M0 at 1.89 mm/year. The pH values of SBF solution at different immersion time were collected, as shown in Figure 7b. The pH the M0 group initially increased rapidly, then increased slowly and finally stabilized at 10.9. In contrast, the pH levels corresponding to M10, M20, and M30 increased slowly and stabilized at 10.3, 9.7, and 9.3, respectively. The results were ascribed to the generation of brucite and Mg-containing calcium phosphate, which precipitated on the Mg matrix surface, hindering the matrix corrosion [33–36]. Furthermore, the Mg^{2+} and Zn^{2+} ion concentrations after immersion for 168 h are exhibited in Figure 7c. The Mg^{2+} and Zn^{2+} ion concentrations of M10, M20, and M30 groups were evidently lower than the M0 group. Especially for the M30 group, both the Mg^{2+} and Zn^{2+} ion concentrations reached 42.5 and 0.62 μg/mL, respectively, which indicated reduced degradation rates.

The typical corrosion surfaces for all samples after immersion for 7 d are exhibited in Figure 8a. It can be seen that a large amount of corrosion products were deposited on the M0 matrix, indicating intense corrosion. The corrosion products obviously decreased for M10, M20, and M30. Only a small amount of agglomerations appeared on the surface of M20 and M30, which was attributed to the change of the surface chemical composition. Analyzing the EDS results detected for area S1 in M20 and area S2 in M30, the agglomerations mainly included large amounts of O, Mg, P, and Ca, as well as small amount of Zn for M20, while almost no additional Zn was detected for M30. Furthermore, massive continuous and wide cracks existed in the corrosion films grown on M20. In contrast, discontinuous and narrow cracks were observed for M30, which was affected by the dehydrogenation process.

To further analyze the degradation mechanism, the surface morphologies after removal of corrosion products are displayed in Figure 8b. Obviously, the precipitated phase remained unattacked while the adjacent Mg matrix seriously dissolved from the matrix surface to the center, which led to numerous deep pits for the M0 sample. Some large cavities were even formed owing to the strong reactions of galvanic corrosion. The precipitated phase increased the potential difference of the corrosion couple with the Mg matrix, thereby accelerating the degradation [37–39]. Comparatively, the corrosion degree corresponding to M10, M20, and M30 samples gradually decreased. For M30, only a few shallow and small corrosion pits were present on the matrix surface, which indicated significantly enhanced corrosion resistance and a uniform degradation mode.

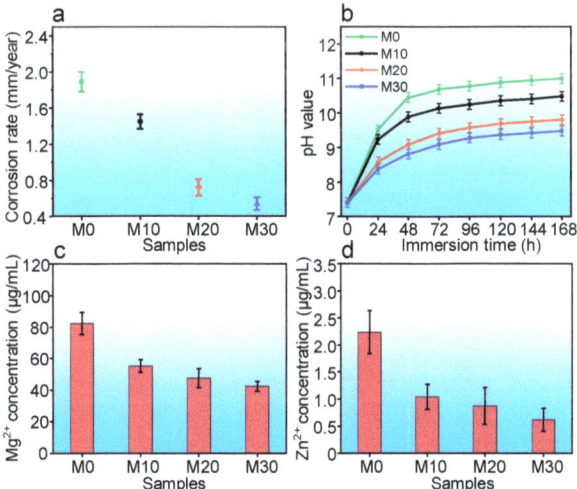

Figure 7. (**a**) Corrosion rates, (**b**) pH values, and (**c**) Mg^{2+} and (**d**) Zn^{2+} ion concentrations for all sample extracts.

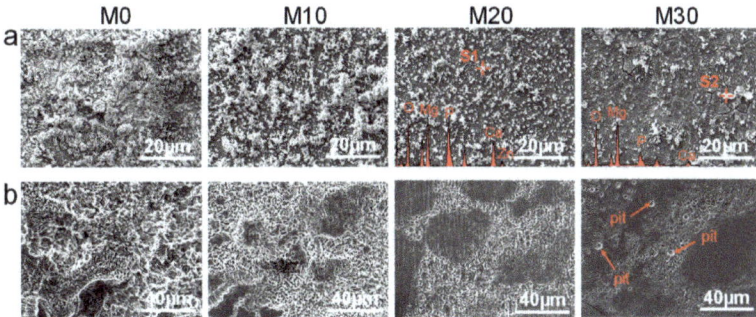

Figure 8. Surface morphologies of laser-sintered parts: (**a**) before and (**b**) after removing corrosion products.

3.5. Cytocompatibility

To evaluate the proliferation of MG-63 cells cultured in the extracts of M0 and M30, cells were stained after culturing for 1, 3, and 5 d. The fluorescence image in Figure 9a shows live cells as green and dead cells as red, respectively. It can be seen that the cell densities for both M0 and M30 groups increased with increasing culture time, indicating the favorable biocompatibility of the Mg-Zn alloy, despite the morphologies of MG-63 cells showing round and unhealthy shapes in M0 and M30 after culturing for 1 d. However, some spindle-shaped cells with filopodia were observed after 5 d. Notably, there were more live cells in the M30 group than in the M0 group, implying better cytocompatibility of M30. CCK-8 assay was carried out and the results are shown in Figure 9b. Overall, the viability of M30 was higher than that of M0. In detail, the cell viability levels for M0 were 57% at day 1, 62% at day 3, and 70% at day 5, respectively, indicating relatively low cell viability. This was attributed to the fast degradation rate leading to increased pH, which was harmful to cell growth [40–42]. Regarding M30, the cell viability reached 62% at day 1, 75% at day 3, and 88% at day 5. It could be concluded that M30 showed better cytocompatibility for cell growth in comparison to M0. The reason was the improvement of the solid solubility of Zn in Mg, which contributed to better corrosion resistance, resulting in a mild environment for cell growth.

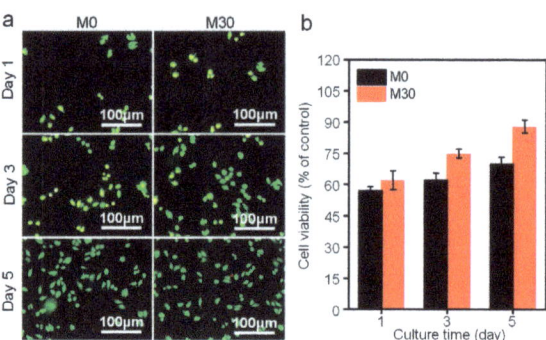

Figure 9. (**a**) Cell morphology and (**b**) cell viability results for MG-63 cells after 1, 3, and 5 d culture in M0 and M30 extracts.

4. Conclusions

In the present study, Mg-Zn solid solution powders were prepared using MA. During MA, the continuous and mandatory mechanical grinding introduced numerous lattice defects and the diffusion channel of Zn into the Mg matrix, resulting in supersaturated solid solution powders after milling for 30 h. The as-prepared powders were then developed into parts using laser sintering, which was a typical feature of the Mg-Zn supersaturated solid solution due to the fast cooling rate. The results showed that the laser-sintered Mg-Zn parts presented relatively high corrosion potential and anti-charge transfer ability. The corrosion rate was improved to 0.54 mm/year. The parts also showed good cytocompatibility.

Author Contributions: Conceptualization, Y.Y. (Youwen Yang); investigation, W.W., M.Y. and Y.Y. (Yingxin Yang); resources, D.W.; writing—original draft preparation, Y.Y. (Youwen Yang); supervision, Z.L. and C.S. All authors have read and agreed to the published version of the manuscript.

Funding: This research was funded by; (1) National Natural Science Foundation of China (51935014, 52165043, 82072084, 81871498); (2) JiangXi Provincial Natural Science Foundation of China (2020ACB214004); (3) The Provincial Key R&D Projects of Jiangxi (20201BBE51012); (4) The Project of State Key Laboratory of High Performance Complex Manufacturing; (5) China Postdoctoral Science Foundation (2020M682114); (6) Key Laboratory of Construction Hydraulic Robots of Anhui Higher Education Institutes, Tongling University (TLXYCHR-O-21YB01); (7) JiangXi Provincial Postgraduate Innovation Fund (YC2021-S604).

Institutional Review Board Statement: Not applicable.

Informed Consent Statement: Not applicable.

Data Availability Statement: The data presented in this study are available on request from the corresponding author.

Conflicts of Interest: The authors declare no conflict of interest.

References

1. Xu, L.; Zhang, E.; Yin, D.; Zeng, S.; Yang, K. In vitro corrosion behaviour of Mg alloys in a phosphate buffered solution for bone implant application. *J. Mater. Sci. Mater. Med.* **2008**, *19*, 1017–1025. [CrossRef]
2. Yu, Y.; Lu, H.; Sun, J. Long-term in vivo evolution of high-purity Mg screw degradation—Local and systemic effects of Mg degradation products. *Acta Biomater.* **2018**, *71*, 215–224. [CrossRef] [PubMed]

3. He, R.; Liu, R.; Chen, Q.; Zhang, H.; Wang, J.; Guo, S. In vitro degradation behavior and cytocompatibility of Mg-6Zn-Mn alloy. *Mater. Lett.* **2018**, *228*, 77–80. [CrossRef]
4. Zeng, R.-C.; Li, X.-T.; Li, S.-Q.; Zhang, F.; Han, E.-H. In vitro degradation of pure Mg in response to glucose. *Sci. Rep.* **2015**, *5*, 13026. [CrossRef] [PubMed]
5. Myrissa, A.; Agha, N.A.; Lu, Y.; Martinelli, E.; Eichler, J.; Szakács, G.; Kleinhans, C.; Willumeit-Römer, R.; Schäfer, U.; Weinberg, A.-M. In vitro and in vivo comparison of binary Mg alloys and pure Mg. *Mater. Sci. Eng. C* **2016**, *61*, 865–874. [CrossRef] [PubMed]
6. Li, H.; Wen, J.; He, J.; Shi, H.; Liu, Y. Effects of Dy Addition on the Mechanical and Degradation Properties of Mg–2Zn–0.5 Zr Alloy. *Adv. Eng. Mater.* **2020**, *22*, 1901360. [CrossRef]
7. Feng, P.; Kong, Y.; Liu, M.; Peng, S.; Shuai, C. Dispersion strategies for low-dimensional nanomaterials and their application in biopolymer implants. *Mater. Today Nano* **2021**, *15*, 100127. [CrossRef]
8. Shuai, C.; He, C.; Qian, G.; Min, A.; Deng, Y.; Yang, W.; Zang, X. Mechanically driving supersaturated Fe–Mg solid solution for bone implant: Preparation, solubility and degradation. *Compos. Part Eng.* **2021**, *207*, 108564. [CrossRef]
9. Yang, Y.; Lu, C.; Peng, S.; Shen, L.; Wang, D.; Qi, F.; Shuai, C. Laser additive manufacturing of Mg-based composite with improved degradation behaviour. *Virtual Phys. Prototyp.* **2020**, *15*, 278–293. [CrossRef]
10. Li, H.; Peng, Q.; Li, X.; Li, K.; Han, Z.; Fang, D. Design, Microstructures, mechanical and cytocompatibility of degradable Mg–Zn based orthopedic biomaterials. *Mater. Des.* **2014**, *58*, 43–51. [CrossRef]
11. Qin, H.; Zhao, Y.; An, Z.; Cheng, M.; Wang, Q.; Cheng, T.; Wang, Q.; Wang, J.; Jiang, Y.; Zhang, X.; et al. Enhanced antibacterial properties, biocompatibility, and corrosion resistance of degradable Mg-Nd-Zn-Zr alloy. *Biomaterials* **2015**, *53*, 211–220. [CrossRef] [PubMed]
12. Qiao, Y.; Zhang, W.; Tian, P.; Meng, F.; Zhu, H.; Jiang, X.; Liu, X.; Chu, P. Stimulation of bone growth following zinc incorporation into biomaterials. *Biomaterials* **2014**, *35*, 6882–6897. [CrossRef] [PubMed]
13. Kubasek, J.; Vojtech, D. Structural characteristics and corrosion behavior of biodegradable Mg-Zn, Mg-Zn-Gd alloys. *J. Mater. Sci. Mater. Med.* **2013**, *24*, 1615–1626. [CrossRef] [PubMed]
14. Suryanarayana, C. Phase formation under non-equilibrium processing conditions: Rapid solidification processing and mechanical alloying. *J. Mater. Sci.* **2018**, *53*, 13364–13379. [CrossRef]
15. Varalakshmi, S.; Kamaraj, M.; Murty, B.S. Processing and properties of nanocrystalline CuNiCoZnAlTi high entropy alloys by mechanical alloying. *Mater. Sci. Eng.* **2010**, *527*, 1027–1030. [CrossRef]
16. Othman, A.R.; Sardarinejad, A.; Masrom, A.K. Masrom, Effect of milling parameters on mechanical alloying of aluminum powders. *Int. J. Adv. Manuf. Technol.* **2015**, *76*, 1319–1332. [CrossRef]
17. Amram, D.; Schuh, C.A. Mechanical alloying produces grain boundary segregation in Fe–Mg powders. *Scr. Mater.* **2020**, *180*, 57–61. [CrossRef]
18. Lu, L.; Lai, M.O.; Zhang, S. Diffusion in mechanical alloying. *J. Mater. Process. Technol.* **1997**, *67*, 100–104. [CrossRef]
19. Yap, C.Y.; Chua, C.K.; Dong, Z.L.; Liu, Z.H.; Zhang, D.Q.; Loh, L.E.; Sing, S.L. Review of selective laser melting: Materials and applications. *Appl. Phys. Rev.* **2015**, *2*, 041101. [CrossRef]
20. Bremen, S.; Meiners, W.; Diatlov, A. Selective laser melting: A manufacturing technology for the future? *Laser Tech. J.* **2012**, *9*, 33–38. [CrossRef]
21. Wang, D.; Ye, G.; Dou, W.; Zhang, M.; Yang, Y.; Mai, S.; Liu, Y. Influence of spatter particles contamination on densification behavior and tensile properties of CoCrW manufactured by selective laser melting. *Opt. Laser Tech.* **2020**, *121*, 105678. [CrossRef]
22. Sun, S.; Liu, P.; Hu, J.; Hong, C.; Qiao, X.; Liu, S.; Zhang, R.; Wu, C. Effect of solid solution plus double aging on microstructural characterization of 7075 Al alloys fabricated by selective laser melting (SLM). *Opt. Laser Technol.* **2019**, *114*, 158–163. [CrossRef]
23. Kumar, P.P.; Bharat, A.R.; Sai, B.S.; Sarath, R.P.; Akhil, P.; Reddy, G.P.K.; Kondaiah, V.; Sunil, B.R. Role of microstructure and secondary phase on corrosion behavior of heat treated AZ series magnesium alloys. *Mater. Today: Proc.* **2019**, *18*, 175–181. [CrossRef]
24. Sung, Y.M.; Lee, Y.J.; Park, K.S. Kinetic analysis for formation of Cd1-x Zn x Se solid-solution nanocrystals. *J. Am. Chem. Soc.* **2006**, *128*, 9002–9003. [CrossRef] [PubMed]
25. Yeh, J.W.; Chang, S.Y.; Hong, Y.D.; Chen, S.K.; Lin, S.J. Anomalous decrease in X-ray diffraction intensities of Cu–Ni–Al–Co–Cr–Fe–Si alloy systems with multi-principal elements. *Mater. Chem. Phys.* **2007**, *103*, 41–46. [CrossRef]
26. Shang, C.; Axinte, E.; Sun, J.; Li, X.; Li, P.; Du, J.; Wang, Y. CoCrFeNi(W1−xMox) high-entropy alloy coatings with excellent mechanical properties and corrosion resistance prepared by mechanical alloying and hot pressing sintering. *Mater. Des.* **2017**, *117*, 193–202. [CrossRef]
27. Regev, M.; Rosen, A.; Bamberger, M. Qualitative model for creep of AZ91D magnesium alloy. *Metall. Mater. Trans.* **2001**, *32*, 1335–1345. [CrossRef]
28. Maeshima, T.; Oh-Ishi, K. Solute clustering and supersaturated solid solution of AlSi10Mg alloy fabricated by selective laser melting. *Heliyon* **2019**, *5*, e01186. [CrossRef]
29. Yang, Y.; Lu, C.; Shen, L.; Zhao, Z.; Peng, S.; Shuai, C. In-situ deposition of apatite layer to protect Mg-based composite fabricated via laser additive manufacturing. *J. Magnes. Alloy.* **2021**, in press.

30. Li, J.; Qiu, Y.; Yang, J.; Sheng, Y.; Yi, Y.; Zeng, X.; Chen, L.; Yin, F.; Su, J.; Zhang, T.; et al. Effect of grain refinement induced by wire and arc additive manufacture (WAAM) on the corrosion behaviors of AZ31 magnesium alloy in NaCl solution. *J. Magnes. Alloy.* **2021**, in press. [CrossRef]
31. Xia, Y.H.; Zhang, B.P.; Lu, C.X.; Geng, L. Improving the corrosion resistance of Mg-4.0Zn-0.2Ca alloy by micro-arc oxidation. *Mater. Sci. Eng.* **2013**, *33*, 5044–5050. [CrossRef]
32. Tamilselvi, S.; Rajendran, N. In vitro corrosion behaviour of Ti-5Al-2Nb-1Ta alloy in Hanks solution. *Mater. Corros.* **2007**, *58*, 285–289. [CrossRef]
33. Klein, F.; Bach, W.; Jöns, N.; McCollom, T.; Moskowitz, B.; Berquó, T. Iron partitioning and hydrogen generation during serpentinization of abyssal peridotites from 15° N on the Mid-Atlantic Ridge. *Geochim. Cosmochim. Acta* **2009**, *73*, 6868–6893. [CrossRef]
34. Yang, Y.; He, C.; Dianyu, E.; Yang, W.; Qi, F.; Xie, D.; Shuai, C. Mg bone implant: Features, developments and perspectives. *Mater. Des.* **2020**, *185*, 108259. [CrossRef]
35. Atrens, A.; Johnston, S.; Shi, Z.; Dargusch, M. Viewpoint-Understanding Mg corrosion in the body for biodegradable medical implants. *Scr. Mater.* **2018**, *154*, 92–100. [CrossRef]
36. Qian, G.; Zhang, L.; Wang, G.; Zhao, Z.; Peng, S.; Shuai, C. 3D Printed Zn-doped Mesoporous Silica-incorporated Poly-L-lactic Acid Scaffolds for Bone Repair. *Int. J. Bioprinting* **2021**, *7*, 346. [CrossRef] [PubMed]
37. Song, M.-S.; Zeng, R.-C.; Ding, Y.-F.; Li, R.; Easton, M.; Cole, I.; Birbilis, N.; Chen, X.-B. Recent advances in biodegradation controls over Mg alloys for bone fracture management: A review. *J. Mater. Sci. Technol.* **2019**, *35*, 535–544. [CrossRef]
38. Zhang, Y.; Li, J.; Li, J. Effects of calcium addition on phase characteristics and corrosion behaviors of Mg-2Zn-0.2 Mn-xCa in simulated body fluid. *J. Alloy. Compd.* **2017**, *728*, 37–46. [CrossRef]
39. Yan, K.; Bai, J.; Liu, H.; Jin, Z.-Y. The precipitation behavior of MgZn2 and Mg4Zn7 phase in Mg-6Zn (wt.%) alloy during equal-channel angular pressing. *J. Magnes. Alloy.* **2017**, *5*, 336–339. [CrossRef]
40. Xiao, C.; Wang, L.; Ren, Y.; Sun, S.; Zhang, E.; Yan, C.; Qin, G. Indirectly extruded biodegradable Zn-0.05 wt% Mg alloy with improved strength and ductility: In vitro and in vivo studies. *J. Mater. Sci. Technol.* **2018**, *34*, 1618–1627. [CrossRef]
41. Qi, F.; Zeng, Z.; Yao, J.; Cai, W.; Zhao, Z.; Peng, S.; Shuai, C. Constructing core-shell structured BaTiO3@ carbon boosts piezoelectric activity and cell response of polymer scaffolds. *Mater. Sci. Eng.* **2021**, *126*, 112129. [CrossRef] [PubMed]
42. Gao, C.; Yao, M.; Peng, S.; Tan, W.; Shuai, C. Pre-oxidation induced in situ interface strengthening in biodegradable Zn/nano-SiC composites prepared by selective laser melting. *J. Adv. Res.* **2021**. In Press. [CrossRef]

Article

Strain Rate Dependence of Compressive Mechanical Properties of Polyamide and Its Composite Fabricated Using Selective Laser Sintering under Saturated-Water Conditions

Xiaodong Zheng [1], Jiahuan Meng [1] and Yang Liu [1,2,*]

1 Faculty of Mechanical Engineering & Mechanics, Ningbo University, Ningbo 315211, China; zhengshangjue@163.com (X.Z.); mengjiahuan2022@163.com (J.M.)
2 Key Laboratory of Impact and Safety Engineering, Ministry of Education, Ningbo University, Ningbo 315211, China
* Correspondence: liuyang1@nbu.edu.cn

Abstract: In this work, polyamide 12 (PA12) and carbon fiber reinforced polyamide 12 (CF/PA12) composites were fabricated using selective laser sintering (SLS), and the coupling effects of the strain rate and hygroscopicity on the compressive mechanical properties were investigated. The results showed that the CF/PA12 had a shorter saturation time and lower saturated water absorption under the same conditions, indicating that the SLS of CF/PA12 had lower hydrophilia and higher water resistance when compared to the SLS of PA12. It was observed that as the strain rate increased, and the ultimate compression strength and the yield strength monotonically increased with almost the same slope, indicating that the strain rate had the same positive correlation with the compressive strength of the SLS of PA12 and CF/PA12. The water immersion results showed a significant reduction of 15% in the yield strength of SLS of PA12, but not very significant in CF/PA12. This indicated that the carbon fiber was favorable for maintaining the mechanical properties of polyamide 12 after absorbing water. The findings in this work provide a basic knowledge of the mechanical properties of SLS polyamide under different loading and saturated-water conditions and thus is helpful to widen the application of SLS products in harsh environments.

Keywords: selective laser sintering; carbon fiber reinforced polyamide composites; strain rate; hygroscopicity

Citation: Zheng, X.; Meng, J.; Liu, Y. Strain Rate Dependence of Compressive Mechanical Properties of Polyamide and Its Composite Fabricated Using Selective Laser Sintering under Saturated-Water Conditions. *Micromachines* 2022, 13, 1041. https://doi.org/10.3390/mi13071041

Academic Editor: Cyril Mauclair

Received: 14 June 2022
Accepted: 27 June 2022
Published: 30 June 2022

Publisher's Note: MDPI stays neutral with regard to jurisdictional claims in published maps and institutional affiliations.

Copyright: © 2022 by the authors. Licensee MDPI, Basel, Switzerland. This article is an open access article distributed under the terms and conditions of the Creative Commons Attribution (CC BY) license (https://creativecommons.org/licenses/by/4.0/).

1. Introduction

Selective laser sintering (SLS) is an important branch of laser additive manufacturing technology and an important processing method used for the fabrication of polymer products [1,2]. Compared with the traditional plastic molding processes, such as extrusion molding [3] or injection molding [4], SLS is capable of fabricating complex and fine plastic workpieces with less loss of raw material, because the recycled raw materials can be used in the next fabrication. Thus, it can reduce production costs significantly.

Polyamide, also known as nylon, along with its composites has been the most widely used engineering plastics in recent decades due to its excellent performance, such as heat resistance, impact resistance, high strength, anti-seismic, etc. [5]. The components made from nylon are light, non-rusting, and post-maintenance, which has caused nylon parts to gradually replace some of the metal parts in automobile and consumer electronics industries [6]. As a result, the mechanical performance of polyamide became the focus of attention. Connor et al. [7] comparatively studied the tensile properties of the polyamide 12 and glass bead reinforced polyamide 12 composites, and found that the addition of glass beads increased the tensile and bending strength by 39% and 15%, respectively. Cai et al. [8] compared the tensile strength of SLS polyamide 12 in different directions and found that the difference of strength in X- and Y-directions was very small, and it exhibited approximate isotropy.

As polyamide and its composites are widely used in complex service conditions, the influence of the load strain rate on its properties is focused [9,10]. Wang et al. [11] studied the influence of the tensile strain rate on the elastoplastic deformation and failure behavior of polyamide composites. Although the strain rate had a limited effect on the deformation and failure characteristics, Young's modulus increased significantly upon the increasing strain rate, indicating that loading strain rate had a significant influence on the mechanical properties of polyamide 12 and its composites. Sagradov et al. [12] proposed a method to analyze the strain rate-related material and damage behavior of polyamide 12 by SLS. In this work, two different situations were considered: multiple tensile tests where the strain rate changed during the tensile load; multiple relaxation tests where the strain rate changed at the same time each test.

Polyamide is a semi-crystalline thermoplastic with polar amide groups. When exposed to hydrothermal conditioning, the absorbed water molecule replaces the existing inter-chain amide-amide bonding with amide-water bonding [13]. This highly impacts the mechanical properties. There have been several studies regarding the hygrothermal behavior of glass fiber reinforced polyamides (GF/PA) composites on water diffusion and mechanical properties. Li et al. [14] found that the tensile, bending, and interlaminar shear strength of CF/PA6 composite are decreased by 35%, 53%, and 5%, respectively, after being immersed for 40 h. Lin et al. [15] immersed carbon fiber reinforced polyamide 6 samples into the water at different temperatures (20 °C, 40 °C, and 60 °C), and found that tensile strength, Young's modulus, and impact strength decrease monotonously with the increasing temperature. Chaichanawong et al. [16] found that the mechanical properties of glass fiber reinforced polyamide after being saturated with water were highly related to the immersion time. Within the initial 35 days, the tensile strength decreased mildly and then decreased sharply as prolonging immersion time. Do et al. [17] comparatively studied the mechanical properties of polyamide-6 and polypropylene after being saturated with water. It was found that the polyamide 6 had better ultimate tensile strength, elastic modulus, and elongation than those of polypropylene.

In the research of fabric reinforced composites, SHPB (Split Hopkinson Pressure Bar) apparatus was often used for the experimental determination of dynamic mechanical properties. Yang et al. [18] analyzed the stress uniformity of split Hopkinson bar specimens. Song et al. [19] studied the compression behavior of braided carbon/epoxy laminate composites under in-plane and out-of-plane loads using the SHPB device. The results showed that the stress–strain curve, maximum compressive stress and strain all change with the strain rate.

From the literature, it is observed that the strain rate and hygroscopicity have a great influence on the mechanical properties of polyamide. However, it is observed that limited effort is made to the coupling effect of these two factors. In this work, polyamide 12 and the carbon fiber reinforced polyamide 12 composites were prepared using selective laser sintering, and then a series of compressive tests of under different strain rates (10^{-4} to 2000 /s) were conducted. Before testing, these samples were immersed into the water until they were saturated, then the influence of strain rate and hygroscopicity on the compressive properties were comparatively studied.

2. Experimental Detail

2.1. Materials

In this study, polyamide 12 (PA12) powder with spherical shape and mean particle size of 120 μm is used in this work, the apparent density is 0.48 g/cm^3. The composite powder is prepared by mixing 20 wt.% of short carbon fiber into the polyamide 12 powder (CF/PA12) and is mixed uniformly by ball milling, the composite powder is gray-black powder with density of 0.52 g/cm^3. A selective laser sintering apparatus (HT252P, Hunan Farsoon High-Technology Co., Ltd., Hunan, China) was employed to prepare the samples.

The schematic diagram of SLS is shown in Figure 1, which is equipped with a 60 W carbon dioxide laser with a focal laser beam diameter of ≤0.5 mm. The scanning system used was a dual-axis mirror positioning system and a galvanometer optical scanner, which directs the laser beam in the X and Y axes through the F-theta lens. The building envelope was 250 × 250 × 300 mm³. A heater was equipped to preheat the raw powder material, which could provide a maximum temperature of up to 225 °C. During the process, high purity nitrogen was filled into the chamber to protect the sample from oxidation.

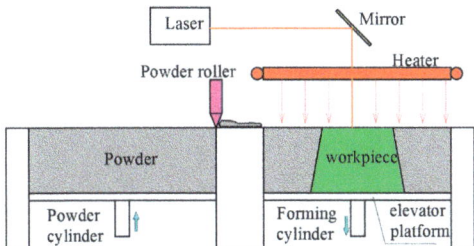

Figure 1. Schematic diagram of the selective laser sintering.

Cylinder samples with dimensions of φ8 × 8 mm are prepared by SLS, the processing parameters are determined as laser power of 45 W, laser scanning speed of 10 m/s, layer thickness of 0.1 mm. In order to investigate the influence of hygroscopicity on the mechanical properties of SLS PA12 and its composites, the as-built PA12 and CF/PA12 samples are immersed into distilled water at room temperature for 72 h, and one group of the as-built samples is used as counterpart. Then quasi-static compression and SHPB tests are performed on immersed and unimmersed PA12 and CF/PA12 samples.

2.2. Differential Scanning Calorimetry (DSC)

The thermal analysis of CF/PA12 and PA12 samples is carried out by differential scanning calorimeter. A sample weighing approximately 410 mg is heated from room temperature to 450 °C at a rate of 10 °C/min using argon as a protective gas. The melting temperature is determined at the maximum heat capacity and temperature. Degrees of crystallinity Xc are determined using DSC on Q200 equipment. In composite materials, the degree of crystallinity Xc is determined as follows, Equation (1):

$$\Delta X_c = \frac{\Delta H_f}{\Delta H_f^0 \times \left(1 - W_f\right)} \qquad (1)$$

where ΔH_f is the enthalpy of fusion of the tested polymer and ΔH_f^0 the theoretical enthalpy of fusion for a 100% crystalline material and W_f the fibre weight fraction. The latter was determined by TGA from matrix burn off tests at 450 °C for 1.5 h under argon (Ar).

2.3. Compression Tests

Quasi-static compression tests are carried out on Instron 5966 servo hydraulic material test machine with compression strain rates of 0.0001 and 0.1, respectively. At least three specimens are tested and average compressive stress–strain curves were obtained. Compression is carried out along the out-of-plane direction of the composite sample. As a contrast test, the quasi-static sample has the same geometry and size as the dynamic compression sample.

Out-of-plane compression tests are carried out on samples at high strain rates using a SHPB (Φ7, Key Laboratory of Impact and Safety Engineering, Ningbo University) apparatus. The detail principle for SHPB can refer to our previous work [20]. Different strain rates are obtained by changing the chamber pressure from 0.15 MPa to 0.45 MPa.

3. Results and Discussion

3.1. Water Absorption of SLS PA12 and CF/PA12

Under the same environmental conditions, an electronic scale with a measurement accuracy of 0.0001 g was used to test the weight of the two group samples before and after the immersion, and the water absorption rate is expressed as follows [18]:

$$\Delta m = \frac{m_1 - m_0}{m_0} \times 100\% \tag{2}$$

$$\Delta M = m_0 - m_2 \tag{3}$$

In Equations (2) and (3), Δm represents the water absorption rate of material, m_0 and m_1 represent the weight of the sample before and after water immersion respectively. And m_2 represents the weight of the sample after drying. ΔM represents the weight of sample hydrolysis. The above formula was used to analyze and calculate the corresponding relationship between the average water absorption of the two materials.

From Figure 2, it is observed that Δm of PA12 sample increases sharply in the first 36 h as the samples were immersed into water. With further prolonged immersed time, the Δm increases slowly and reaches a saturation state with Δm of 5.47%. In contrast, the Δm of CF/PA12 sample shows a faster upward trend in the first 24 h, and then it reaches a saturation state with Δm of 4.6% once the immersion time exceeds 24 h, indicating that shorter saturation time is needed to achieve a saturation state for the CF/PA12. This is because the carbon fiber has a high specific surface area and is uniformly dispersed in the PA12 matrix, which is able to decrease the diffusion distance of water molecules in the composite.

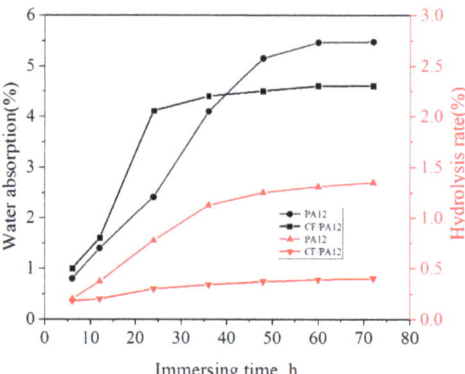

Figure 2. Relationship among the water absorption rate, hydrolysis rate, and immersion time.

In general, the SLS of CF/PA12 has lower hydrophilia and higher water resistance compared with PA12. This is due to the fact that, due to the higher content of the amide group and lower crystallinity, the water absorption of polyamide is better. Although the SLS of PA12 contains a lot of amide groups, the addition of carbon fiber reduces the composition, amide group content, and crystallization property of PA12 [21,22], thus reducing the water absorption rate in CF/PA12.

Figure 3a shows the change of melting peak at different soaking times. The results show that the enthalpy of melting peak increases with the increase of immersion time. In addition, there is no change in the form of the melting peak. Then, in Figure 3b, we found that after 3 days of immersion, the crystallinity ratio of PA12 increased from 9.3% to 10.2%, and the crystallinity ratio of CF/PA12 increased from 10.7% to 11.7%. This process is associated with the phenomenon of chemical crystallization, which is well known in the literature. When chain breaking occurs, the amorphous chain in the polymer regains sufficient fluidity to form new microcrystals. The increase of crystallinity ratio has a significant effect on the mechanical properties of nylon materials. As can be seen from Figure 3b, after 3 days of aging, the crystallinity ratio increased from about 9% to about 11%. This process is associated with the chemi-crystallization phenomenon, well known in the literature [22,23]. Thus, the increase of crystallinity results in the decrease of ductility [23]. This is highly consistent with the test results in Figure 3b.

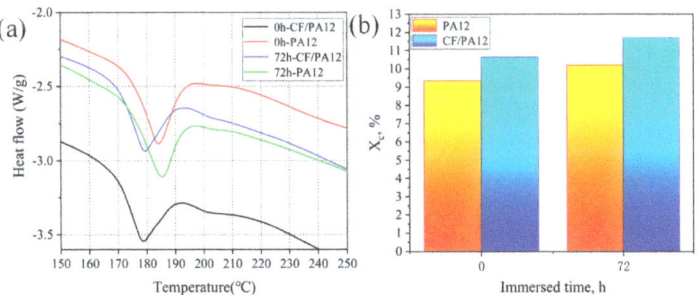

Figure 3. Effect of aging at water on (**a**) the melting peak and (**b**) the crystallinity ratio.

Figure 4 shows the surface morphology of PA12 and CF/PA12 before and after immersion. Figure 4a,c show that the pore size of the nylon sample increases slightly after immersion, and there are two large holes. Then, the surface of CF/PA12 in immersion quality have no obvious change, only small pore space.

Figure 4. Surface morphology of CF/PA12 and PA12 before and after immersion: (**a**) 0 h-PA12; (**b**) 0 h-CF/PA12; (**c**) 72 h-PA12; and (**d**) 72 h-CF/PA12.

3.2. Compressive Mechanical Properties

3.2.1. Influence of Strain Rate

Figure 5 illustrates the compressive stress–strain curves of PA12 and CF/PA12 under different strain rates (10^{-4} to 2000/s). From the figure, it is observed that within the quasi-static loading range, the stress–strain curves have the same shape, and it is easy to distinguish the boundary point between the elastic stage (~5%) according to the curves. Under impact load, the elastic strain of PA12 and CF/PA12 nylon samples does not exceed 2% and 3%, respectively. Further, the stress–strain curves depend on the strain rate, and, therefore, the ultimate compression strength (UCS) and yield strength (YS) monotonically increase as the strain rate increases from 10^{-4} to 2000/s. The maximum difference of YS within the strain rate range was up to 62 MPa. Moreover, the flow stress of PA12 and CP/PA12 within the plastic stage increase with increasing strain, indicating that the PA12 and CP/PA12 have a strain strengthening effect. This kind of work-hardening ability is advantageous in structural applications to guarantee a large safety margin before fracture.

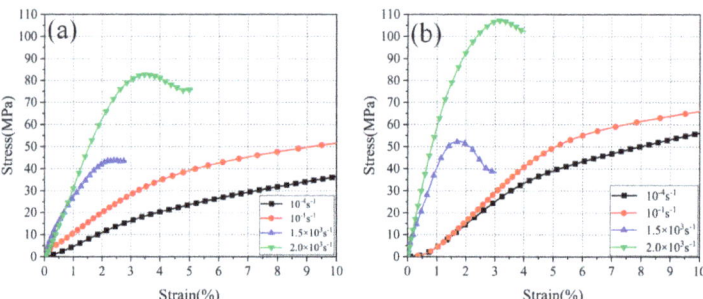

Figure 5. Compressive stress–strain curves of the (**a**) PA12 and (**b**) CF/PA12 under different strain rates.

The addition of carbon fiber poses a great influence on the compressive mechanical properties of SLS PA12. The yield stress of the two groups of samples was extracted from the stress–strain curves and is presented in Figure 6. The yield stress of two groups of material almost increased linearly with logarithmic strain rate, illustrating the obvious strain rate hardening effect. Moreover, the slope of the lines is found to be almost the same, indicating that strain rate has the same influence on the PA12 and CF/PA12.

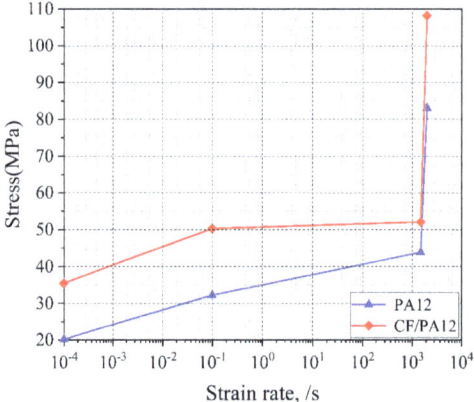

Figure 6. Effect of strain rate on the compressive mechanical properties of PA12 and CF/PA12 samples.

Moreover, it is also observed that the yield stress of CF/PA12 is much higher than that of PA12 under the same strain rate (the former is 15–25 MPa higher than the latter). This is due to the reinforcement caused by carbon fiber, and the reinforcement mechanisms will be discussed later.

It is observed that as the strain rate decreased, the reinforcing effect on yield strength and strain of carbon fiber was enhanced monotonously. Moreover, the reinforcing effect of yield strength was much larger than that of yield strain.

It is observed that the slope of the curve in Figure 7 steepens as the strain rate increases. As can be seen from Figure 7, with the increase of strain rate, the peak stress and modulus of composite material increase significantly. Although the peak strain decreases with the increase of strain rate, the dynamic failure strain and failure stress are much lower than the quasi-static failure strain.

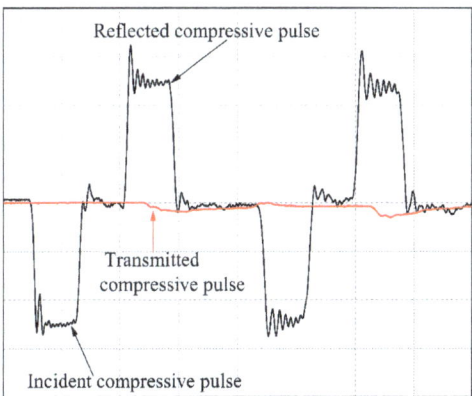

Figure 7. Typical dynamic compressive responses from strain gages.

3.2.2. Influence of Water Immersion

Figure 8 depicts the compressive stress–strain curves of the PA12 at different strain rates (10^{-4} to 2000/s). It is found that in the elastic deformation stage, the influence of water immersion on the elasticity modulus is negligible. This finding is different from the melting pultrusion impregnation of PA6 [24]. Nevertheless, after yielding, the yield strength of immersed PA12 is found to be smaller than that of the unimmersed PA12 at each strain rate. Normally, the water immersion causes the compressive strength to reduce by more than 15% at different strain rates (as illustrated in Figure 9a). This result is consistent with the finding reported by [25]. This also indicates that immersion water has a negative influence on the mechanical properties of SLS PA12. The reasons may be as follows:

The reversible hydrolysis reaction of the molecular chain takes place after absorbing water and leads to the reduction of the molecular weight of PA12 polymer. As consequence, the compressive strength of PA12 is also decreased after water immersion. Further, the amide groups occur repeatedly in the PA12 molecular chain belonging to the polar group. When the molecular chain does not contain a water molecule, the hydrogen atom on the amide group combined with the carbonyl group on another amide group to form a hydrogen bond thus increasing the crystallinity of the PA12. Meanwhile, the intermolecular force is strengthened simultaneously. However, after PA12 was soaked, a certain number of water molecules were stored in the pores of the sample. The water molecules cause the carbonyl group in the nylon molecular chain to dissociate from the hydrogen in the amide group, instead, forming a closer hydrogen bond with the water molecule. Thus, the interaction force between nylon molecules is reduced. In combination with the decreasing molecular weight caused by hydrolysis which reduces the compressive strength [24].

With regard to the SLS of CF/PA12, as shown in Figure 9, it is observed that the difference of compressive strength between immersed and unimmersed CF/PA12 is less than 4 MPa (8%), and the yield strain changes slightly, indicating that the influence of water immersion is relatively smaller than that of SLS of PA12. The reason is that after the nylon CF/PA12 samples were immersed in water, the water molecules absorbed by the sample were tightly locked by the carbon fiber, as it has strong water absorption and locking performance. Therefore, the hydrolysis reaction between the water molecules and polymer molecular chains is prevented effectively. As a consequence, the compressive mechanical properties of CF/PA12 immersed in water are slightly influenced. This indicates that the carbon fiber reinforced nylon can maintain its mechanical properties in humid environments or after being immersed in water.

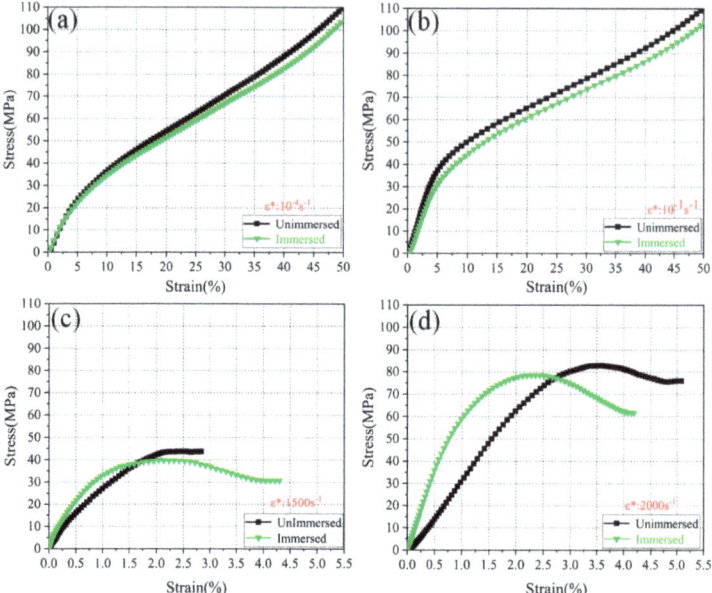

Figure 8. The stress–strain curves of SLS of PA12 under different strain rates: (**a**) 10^{-4} s^{-1}; (**b**) 10^{-1} s^{-1}; (**c**) 1500 s^{-1}; (**d**) 2000 s^{-1}.

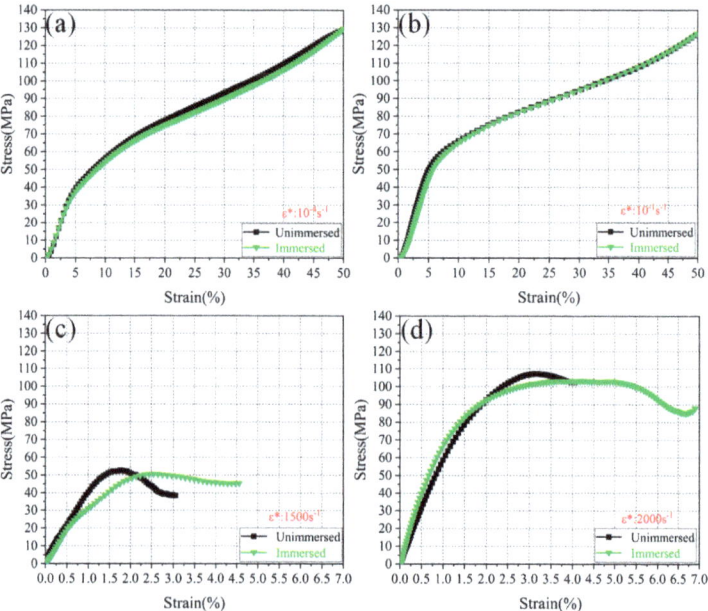

Figure 9. The stress–strain curves of SLS of CF/PA12 under different strain rates: (**a**) 10^{-4} s^{-1}; (**b**) 10^{-1} s^{-1}; (**c**) 1500 s^{-1}; (**d**) 2000 s^{-1}.

Figure 10 shows the change of yield strength with the increase of strain before and after immersing of PA12 and CF/PA12. Figure 10a shows that the yield strength of PA12 before immersing is stronger than after immersing. Figure 10b shows that the yield strength of CF/PA12 is almost unaffected by the immersing factor as the strain rate increases.

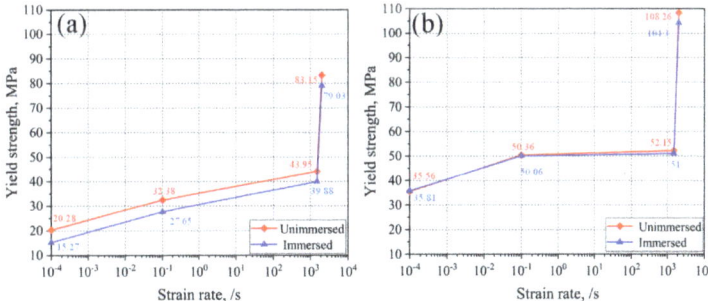

Figure 10. Comparison of yield strength versus strain rate of (**a**) PA12 and (**b**) CF/PA12.

Moreover, it is found that both the strain hardening rate of PA12 and CF/PA12 decrease slightly after water immersion, and the reasons are as follows: the strain hardening behavior of polyamide after yielding is due to the orientation and crystallization of chain segments under an external force. However, the regularity of molecular chains is distorted after immersion, and, thus, causes a smaller inter-chain force between the molecular chains. Further, the orientation of chains is easy to take place, thus causing a lower strain hardening rate [13].

3.3. Fracture Surface

In the quasi-static compression experiment, only CF/PA12 samples were broken, so the cross-section was photographed by SEM to reveal the cause of fracture.

As shown in Figure 11a, the PA12 samples did not break under quasi-static compression. The CF/PA12 samples tend to contract inward in the vertical direction and expand outward in the horizontal direction under compression loading. Moreover, the rates of contraction or expansion are found to increase upon the increasing strain rate. This could be due to the fact that compression usually leads to inhomogeneous deformation along the compression direction and the radial direction due to the frictional force between the sample and the compression anvils [26].

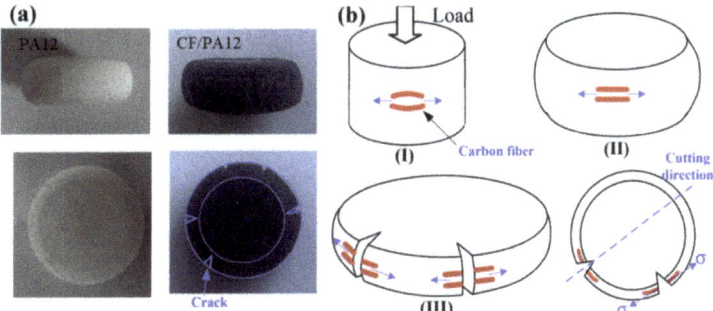

Figure 11. Schematic diagram for the cracking process of CF/PA12 and PA12 Sample: (**a**) PA12 and CF/PA12; (**b**) CF/PA12.

Further, as stated before, the carbon fibers were stretched to fracture, the reasons are as follows: Under compressive deformation, the samples expand outward in the horizontal direction; however, the carbon fibers embedded in the sample were fixed by the matrix material, thus causing the fibers to stretch along the radial direction of the circle (stage II in Figure 11b). It is also observed that the core area of a compressed sample experiences the smallest deformation strain while the outer edge area is deformed most heavily [27], which causes cracks on the edge initially, thus causing the carbon fibers to fracture along the tangent of the circle ultimately (stage III in Figure 11b). It should be noted that the fracture surfaces shown in Figure 11 were prepared by cutting the fractured samples along the cracks; thus, the fracture surfaces of carbon fibers were mainly perpendicular to the observation plane.

After compressive loading, the PA12 samples were compressed into a drum shape, but not fractured at all, while CF/PA12 samples fractured, their surfaces could give valuable information on the fracture behavior and modes, as illustrated in Figure 12. The samples did not break under high-speed impact. Therefore, what we showed is the samples broken under quasi-static condition. From the figure, it is observed that within the quasi-static loading range, the fracture surfaces are rough and uneven. This kind of morphology reflects tensile rather than compression deformation characteristics of polyamide [28]. In addition, the fracture surfaces of carbon fiber also exhibit tensile characteristics, and the surface becomes more flat under a higher strain rate. Moreover, cracks always exist close to the carbon fiber, indicating that the carbon fibers were stretched relative to the matrix during compression.

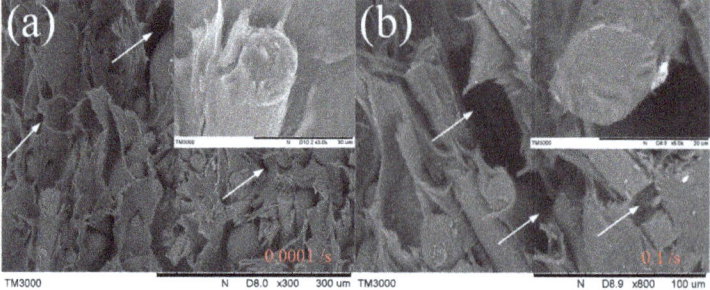

Figure 12. Fracture surfaces of CF/PA12 under different strain rates: (**a**) 10^{-4} s^{-1}; (**b**) 10^{-1} s^{-1}.

4. Conclusions

In this work, the influences of strain rate and hygroscopicity on the compressive properties of selective laser sintering (SLS) of polyamide 12 (PA12) and the carbon fiber reinforced polyamide 12 (CF/PA12) composites were comparatively studied, and the following conclusions are drawn.

The CF/PA12 had shorter saturation time and lower saturated water absorption under the same conditions, indicating that the SLS of CF/PA12 had lower hydrophilia and higher water resistance when compared with that of the SLS of PA12.

With the increasing strain rate, the ultimate compression strength and yield strength monotonically increased with almost the same slope, indicating that the strain rate had the same positive correlation with the compressive strength of SLS of PA12 and CF/PA12.

Compared with quasi-static state, PA12 and CF/PA12 can withstand nearly twice the yield strength under impacting load thanks to their good plasticity. Therefore, these two materials are better able to resist impact loading.

Water immersion resulted in a significant reduction of 15% in yield strength of SLS of PA12, but not so much to the CF/PA12, indicating that the carbon fibers favor for maintaining mechanical properties of polyamide 12 after absorbing water.

Author Contributions: Conceptualization, Y.L.; Data curation, X.Z. and J.M.; Formal analysis, X.Z.; Funding acquisition, Y.L.; Investigation, X.Z.; Methodology, J.M.; Project administration, Y.L.; Writing–original draft, X.Z. All authors have read and agreed to the published version of the manuscript.

Funding: This research received no external funding.

Conflicts of Interest: The authors declare no conflict of interest.

References

1. Olakanmi, E.O.; Cochrane, R.F.; Dalgarno, K.W. A Review on Selective Laser Sintering/Melting (SLS/SLM) of Aluminium Alloy Powders: Processing, Microstructure, and Properties. *Prog. Mater. Sci.* **2015**, *74*, 401–477. [CrossRef]
2. Yuan, S.Q.; Shen, F.; Chua, C.K.; Zhou, K. Polymeric Composites for Powder-Based Additive Manufacturing: Materials and Applications. *Prog. Polym. Sci.* **2019**, *91*, 141–168. [CrossRef]
3. Carneiro, O.S.; Silva, A.F.; Gomes, R. Fused Deposition Modeling with Polypropylene. *Mater. Des.* **2015**, *83*, 768–776. [CrossRef]
4. Pappu, A.; Pickering, K.L.; Thakur, V.K. Manufacturing and Characterization of Sustainable Hybrid Composites Using Sisal and Hemp Fibres as Reinforcement of Poly (Lactic Acid) via Injection Moulding. *Ind. Crop. Prod.* **2019**, *137*, 260–269. [CrossRef]
5. Yuan, M.Q.; Diller, T.T.; Bourell, D.; Beaman, J. Thermal Conductivity of Polyamide 12 Powder for Use in Laser Sintering. *Rapid Prototyp. J.* **2013**, *19*, 437–445. [CrossRef]
6. Güler, T.; Demirci, E.; Yildiz, A.R.; Yavuz, U. Lightweight Design of an Automobile Hinge Component Using Glass Fiber Polyamide Composites. *Mater. Test.* **2018**, *60*, 306–310. [CrossRef]
7. O'Connor, H.J.; Dowling, D.P. Comparison between the Properties of Polyamide 12 and Glass Bead Filled Polyamide 12 Using the Multi Jet Fusion Printing Process. *Addit. Manuf.* **2020**, *31*, 2214–8604.
8. Cai, C.; Tey, W.S.; Chen, J.Y.; Zhu, W.; Liu, X.J.; Liu, T.; Zhao, L.H.; Zhou, K. Comparative Study on 3D Printing of Polyamide 12 by Selective Laser Sintering and Multi jet Fusion. *J. Mater. Process. Technol.* **2021**, *288*, 116882. [CrossRef]
9. Crespo, M.; Gómez-del, R.M.T.; Rodríguez, J. Failure of SLS Polyamide 12 Notched Samples at High Loading Rates. *Theor. Appl. Fract. Mec.* **2017**, *92*, 233–239. [CrossRef]
10. Todo, M.; Takahashi, K.; Be´guelin, P.; Kausch, H.H. Strain-Rate Dependence of the Tensile Fracture Behaviour of Woven-Cloth Reinforced Polyamide Composites. *Compos. Sci. Technol.* **2000**, *60*, 763–771. [CrossRef]
11. Wang, K.; Xie, X.; Wang, J.; Zhao, A.D.; Peng, Y.; Rao, Y.N. Effects of Infill Characteristics and Strain Rate on the Deformation and Failure Properties of Additively Manufactured Polyamide-Based Composite Structures. *Results Phys.* **2020**, *18*, 103346. [CrossRef]
12. Sagradov, I.; Schob, D.; Roszak, R.; Maasch, P.; Sparr, H.; Ziegenhorn, M. Experimental Investigation and Numerical Modelling of 3D Printed Polyamide 12 with Viscoplasticity and a Crack Model at Different Strain Rates. *Mater. Today Commun.* **2020**, *25*, 101542. [CrossRef]
13. Li, H.F.; Wang, Y.; Zhang, C.W.; Zhang, B.M. Effects of Thermal Histories on Interfacial Properties of Carbon Fiber/Polyamide 6 Composites:Thickness, Modulus, Adhesion and Shear Strength. *Compos. Part A. Appl. Sci. Manuf.* **2016**, *85*, 31–39. [CrossRef]
14. Li, H.F.; Wang, S.F.; Sun, H.X.; Wang, Y.; Zhang, B.M. Moisture Absorption and Mechanical Properties of Continuous Carbon Fiber/Nylon 6 Thermoplastic Composites. *J. Compos. Mater.* **2019**, *36*, 114–121.
15. Sang, L.; Wang, C.; Wang, Y.Y.; Hou, W.B. Effects of Hydrothermal Aging on Moisture Absorption and Property Prediction of Short Carbon Fiber Reinforced Polyamide 6 Composites. *Compos. Part B. Eng.* **2018**, *153*, 306–314. [CrossRef]
16. Chaichanawong, J.; Thongchuea, C.; Areerat, S. Effect of Moisture on the Mechanical Properties of Glass Fiber Reinforced Polyamide Composites. *Adv. Powder Technol.* **2016**, *27*, 898–902. [CrossRef]
17. Do, V.T.; Tran, H.D.N.; Chun, D.M. Effect of Polypropylene on the Mechanical Properties and Water Absorption of Carbon Fiber Reinforced Polyamide-6/Polypropylene Composite. *Compos. Struct.* **2016**, *150*, 240–245. [CrossRef]
18. Yang, L.M.; Shim, V.P.W. An Analysis of Stress Uniformity in Split Hopkinson Bar Test Specimens. *Int. J. Impact Eng.* **2005**, *31*, 129–150. [CrossRef]
19. Carrillo, J.G.; Gamboa, R.A.; Flores-Johnson, E.A.; Gonzalez-Chi, P.I. Ballistic Performance of Thermoplastic Composite Laminates Made from Aramid Woven Fabric and Polypropylene Matrix. *Polym. Test.* **2012**, *31*, 512–519. [CrossRef]
20. Liu, Y.; Meng, J.; Zhu, L.; Chen, H.; Li, Z.; Li, S.; Wang, D.; Wang, Y.; Kosiba, K. Dynamic Compressive Properties and Underlying Failure Mechanisms of Selective Laser Melted Ti-6Al-4V Alloy under High Temperature and Strain Rate Conditions. *Addit. Manuf.* **2022**, *54*, 102772. [CrossRef]
21. Hocker, S.J.; Kim, W.T.; Schniepp, H.C.; Kranbuehl, D.E. Polymer Crystallinity and the Ductile to Brittle Transition. *Polymer* **2018**, *158*, 72–76. [CrossRef]
22. Deshoulles, Q.; Le Gall, M.; Dreanno, C.; Arhant, M.; Priour, D.; Le Gac, P.-Y. Modelling Pure Polyamide 6 Hydrolysis: Influence of Water Content in the Amorphous Phase. *Polym. Degrad. Stab.* **2021**, *183*, 109435. [CrossRef]
23. Rabello, M.S.; White, J.R. Crystallization and Melting Behaviour of Photodegraded Polypropylene I. Chemi-Crystallization. *Polymer* **1997**, *38*, 6379–6387. [CrossRef]
24. Das, V.; Kumar, V.; Singh, A.; Gautam, S.S.; Pandey, A.K. Compatibilization Efficacy of LLDPE-g-MA on Mechanical, Thermal, Morphological and Water Absorption Properties of Nylon-6/ LLDPE Blends. *Polym. Plast. Technol.* **2012**, *51*, 446–454. [CrossRef]
25. Dhakal, H.N.; Zhang, Z.Y.; Richardson, M.O.W. Effect of Water Absorption on the Mechanical Properties of Hemp Fiber Reinforced Unsaturated Polyester Composites. *Compos. Sci. Technol.* **2007**, *67*, 1674–1683. [CrossRef]
26. Sun, J.L.; Trimby, P.W.; Yan, F.K.; Liao, X.Z.; Tao, N.R.; Wang, J.T. Shear Banding in Commercial Pure Titanium Deformed by Dynamic Compression. *Acta Mater.* **2014**, *79*, 47–58. [CrossRef]

27. Yang, D.K.; Cizek, P.; Hodgson, P.D.; Wen, C.E. Microstructure Evolution and Nanograin Formation during Shear Localization in Cold-Rolled Titanium. *Acta Mater.* **2010**, *58*, 4536–4548. [CrossRef]
28. Kalinka, G. Effect of Transcrystallization in Carbon Fiber Reinforced Poly (Phenylene Sulfide) Composites on the Interfacial Shear Strength Investigatied with the Single Fiber Pull-Out Test. *J. Macromol. Sci. B.* **1996**, *35*, 527–546.

Article

Four-Dimensional Stimuli-Responsive Hydrogels Micro-Structured via Femtosecond Laser Additive Manufacturing

Yufeng Tao [1,2,*], Chengchangfeng Lu [3], Chunsan Deng [2], Jing Long [2], Yunpeng Ren [1], Zijie Dai [1,*], Zhaopeng Tong [1], Xuejiao Wang [1], Shuai Meng [1], Wenguang Zhang [2], Yinuo Xu [2] and Linlin Zhou [2]

1. Institute of Micro-Nano Optoelectronics and Terahertz Technology, Jiangsu University, Zhenjiang 212013, China; renyp@ujs.edu.cn (Y.R.); Tongzp@ujs.edu.cn (Z.T.); wangxj0122@hnu.edu.cn (X.W.); 2212003077@stmail.ujs.edu.cn (S.M.)
2. Wuhan National Laboratory for Optoelectronics, Huazhong University of Science and Technology, Wuhan 430074, China; chunsan.deng@foxmail.com (C.D.); D201780704@hust.edu.cn (J.L.); zhangwg@hust.edu.cn (W.Z.); xvyinuo@hust.edu.cn (Y.X.); m201772798@hust.edu.cn (L.Z.)
3. Whiting School of Engineering, Johns Hopkins University, Baltimore, MD 21218, USA; c9lu@ucsd.edu
* Correspondence: taoyufeng@ujs.edu.cn (Y.T.); Daizijie@ujs.edu.cn (Z.D.)

Abstract: Rapid fabricating and harnessing stimuli-responsive behaviors of microscale bio-compatible hydrogels are of great interest to the emerging micro-mechanics, drug delivery, artificial scaffolds, nano-robotics, and lab chips. Herein, we demonstrate a novel femtosecond laser additive manufacturing process with smart materials for soft interactive hydrogel micro-machines. Bio-compatible hyaluronic acid methacryloyl was polymerized with hydrophilic diacrylate into an absorbent hydrogel matrix under a tight topological control through a 532 nm green femtosecond laser beam. The proposed hetero-scanning strategy modifies the hierarchical polymeric degrees inside the hydrogel matrix, leading to a controllable surface tension mismatch. Strikingly, these programmable stimuli-responsive matrices mechanized hydrogels into robotic applications at the micro/nanoscale (<300 × 300 × 100 μm^3). Reverse high-freedom shape mutations of diversified microstructures were created from simple initial shapes and identified without evident fatigue. We further confirmed the biocompatibility, cell adhesion, and tunable mechanics of the as-prepared hydrogels. Benefiting from the high-efficiency two-photon polymerization (TPP), nanometer feature size (<200 nm), and flexible digitalized modeling technique, many more micro/nanoscale hydrogel robots or machines have become obtainable in respect of future interdisciplinary applications.

Keywords: femtosecond laser; additive manufacturing; hyaluronic acid methacryloyl; polyethylene glycol diacrylate; stimuli-responsiveness

1. Introduction

The development of modern robotics/sensors [1], micro/nanomechanics [2], tissue engineering [3], and drug delivery [4] introduces an urgent demand for smart micro- or nanostructured machines or robots [5]. However, conventional 3D-printed constructs have fallen short of expectations, mainly due to their bulky volume and inability to mimic the dynamic human tissues. Thereby, the shape-reconfigurable hydrogels emerge as a new scientific frontier for cancer treatment [6], wound healing [7], or biomimetic applications [8–15] acting similar to artificial muscles [11], grippers [12], actuators, active origami or machines [13], swellable scaffolds [14], organic electronics [15], microneedles [16,17], etc.

These smart hydrogel devices would ideally possess the embeddable volume and the reconfigurable morphology to be applied [9,10]. The composition of smart hydrogels re-defines the environment-to-hydrogel interactions, the controllability of responsive behavior [13], and the dynamic mechanical properties to a large extent. In the context of biomimetic applications, hydrogel-based materials have showcased certain advantages such as softness [14], biocompatibility and degradability [18], and massive potential for cell adhesion and proliferation.

Many additive manufacturing methods have been demonstrated with the four-dimensional (4D) time-dependent shape reconfiguration to obtain controllable stimuli-responsive behavior [19]. Four-dimensional printing has revolutionized traditional three-dimensional printing manufacturing products worldwide. In literature reviews, the state-of-the-artwork of 4D printing generally integrates the time-dependent behavior of stimulus-responsive materials [20], the associated materials interacting with various stimuli (physical, chemical, and even biological signals) for pre-designed motion or actuation [21,22]. The already-known 4D examples can be found from two-photon stereolithography [23] to 3D printing [24,25], extrusion fabrication, or ink-writing methods [26,27].

Among the existing macroscopic 4D products and fabrications with a relative millimeter or sub-millimeter resolution, the TPP using the femtosecond laser as a light source captures roaring attention. As is known, the primary-stage products of TPP are mostly stationary without controllable stimuli-inspired properties or artificial shape morphing. We recently reported a carbon nanotube-doped hydrogel with swelling-to-shrinkage behavior using two-photon polymerization (TPP) [15], where the swellable hydrogel scaffold absorbs functional materials for semiconductor applications. Its confinement of nonlinear two-photon absorption (TPA) within the submicron focal volume provided an ultra-fine spatial resolution [28]. Flexible laser parameters and a computer-aided technique [29] promised a tunable formation quality in fabricating quasi-arbitrary three-dimensional (3D) devices. In a nutshell, TPP will be a predominant tool [14,23,28] for fast fabricating core devices in interdisciplinary research due to its structural diversity and selective resolution.

Micro/nanostructures fabricated via TPP are formed through high-density covalent bonding networks. The solidified structures, generally, cannot undergo large transformations such as soft/elastomer materials. Recently, the smart properties of hydrogels have been improved. For example, the traditional actuation uses a residual stress-driven method for thermal shape deformation [30,31]. Some researchers use a unique laser writing process for pre-designed shape morphing [32], or a doping carbon nanotube to enhance the light-responsive behavior [33,34], or use a bio-environment to tune the optical properties [35,36]. Although having achieved tremendous progress [29–36], the stimuli-responsive behavior still deserves further investigation for complex high-freedom shape reconfiguration. The material limitation and monotonous laser scanning methods are now restraining this micro/nanoscale additive manufacturing from innovative development and broadband applications.

Herein, to obtain direct fabricating and temporally controlling micro/nanoscale quasi-arbitrary 3D geometries without filling other functional particles, we tentatively perform the modified TPP (Figure 1) on the bio-compatible hyaluronic acid methacryloyl (HAMA, Figure 1a) [7,17,23] with polyethylene glycol diacrylate [37] for responsiveness (Figure 1a). The shifted laser-scanning space creates a hetero-distribution of polymeric degrees through the formed hierarchical micro/nanostructures (Figure 1c), leading to a controllable surface tension mismatch. Following the programmable shape morphing, we display and identify several reconfigurable micro-scale structures by applying external stimuli for the first time. Shown by the cell loading experiment (Figures 1c, S1 and S2), the hydrogels allow fibril cells to crawl on the surface freely after one week, and the summarized cell viability exceeds 98%, implying the as-prepared hydrogels to be ready for cell culture [6,38] or tissue engineering.

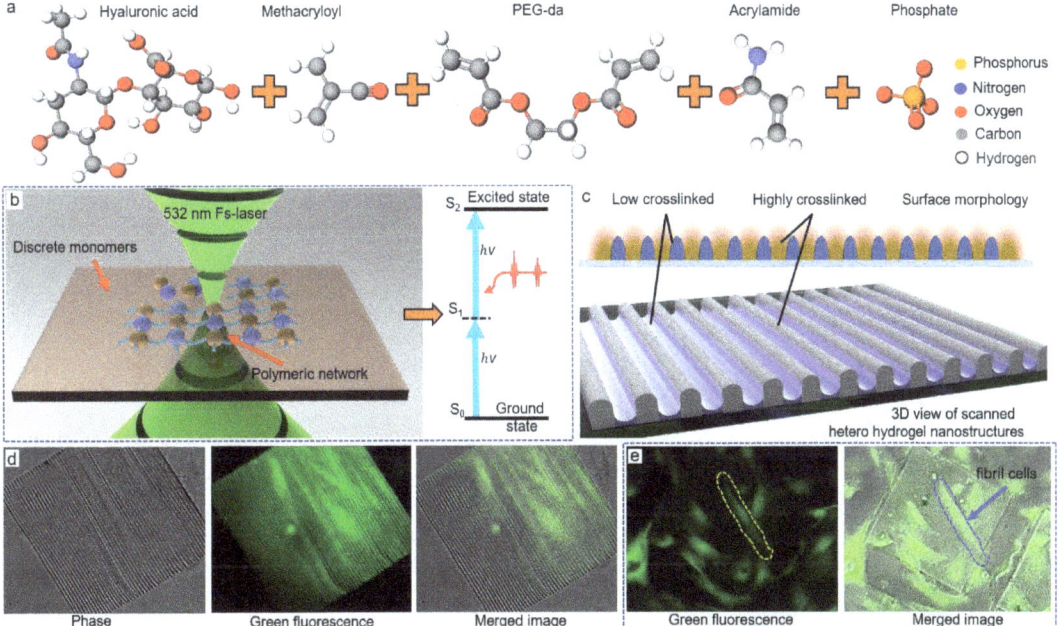

Figure 1. The schematic procedure of TPP incorporating stimuli-responsive hydrogels: (**a**) the main components used in the photoresist containing the methacryloyl-modified hyaluronic acid, PEG-da, and acrylamide as structural or functional materials rendering the hydrogel with stimuli-responsive ability; (**b**) illustration of the laser beam focused on the interface between material and substrate, the laser beam assembles the discrete functional polymers into a polymerized matrix, and the two-photon absorption process; (**c**) description of the used hetero-scanning TPP strategy, where the laser-scanned path forms highly crosslinked nanowires, while the interconnecting spacing is a low-degree crosslinked area, and the 3D view illustrates an interlocking morphology; (**d**) phase image, fluorescence image, and the merged images of typical grating structure created by our proposed method; (**e**) the fluorescence and images of loading fibril cells on the prepared square hydrogels for 2 weeks, where the active cells crawled on the surface of the structure after one week, indicating a desirable adhesion and bio-compatibility for cell culture.

Interestingly, the optical double-frequency technique is adopted here for generating the 532 nm green femtosecond laser beam during TPP fabrication for the first time. The double-frequency crystal is inserted in the path to decrease the optical power and wavelength simultaneously. We also use a micro-mechanics platform to characterize the mechanical behaviors of the TPP-fabricated hydrogels. A surface tension analysis expounds the mechanics' theoretical rationales during the controllable shape morphing.

2. Materials and Methods

2.1. Material Preparation

Hyaluronic acid methacryloyl (HAMA, 30 wt%), 2-hydroxy-2-methylpropiophenone (2 wt%, molecular structure seen in Supplementary Materials, Figure S3) [39,40], acrylamide (10%, Figure 1c), and poly(2-ethyl-2-oxazoline) diacrylates (PEG-da 475, 55 wt%) were mixed in phosphate-buffered saline (PBS, 4.8 wt%) for the responsive photoresist. Then, the mixture was pre-processed under 30 min ultra-sonication for dispersion and then was magnetically stirred at 800 rpm for 8 h. We purchased HAMA from Aladdin (Shanghai, China), and PBS from HyClone (Logan, UT, USA). All other reagents were purchased from Sigma-Aldrich (St. Louis, MO, USA) and were not purified before usage. The whole fabrication procedure (including development and applying external stimuli) was carried out without light illumination. In the following experimental section, a family of the photoresists in different weight ratios of two monomers were prepared in the same procedure.

2.2. Laser System and TPP Additive Manufacturing

A barium metaborate crystal ($Ba(BO_2)_2$) transformed the 1064 nm wavelength femtosecond laser beam from a Ti:Sapphire femtosecond laser (Chameleon-Discovery, Coherent, CA, USA) into a green 532 nm beam (Figure 2) based on the double-frequency effect. Optical power deceased to mW level in the optical path. An expander and an acoustic-optics modulator (AOM), a half-wave plate, a Glan mirror, and an aperture slot were placed in the optical path following our previous femtosecond laser direct writing system [41] (Figure 2a). The terminal biological microscope (IX83, Olympus, Tokyo, Japan) contained a two-dimensional nanometer-step moving platform, in situ charge-coupled device (CCD), and dichroic mirror, where three 20×, 40×, and 100× oil-immersed objectives were optional for different focusing lengths. The normal data slicing technique processed digitalized models such as 3D additive manufacturing [42,43]. The linearly polarized, sequential, near-infrared laser pulse propagated along the bottom-to-top path into the photon-sensitive photoresist without masks for 3D fabrication [44].

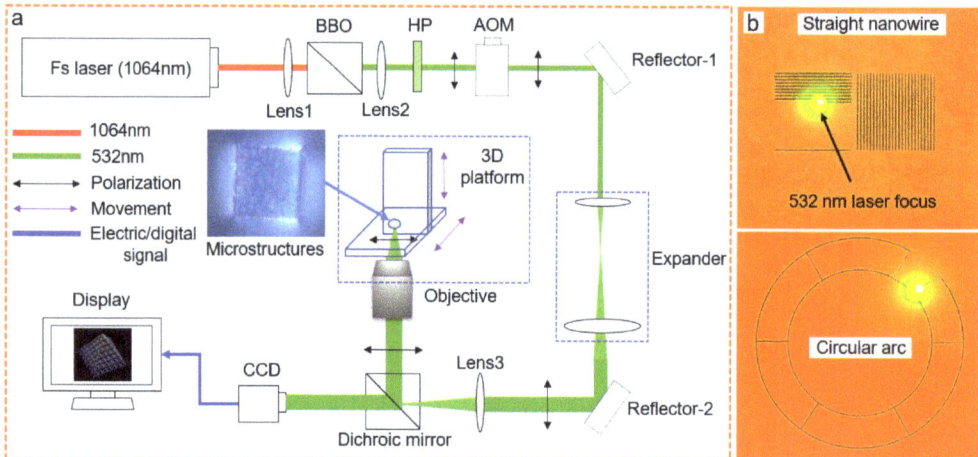

Figure 2. (**a**) The optical configuration of the double-frequency technique in ultrafast laser system for green laser beam; (**b**) the scanned straightforward nanowires and a pattern of curvy lines using the 532 nm femtosecond laser.

During the TPP process [29,44], we carefully adjusted laser power for resonant two-photon absorption ranging from 1.4 to 5 mW to avoid carbonization or incomplete photopolymerization. The repetition rate of the ultrafast pulsed laser beam was 78 MHz with an approximate 100 fs pulse width. Subsequently, we three-dimensionally moved laser focus inside photoresist tightly on the substrate to form solid gelation. The available moving speeds of x- and y-axis were both 50 μm/s following the pre-designed straight or curvilinear path (Seen in the Video S1). After TPP, the unsolidified photoresist was washed away by rinsing in alcohol (purity > 99%) for over 5 min.

2.3. Measurement

Substrates to be observed by Nova Nano SEM field emission electron microscope (SEM) with acceleration voltages of 5 KV were pre-coated with indium tin oxide semiconductor (ITO) film for electric conduction. The SEM software analyzed the size and spatial resolution. The imaging spectrometer (island-320, Teledyne Princeton Instruments, Princeton, NJ, USA) reflected the fluorescence images of hydrogel structures with a high-resolution scientific complementary metal oxide semiconductor (sCMOS, KURO, Teledyne Princeton Instruments, Princeton, NJ, USA). The software of island-320 set the color of fluorescence images. A CCD camera installed on the digitalized inverted microscope (ix83, Olympus, Tokyo, Japan) recorded the dynamic responsiveness. The minimum time interval of each video was a single sec. The external water stimuli were realized by dropping deionized water onto the sample or blowing moisture from a humidifier to investigate the responsive behavior.

The pH variation was realized by adding the diluted hydrochloric acid into in-solution hydrogels (Figure S4). A light stimulus was applied using the same optical system to the fabrication. The heating condition was realized by placing the sample on the thermoelectric cooler with a digital control step size of 0.2 °C at a range from 20 to 40 °C. The solidified samples' heat/water micro-forces that occurred were measured by FT-MTA2 of FemtoTools (Switzerland) with FT-S10000-TP tungsten probes of 50×50 μm^2 tip radius and 5 nN resolution. Young's Modulus of the solidified samples was measured using FT-S100-TP of 2×2 μm^2 tip radius and at the same resolution.

3. Results and Discussion

3.1. Fabricating Hollow 3D Structures with Selected Spatial Resolution

By laser scanning inside the photoresist (Figure 2b), the smart properties of materials provided fundamental responsiveness as beneficial advantages in devices. For example, the hydrogel-based microlenses (Figure 3a) exhibited the ability to change light facula similar to a dynamic focus lens (Video S2). Seen in Figure 3a, the diameter of the lens changed from about 20 μm to 29 μm, meaning the swelling area in the X–Y plane increased by at least 100% (seen in the Video S3). Subsequently, we changed the volume of the hydrogels and measured the volume alternation to confirm an approximate swelling ratio of >210% (seen in Figure S5). Moreover, the as-prepared hydrogel inherited the pH-responsive ability such as the previously studied bio-materials [6,45,46]. By slowly changing the pH value to an acidic environment, the in-solution hydrogel further swelled and stretched itself out (seen in Video S4). With the micro/nanoscale deformable structures, these smart devices promise broadband applications for embedding bio-conditions.

The experimental observation confirmed the desirable mobility of the photoresist and highly effective two-photon absorption at a mild laser power. The photon-induced cross-linking reaction was confined at the submicron voxel (Figure S5). All complex 3D scaffolds self-stood on substrates in the absence of supportive tools. Both the minimum line width and minimum height could reach 150 nm (Figure S6). The experiment generally concluded a suitable scanning speed from 30 to 140 μm/s with an average optical power distributed from 2 to around 20 mW. To check the formation quality for complex hollow structures widely applied for cell proliferation [9], micro machinery [47] or microfluidic [48], a batch of scaffolds consisting of specific tetrahedrons and cubes was firstly fabricated and characterized (Figure 3b,e).

An ultrafine feature size was observable in the CCD or SEM images (tetrahedron in Figure 3b,c, cubes in Figure 3d,e). By adjusting the scanning speed, the equivalent power exposure dose affected the volume of the cross-linking degree a lot. For example, the spatial resolution of Figure 3b scanned at 10 μm/s was better than that of Figure 3c scanned at 1 μm/s. The same selective ability was manifested again in the shape of the cubes. A higher exposure dose of the pulsed laser beam triggered a higher-level polymerization. Based on this factor, the optical parameters could determine the feature size.

Furthermore, the spectrometer-reconstructed fluorescence images of structures (Figure 3c,e) matched well with the SEM images. The highlighted fluorescence implied that the freestanding hollow structures used pure organic hydrogel materials without hard metal. Due to the bio-compatibility, desirable adhesion, and structural complexity of HAMA, and PEG-da [31], the demonstrated scaffolds promised more practical cell applications. The tests of cell viability and adhesion of hydrogels are ongoing, and will be reported on soon.

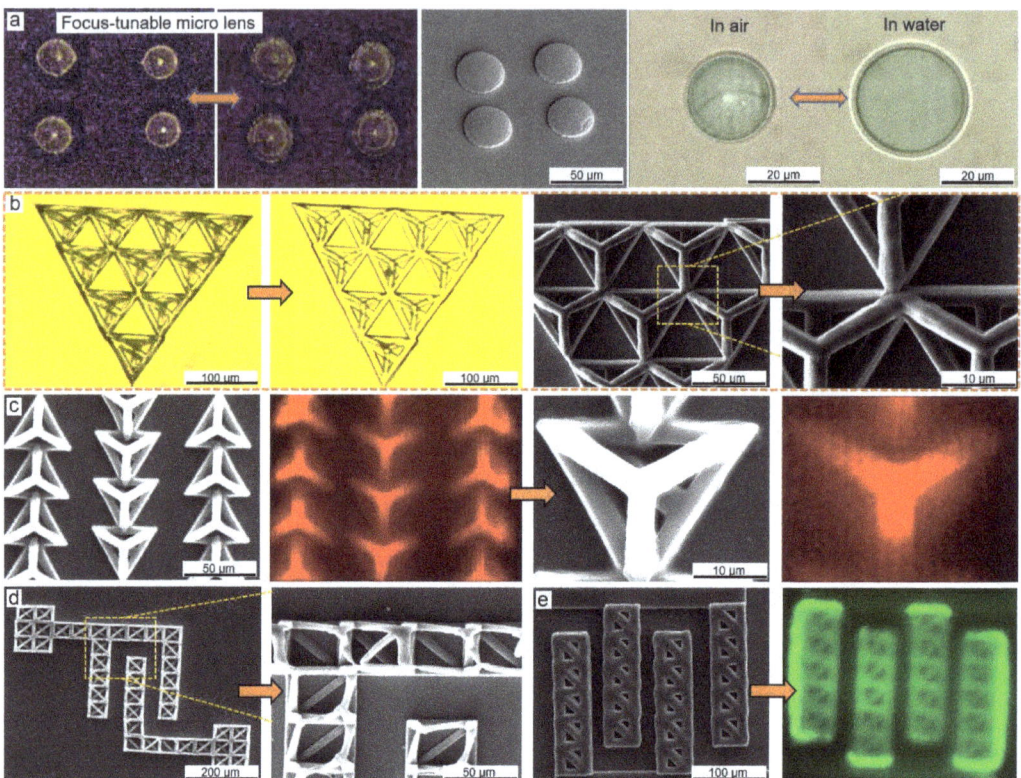

Figure 3. (**a**) A group of dynamic focus lenses using the proposed hydrogel materials, which tuned the size of facula by swelling-to-shrinkage on background light in a dark room; additionally, a comparison of the single hydrogel lens before and after swelling is shown; (**b**–**e**) the fabricated bio-scaffolds of tetrahedrons and cubes, respectively. (**b**) CCD, SEM, and zoomed-in SEM images of a triangle array of tetrahedrons is contained in the first panel; (**c**) SEM and fluorescence images (red) of tetrahedrons scanned at slow scanning speed about 5 µm/s; (**d**,**e**) observation of two arrays of cubes. Zoomed-in view shows a micrometer-level resolution (approximately 2 µm in (**d**), and 16 µm in (**e**)) to self-support the complex cube-stacked structures.

Additionally, TPP is typically an additive manufacturing process, where voxels stack up every layer, so the resolution of the laser voxels also plays a vital role in precise control. A smaller laser voxel determines a better spatial arrangement in the same exposure dose. By changing the magnifying ability of objectives, for example, N.A. from 1.2 to 1.43, the fine voxel generated an ultrafine resolution. As compared in Figure 3d,e, samples were scanned with N.A. of objectives = 1.43 and 1.2, respectively. The high accuracy implied the use of small voxels and a tight arrangement. All the demonstrated hydrogels here closely followed the design, although not ideally, as some geometric variations resulted from the intrinsic material properties, development, and observation method.

In addition to the selective spatial resolution and biocompatibility, this 532 nm TPP utilized optical parameters and a material ratio to modulate the mechanical properties in a wide range. As tested using our previously reported micro-mechanics technique [10], a Young's modulus of as-prepared hydrogels presented a wide range from KPa to MPa

(Figure 4a), covering the general requirements from tissue engineering to mechanics. These tunable mechanics proved that hydrogels work as a structural material and functional material simultaneously.

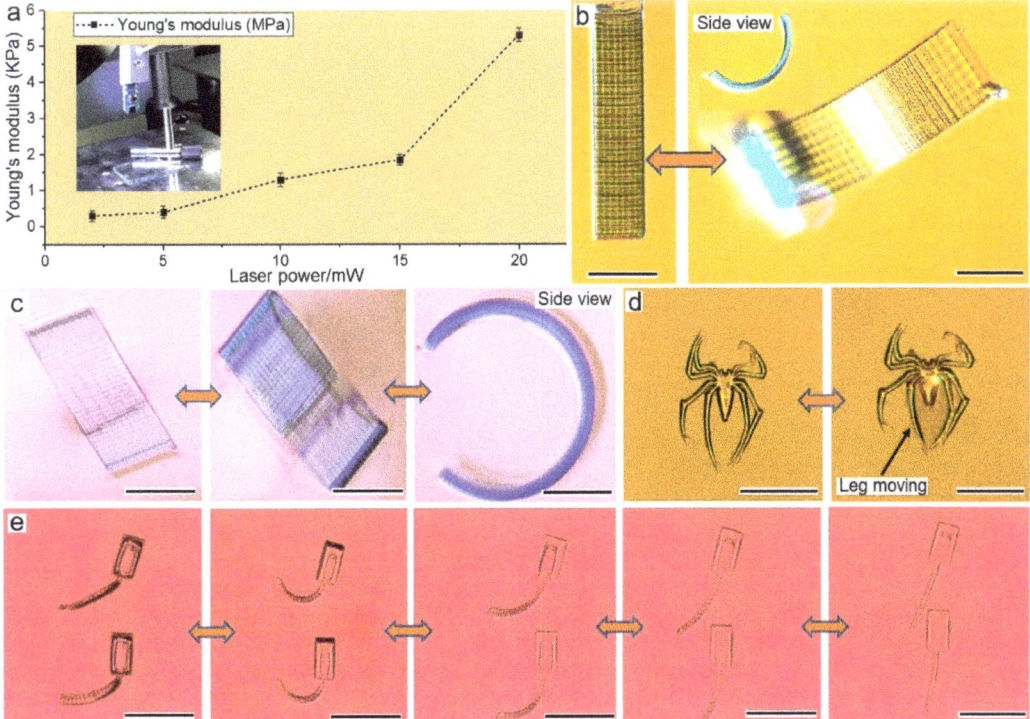

Figure 4. (**a**) Young's modulus of hydrogels; (**b**) water reversely bent a planar hydrogel from the initial planar to a bent shape, and the scale bar is 50 µm; (**c**) ethanol bent a planar hydrogel to a bent shape as well, and the scale bar is 50 µm; (**d**) a spider-shaped hydrogel was fabricated using our smart materials and then scanned by a laser beam for local actuation, the area projected by laser absorbed photon energy, and thermally swelled to commence shape reconfiguration, and the scale bar is 50 µm; (**e**) the tadpole-shaped hydrogels stretched their tails by swelling, and the scale bar of CCD images is 50 µm.

3.2. Humidity and Light-Triggered Reverse Shape Morphing

Traditionally, researchers have often combined soft active hydrogels with hard inert materials in dual-layer designs for actuation in multi-step fabrication. Utilizing the different swelling-to-shrinkage degrees of different materials, subject to the environment, for example, the temperature results in self-folding machines. However, TPP incorporating smart hydrogels enables the macroscopic stationary structure to reach a micron-to-nanometer level 4D function using single materials in a single step. The molecular interactive force between the functional groups and applied stimuli (polar solvent, water, acid, or alkali solvent) contributed to the stimuli responsiveness. Therefore, we changed the optical power and spacing width formed during scanning to display the resilient shape deformation ability.

In nature, many plants use water sorption and desorption for motion or reversible shape morphing. To mimic this behavior at a microscale, we fabricated the water-swelling hydrogel. The polymeric matrix consisted of permanent covalent carbon bonds in polymer materials, and various chemical functional groups could collaborate with outside-applied stimuli for judicious motion. As seen in Figure 4b,c, the humidity (or water) reversed the single-layer planar-like hydrogel, and the initial plane changed into a C type. No matter how frequently it is immersed into water or heated, the basic frame of the sample stayed unchanged, which denoted the cross-linked network's existence as a skeleton for structural integrity.

Here, the water molecules worked as the triggering condition, and the molecular force captured water to swell or shrink by heating to recover. The volume ratio shrank over 200% in evaporation, demonstrating a high water retention (Video S5). The reproducible volume changing meant that incredibly soft materials with a high liquid content are applicable to various biological and clinical research areas, from osteoporosis through tissue regeneration to hemorrhage control.

Figure 4d illustrates another kind of actuating method, where the hydrogel absorbed light energy and caused a local shape deformation, causing the spider-shaped hydrogel to activate, corresponding to the applied laser beam. In the light-fueled reconfiguration, the formed matrix absorbed photon energy and converted it into mechanical properties. The amplitude, location, frequency, and speed of the shape-changing properties passively depended on the applied laser beam (seen in Video S6). Here, both the laser pressure [49] and osmotic pressure in the water [50] contributed to the local shape morphing. Without the osmotic pressure (we evaporated the water off), the responsive activity of the spider hydrogel decreased significantly (seen in Video S7).

Furthermore, we fabricated the tadpole-shaped hydrogel, which swung its tail shape using the swelling effect (seen in Video S8), where the in-plane tail bent in air but straighten in water (Figure 4e). As a typical reverse process, we could prove the shrinkage of the tail to the initial state (seen in Video S9). In the discussion, the critical factor, besides the material affecting the bending and stretching, was the groove depth in the tail, which has previously been explored as a mechanism for shape deformation using a self-folding theory based on Timoshenko's theory [51].

Then, we fabricated a smart two-layer structure (Figure 5a), and the swelling happened out-plane in a perpendicular direction. The planar hydrogel bent upward reversely, and a part of the hydrogel relocated on the substrate due to intrinsic adhesion. The trick for the reconfigurable two-layer structure was the uneven scanning space of two layers. Therefore, the densities of two layer (seen in the SEM image of Figure 5a) varied a lot, leading to an uneven swelling or shrinkage degree and inducing shape morphing on the upper layer. The interface between the two layers was linked by smooth covalent bonding, with no mismatch of the traditional dual-layer design for actuation. Notably, the bending direction was perpendicular to the substrate (Figure 5b), implying a direction control using a two-layer structure. No fracture or physical damage was found in any of the shape-morphing hygromorphic hydrogels. Reverse programmability also meant that the functional groups were well maintained during and after TPP. The micro-structured hydrogels required only several seconds for shape reconfiguration, outperforming those bulky hydrogels of slow diffusive swelling rates [52,53] due to the micro/nanoscale surface effect, which made them more applicable for various aqueous environments.

3.3. Heat-Induced Shrinkage Behavior

Besides the humidity or light stimuli for responsiveness, the heating process also led to a self-bending action similar to an artificial muscle (Figure 6). The unique features found by heating, provide possibilities for sensing or actuation as well. If heated, the water uptaken by hydrogel would evaporate. Subsequently, the created surface tension changed to form a shrinkage-based 3D structure. The interspacing of adjacent nanowires modulated the bending degree. Illustrated by the flower (Figure 6a), heart (Figure 6b), and grid structure (Figure 6c), heat-induced deformation became predictable and useful. The hydrogel detected the temperature shifting in the ambient environment and changed its surface tension in the macroscope view.

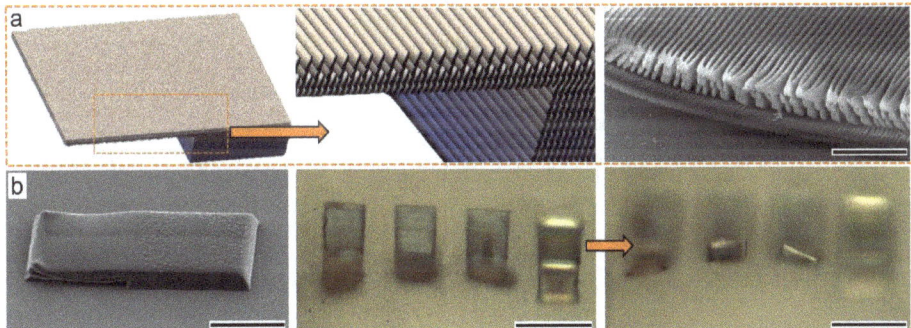

Figure 5. (**a**) The model of dual-layer design, its zoomed-in view, and a side-view SEM image of the differentiated two layers. The upper layer has a higher density of arranged nanowires, and the lower layer has a relatively smooth density; scale bar of the SEM image is 5 µm; (**b**) a two-layer design realizes upward bending out-plane, a group of four two-layer hydrogels demonstrates upward bending observed at height-changed focus. The scale bar of SEM image is 50 µm, scale bar of the CCD image is 100 µm.

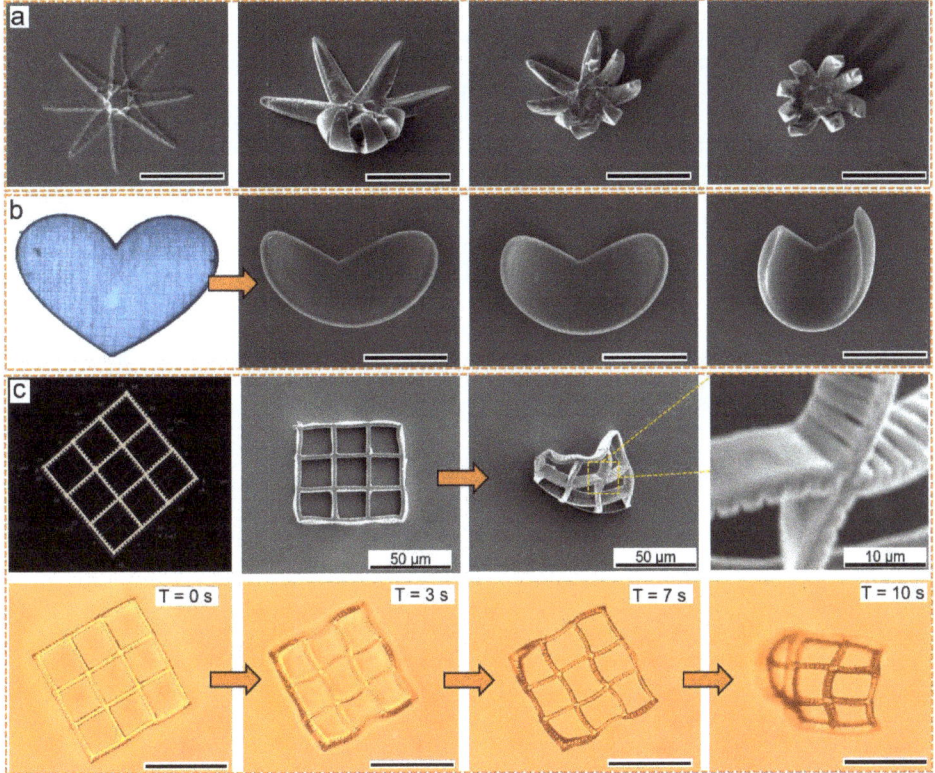

Figure 6. TPP-fabricated thermal-responsive hydrogel for temperature-controlled shaping–morphing: (**a**) the closure of flower mimic hydrogel working as a thermal gripper, where the scale bar is 50 µm; (**b**) an unevenly shrunk hydrogel from initial symmetrical heart shape, where scale bar is 50 µm; (**c**) the model, SEM image, and shrinkage process of a grid-shaped hydrogel, the scale bar is 50 µm.

The heterostructure consisted of solidified hydrogel nanowires, and the smooth spacing resulted in a divergence in shrinkage behavior. This divergence caused residual stress at the molecular level and caused the inward-direction contractile surface tension to accumulate. As seen in Figure 6a, the heat-transferring process differentiated in the eight petals, causing a disorderly shrinkage. Subsequently, we fabricated a symmetrical heart shape, where the heating process showed an asymmetrical shrinkage (Figure 6b). As seen in Figure 6c, another grid hydrogel self-folded into an out-plane uneven ball through heating. Conclusively, the uneven distribution of the geometry intensified the self-folding character and decreased the responsive time. The underlying mechanism for the controllable shape morphing could be found in the explanation section on surface tension (Supplementary Materials). The mechanics platform (seen in Figures S7 and S8, Supplementary Materials) further verified the tunable mechanical properties for reverse shape morphing.

4. Conclusions

In this study, we succeeded in developing a composite hydrogel material sensitive to a water/light/heat environment with a 532 nm femtosecond laser TPP. Compared to the mainstream optics/electron beam mask-projected stereolithography, the proposed two-photon polymerization held several advantages. An ultrafine feature size was obtained by staking the nanoscale voxel of the TPP system. The conventional macroscopic signal-triggered patterns or structures were miniaturized to a three-dimensional micron/nanoscale. The nonlinear characteristics of the fabrication processes still offered a sub-micron writing resolution, which is of great interest to micron-robotics, nano-drivers, and wearable sensors. Meanwhile, the stimuli-responsive photoresist contained no metal or alloy to improve biocompatibility. The controllable behaviors of the micro/nanostructures were being fatigue-free, environment-inspired, and quickly responsive, promising broad applications in micron actuators, sensors, micro-robotics, and biomimetic fields.

Supplementary Materials: The following supporting information can be downloaded at: https://www.mdpi.com/article/10.3390/mi13010032/s1, Figure S1: the fluorescence, phase, and merged images of our fabricated hydrogels loading with cells, Figure S2: the summarized cell viability of as-prepared hydrogels in Figure S1, Figure S3: Molecular structure of 2-hydroxy-2-methylpropiophenone used as photon initiator, Figure S4: dimension measurement on the responsive hydrogels before and after saturated swelling using the Nanomeasurer software 1.2, Figure S5: the optical intensity distribution of laser voxel at varied focusing positions, Figure S6: height measurement using advanced laser confocal microscopy, Figure S7: mechanics test platform used for determining the mechanical properties of our hydrogels, Figure S8: compressive and tensile test of micro-probe penetrating or pulling out of as-prepared square hydrogel, Video S1: 532nm laser beam scanning in photoresist, Video S2: Swelled microlenses by dropping water, Video S3: Single swelled microlens, Video S4: pH responsiveness, Video S5: Shrinkage of hydrogel, Video S6: Leg moving of a spider-shaped hydrogel by laser focus, Video S7: No obvious motion when laser scans spider in air, Video S8: Tail swelling of a tadpole-shaped hydrogel, Video S9: Tail shrinkage of a tadpole-shaped hydrogel.

Author Contributions: Conceptualization, Y.T.; methodology, C.L. and C.D.; validation, J.L. and S.M; investigation, W.Z. and L.Z.; resources, Z.D. and Z.T.; writing—original draft preparation, Y.X. and Y.T.; writing—review and editing, X.W. and Y.R. Validation: S.M.; Writing-original draft preparation: Y.X. All authors have read and agreed to the published version of the manuscript.

Funding: This research was financially supported by the National Key R&D Program of China (SQ2018YFB110138), the National Science Youth Fund of China (61805094), the National Natural Science Foundation of China (61774067), the National Science Foundation (CMMI 1265122), and the Fundamental Research Funds for the Central Universities (HUST:2018KFYXKJC027), China Postdoctoral Science Foundation (2017M622417).

Data Availability Statement: Data is contained within the article or Supplementary Materials.

Acknowledgments: The authors gratefully acknowledge Xiong Wei from Huazhong University of Science and Technology and Lu Yongfeng from University of Nebraska-Lincoln for providing the optical system and constructive instructions.

Conflicts of Interest: The authors declare no conflict of interest.

References

1. Kim, H.; Ahn, S.; Mackie, D.; Kwon, J.; Kim, S.; Choi, C.; Moon, Y.; Lee, H.B.; Ko, S. Shape morphing smart 3D actuator materials for micro soft robot. *Mater. Today* **2020**, *41*, 243–269. [CrossRef]
2. Daryadela, S.; Behroozfarb, A.; Minary-Jolandan, M. A microscale additive manufacturing approach for in situ nanomechanics. *Mater. Sci. Eng. C* **2019**, *767*, 138441. [CrossRef]
3. Guimarães, C.F.; Gasperini, L.; Marques, A.P.; Reis, R. The stiffness of living tissues and its implications for tissue engineering. *Nat. Rev. Mater.* **2020**, *5*, 351–370. [CrossRef]
4. Blanco, E.; Shen, H.; Ferrari, M. Principles of nanoparticle design for overcoming biological barriers to drug delivery. *Nat. Biotechnol.* **2015**, *33*, 941–951. [CrossRef]
5. Xin, C.; Jin, D.; Hu, Y.; Yang, L.; Li, R.; Wang, L.; Ren, Z.; Wang, D.; Ji, S.; Hu, K.; et al. Environmentally Adaptive Shape-Morphing Microrobots for Localized Cancer Cell Treatment. *ACS Nano* **2021**, *15*, 18048–18059. [CrossRef] [PubMed]
6. Wang, M.; Hu, H.; Sun, Y.; Qiu, L.; Zhang, J.; Guan, G.; Zhao, X.; Qiao, M.; Cheng, L.; Cheng, L.; et al. A pH-sensitive gene delivery system based on folic acid-PEG-chitosan-PAMAM-plasmid DNA complexes for cancer cell targeting. *Biomaterials* **2013**, *34*, 10120–10132. [CrossRef] [PubMed]
7. Zhang, X.; Chen, G.; Liu, Y.; Sun, L.; Zhao, Y. Black Phosphorus-Loaded Separable Microneedles as Responsive Oxygen Delivery Carriers for Wound Healing. *ACS Nano* **2020**, *14*, 5901–5908. [CrossRef]
8. Do, A.; Worthington, K.S.; Tucker, B.A.; Salem, A.K. Controlled drug delivery from 3D printed two-photon polymerized poly (ethylene glycol) dimethacrylate devices. *Int. J. Pharm.* **2018**, *552*, 217–224. [CrossRef]
9. Feliciano, A.J.; Blitterswijk, C.V.; Moroni, L.; Baker, M.B. Realizing Tissue Integration with Supramolecular Hydrogels. *Acta Biomater.* **2021**, *124*, 1–14. [CrossRef]
10. Wang, J.; Zhang, Y.; Aghda, N.H.; Pillai, A.R.; Thakkar, R.; Nokhodchi, A.; Maniruzzaman, M. Emerging 3D printing technologies for drug delivery devices: Current status and future perspective. *Adv. Drug Deliv. Rev.* **2021**, *174*, 294–316. [CrossRef]
11. Park, N.; Kim, J. Hydrogel-Based Artificial Muscles: Overview and Recent Progress. *Advanced Intelligent Systems. Adv. Intell. Syst.* **2020**, *2*, 1900135. [CrossRef]
12. Liu, Z.; Wang, Y.; Ren, Y.Y.; Jin, G.Q.; Zhang, C.; Chen, W.Y.; Yan, F. Poly(ionic liquid) hydrogel-based anti-freezing ionic skin for a soft robotic gripper. *Mater. Horiz.* **2020**, *7*, 919–927. [CrossRef]
13. Ding, M.; Jing, L.; Yang, H.; Machnicki, C.E.; Fu, X.; Li, K.; Wong, I.; Chen, P.Y. Multifunctional soft machines based on stimuli-responsive hydrogels: From freestanding hydrogels to smart integrated systems. *Mater. Today Adv.* **2020**, *8*, 100088. [CrossRef]
14. Tao, Y.; Wei, C.; Liu, J.; Deng, C.; Cai, S.; Xiong, W. Nanostructured electrically conductive hydrogels via ultrafast laser processing and self-assembly. *Nanoscale* **2019**, *11*, 9176–9184. [CrossRef]
15. Tao, F.; Deng, C.; Long, J.; Liu, J.; Wang, X.; Song, X.; Lu, C.; Yang, J.; Hao, H.; Wang, C.; et al. Multiprocess Laser Lifting-Off for Nanostructured Semiconductive Hydrogels. *Adv. Mater. Inter.* **2021**, 2101250. [CrossRef]
16. Economidou, S.N.; Perea, C.P.P.; Reid, A.; Uddin, M.J.; Windmill, J.F.C.; Lamproud, D.A.; Douroumis, D. 3D printed microneedle patches using stereolithography (SLA) for intradermal insulin delivery. *Mater. Sci. Eng. C* **2019**, 743–755. [CrossRef]
17. Yao, S.; Chi, J.; Wang, Y.; Zhao, Y.; Luo, Y.; Wang, Y. Zn-MOF Encapsulated Antibacterial and Degradable Microneedles Array for Promoting Wound Healing. *Adv. Healthc. Mater.* **2021**, *10*, 2100056. [CrossRef] [PubMed]
18. Freedman, B.; Uzun, O.; Luna, N.; Rock, A.; Clifford, C.; Stoler, E.; Östlund-Sholars, G.; Johnson, C.; Mooney, D. Degradable and Removable Tough Adhesive Hydrogels. *Adv. Mater.* **2021**, *33*, e2008573. [CrossRef]
19. Bernardeschi, I.; Ilyas, M.; Beccai, L. A Review on Active 3D Microstructures via Direct Laser Lithography. *Adv. Intell. Syst.* **2021**, *3*, 2100051. [CrossRef]
20. Nishiguchi, A.; Zhang, H.; Schweizerhof, S.; Schulte, M.F.; Mourran, A.; Möller, M. 4D Printing of a Light-Driven Soft Actuator with Programmed Printing Density. *ACS Appl. Mater. Interfaces* **2020**, *12*, 12176–12185. [CrossRef]
21. Rafiee, M.; Farahani, R.D.; Therriault, D. Multi-Material 3D and 4D Printing: A Survey. *Adv. Sci.* **2020**, *7*, 1902307. [CrossRef] [PubMed]
22. Lui, Y.; Sow, W.; Tan, L.; Wu, Y.; Lai, Y.; Li, H. 4D printing and stimuli-responsive materials in biomedical aspects. *Acta. Biomater.* **2019**, *92*, 19–36. [CrossRef]
23. Kufelt, O.; El-Tamer, A.; Sehring, C.; Schlie-Wolter, S.; Chichkov, B. Hyaluronic acid based materials for scaffolding via two photon polymerization. *Biomacromolecules* **2014**, *10*, 650–659. [CrossRef] [PubMed]
24. Xu, W.; Jambhulkar, S.; Zhu, Y.; Ravichandran, D.; Kakarla, M.; Vernon, B.; Lott, D.G.; Cornella, J.L.; Shefi, O.; Miquelard-Garnier, G.; et al. 3D printing for polymer/particle-based processing: A review. *Compos. B Eng.* **2021**, *223*, 109102. [CrossRef]

25. Ovsianikov, A.; Deiwick, A.; Vlierberghe, S.V.; Dubruel, P.; Möller, L.; Dräger, G.; Chichkov, B. Laser Fabrication of Three-Dimensional CAD Scaffolds from Photosensitive Gelatin for Applications in Tissue Engineering. *Biomacromolecules* **2011**, *12*, 851–858. [CrossRef] [PubMed]
26. Rajabasadi, F.; Schwarz, L.; Medina-Sánchez, M.; Schmidt, O. 3D and 4D lithography of untethered microrobots. *Prog. Mater. Sci.* **2021**, *120*, 100808. [CrossRef]
27. Gladman, A.S.; Matsumoto, E.A.; Nuzzo, R.G.; Mahadevan, L.; Lewis, J.A. Biomimetic 4D printing. *Nat. Mater.* **2016**, *15*, 413–418. [CrossRef]
28. Kawata, S.; Sun, H.B.; Tanaka, T.; Takada, K. Finer features for functional microdevices. *Nature* **2001**, *412*, 697–698. [CrossRef]
29. Malinauskas, M.; Žukauskas, A.; Hasegawa, S.; Hayasaki, Y.; Mizeikis, V.; Buividas, R.; Juodkazis, S. Ultrafast laser processing of materials: From science to industry. *Light-Sci. Appl.* **2016**, *5*, e16133. [CrossRef]
30. Ge, Q.; Qi, H.J.; Dunn, M.L. Active materials by four-dimension printing. *Appl. Phys. Lett.* **2013**, *103*, 131901. [CrossRef]
31. Liu, Y.; Shaw, B.; Dickey, M.D.; Genzer, J. Sequential self-folding of polymer sheets. *Sci. Adv.* **2017**, *3*, e1602417. [CrossRef]
32. Nishiguchi, A.; Mourran, A.; Zhang, H.; Möller, M. In-Gel Direct Laser Writing for 3D-Designed Hydrogel Composites That Undergo Complex Self-Shaping. *Adv. Sci.* **2017**, *5*, 1700038. [CrossRef]
33. Maruo, S.; Ikuta, K.; Korogi, H. Force-controllable, optically driven micromachines fabricated by single-step two-photon micro stereolithography. *J. Microelectromech. Syst.* **2003**, *12*, 533–539. [CrossRef]
34. Xiao, Y.; Lin, J.; Xiao, J.; Weng, M.; Zhang, W.; Zhou, P.; Luo, Z.; Chen, L. A multi-functional light-driven actuator with an integrated temperature-sensing function based on a carbon nanotube composite. *Nanoscale* **2021**, *13*, 6259–6265. [CrossRef]
35. Ceylan, H.; Yasa, I.C.; Yasa, O.; Tabak, A.F.; Giltinan, J.; Sitti, M. 3D-Printed Biodegradable Microswimmer for Theranostic Cargo Delivery and Release. *ACS Nano* **2019**, *13*, 3353–3362. [CrossRef] [PubMed]
36. Yang, Q.; Li, M.; Bian, H.; Yong, J.; Zhang, F.; Hou, X.; Chen, F. Bioinspired Artificial Compound Eyes: Characteristic, Fabrication, and Application. *Adv. Mater. Technol.* **2021**, *6*, 2100091. [CrossRef]
37. Urrios, A.; Parra-Cabrera, C.; Bhattacharjee, N.; Gonzalez-Suarez, A.M.; Rigat-Brugarolas, L.G. 3D-printing of transparent bio-microfluidic devices in PEG-DA. *Lab Chip* **2016**, *16*, 2287–2294. [CrossRef]
38. Czich, S.; Wloka, T.; Rothe, H.; Rost, J.; Penzold, F.; Kleinsteuber, M.; Gottschaldt, M.; Schubert, U.S.; Liefeith, K. Two-Photon Polymerized Poly(2-Ethyl-2-Oxazoline) Hydrogel 3D Microstructures with Tunable Mechanical Properties for Tissue Engineering. *Molecules* **2020**, *25*, 5066. [CrossRef] [PubMed]
39. Yu, H.; Ding, H.; Zhang, Q.; Gu, Z.; Gu, M. Three-Dimensional Direct Laser Writing of PEGda Hydrogel Microstructures with Low Threshold Power using a Green Laser Beam. *Light Adv. Manuf.* **2021**, *2*, 3. [CrossRef]
40. Vinck, E.; Cagnie, B.; Cornelissen, M.; Declercq, H.; Cambier, D. Green light emitting diode irradiation enhances fibroblast growth impaired by high glucose level. *Photomed. Laser Surg.* **2005**, *23*, 167–171. [CrossRef]
41. Tao, Y.; Ren, Y.; Wang, X.; Zhao, R.; Liu, J.; Deng, C.; Wang, C.; Zhang, W.; Hao, H. A femtosecond laser-assembled SnO_2 microbridge on interdigitated Au electrodes for gas sensing. *Mater. Lett.* **2022**, *308*, 131120. [CrossRef]
42. Dai, Z.; Su, Q.; Wang, Y.; Qi, P.; Wang, X.; Liu, W. Fast fabrication of THz devices by femtosecond laser direct writing with a galvanometer scanner. *Laser Phys.* **2019**, *29*, 065301. [CrossRef]
43. Yin, J.; Zhang, W.; Ke, L.; Wei, H.; Wang, D.; Yang, L.; Zhu, H.; Dong, P.; Wang, G.; Zeng, X. Vaporization of alloying elements and explosion behavior during laser powder bed fusion of Cu–10Zn alloy. *Int. J. Mach. Tool. Manuf.* **2021**, *161*, 103686. [CrossRef]
44. Xiong, Z.; Zheng, M.L.; Dong, X.Z.; Chen, W.Q.; Jin, F. Asymmetric microstructure of hydrogel: Two-photon micro fabrication and stimuli-responsive behavior. *Soft Matter* **2011**, *7*, 10353–10359. [CrossRef]
45. Zhou, Y.; Layani, M.; Wang, S.C.; Hu, P.; Ke, Y.J.; Magdassi, S.; Long, Y. Fully Printed Flexible Smart Hybrid Hydrogels. *Adv. Funct. Mater.* **2018**, *28*, 1705365. [CrossRef]
46. Shi, Y.; Ma, C.; Peng, L.; Yu, G. Conductive "Smart" Hybrid Hydrogels with PNIPAM and Nanostructured Conductive Polymers. *Adv. Funct. Mater.* **2015**, *25*, 1219–1225. [CrossRef]
47. Ji, S.; Li, X.; Chen, Q.; Lv, P.; Duan, H. Enhanced Locomotion of Shape Morphing Microrobots by Surface Coating. *Adv. Intell. Syst.* **2021**, *3*, 2000270. [CrossRef]
48. Kuo, A.; Bhattacharjee, N.; Lee, Y.; Castro, K.; Kim, Y.; Folch, A. High-Precision Stereolithography of Biomicrofluidic Devices. *Adv. Mater. Technol.* **2019**, *4*, 1800395. [CrossRef]
49. Zhang, X.; Pint, C.L.; Lee, M.H.; Schubert, B.E.; Jamshidi, A.; Takei, K.; Ko, H.; Gillies, A.; Bardhan, R.; Urban, J.; et al. Optically- and Thermally-Responsive Programmable Materials Based on Carbon Nanotube-Hydrogel Polymer Composites. *Nano Lett.* **2011**, *11*, 3239–3244. [CrossRef]
50. Li, M.; Wang, X.; Dong, B.; Sitti, M. In-air fast response and high speed jumping and rolling of a light-driven hydrogel actuator. *Nat. Commun.* **2020**, *11*, 3988. [CrossRef]
51. Bauhofer, A.A.; Krödel, S.; Rys, J.; Bilal, O.R.; Constantinescu, A.; Daraio, C. Harnessing Photochemical Shrinkage in Direct Laser Writing for Shape Morphing of Polymer Sheets. *Adv. Mater.* **2017**, *29*, 1703024. [CrossRef] [PubMed]
52. Zhang, J.; Guo, Y.; Hu, W.; Soon, R.H.; Davidson, Z.S.; Sitti, M. Liquid Crystal Elastomer-Based Magnetic Composite Films for Reconfigurable Shape-Morphing Soft Miniature Machines. *Adv. Mater.* **2021**, *33*, e2006191. [CrossRef] [PubMed]
53. Guo, W.; Li, W.; Zhou, J. Modeling programmable deformation of self-folding all-polymer structures with temperature-sensitive hydrogels. *Smart Mater. Struct.* **2013**, *22*, 115028. [CrossRef]

Article

Densification, Tailored Microstructure, and Mechanical Properties of Selective Laser Melted Ti–6Al–4V Alloy via Annealing Heat Treatment

Di Wang [1], Han Wang [1], Xiaojun Chen [1], Yang Liu [2,*], Dong Lu [3,4], Xinyu Liu [3,4] and Changjun Han [1,*]

1. School of Mechanical and Automotive Engineering, South China University of Technology, Guangzhou 510641, China; mewdlaser@scut.edu.cn (D.W.); 202020100649@mail.scut.edu.cn (H.W.); xjchan001@163.com (X.C.)
2. Laboratory of Impact and Safety Engineering, Ministry of Education, Ningbo University, Ningbo 315211, China
3. State Key Laboratory of Vanadium and Titanium Resources Comprehensive Utilization, Pangang Group Research Institute Co., Ltd., Panzhihua 617000, China; ludong_1786@163.com (D.L.); cgvermouth2022@163.com (X.L.)
4. Sichuan Advanced Metal Material Additive Manufacturing Engineering Technology Research Center, Chengdu Advanced Metal Materials Industry Technology Research Institute Co., Ltd., Chengdu 610300, China
* Correspondence: liuyang1@nbu.edu.cn (Y.L.); cjhan@scut.edu.cn (C.H.)

Abstract: This work investigated the influence of process parameters on the densification, microstructure, and mechanical properties of a Ti–6Al–4V alloy printed by selective laser melting (SLM), followed by annealing heat treatment. In particular, the evolution mechanisms of the microstructure and mechanical properties of the printed alloy with respect to the annealing temperature near the β phase transition temperature were investigated. The process parameter optimization of SLM can lead to the densification of the printed Ti–6Al–4V alloy with a relative density of 99.51%, accompanied by an ultimate tensile strength of 1204 MPa and elongation of 7.8%. The results show that the microstructure can be tailored by altering the scanning speed and annealing temperature. The SLM-printed Ti–6Al–4V alloy contains epitaxial growth β columnar grains and internal acicular martensitic α' grains, and the width of the β columnar grain decreases with an increase in the scanning speed. Comparatively, the printed alloy after annealing in the range of 750–1050 °C obtains the microstructure consisting of α + β dual phases. In particular, network and Widmanstätten structures are formed at the annealing temperatures of 850 °C and 1050 °C, respectively. The maximum elongation of 14% can be achieved at the annealing temperature of 950 °C, which was 79% higher than that of as-printed samples. Meanwhile, an ultimate tensile strength larger than 1000 MPa can be maintained, which still meets the application requirements of the forged Ti–6Al–4V alloy.

Keywords: additive manufacturing; selective laser melting; laser powder bed fusion; Ti–6Al–4V; heat treatment; annealing

1. Introduction

Ti–6Al–4V alloy has been widely used in the aerospace, energy, biomedical, and automotive sectors [1,2] due to its high strength, low density, high fracture toughness, excellent corrosion resistance, and good biocompatibility [3]. Metal additive manufacturing (AM) has been advancing in the fabrication of geometrically complex metal products, typically including selective laser melting (SLM), directed energy deposition, metal binder jetting, and sheet lamination [4–6]. SLM has been widely applied to manufacture complex titanium parts with short lead time, great design freedom, and comparable product performance to forged counterparts [7], such as aircraft brackets [8], cervical fusion cages [9], bone implants [10], and partial denture clasps [11].

A considerable number of research works have been conducted on the fabrication of the Ti–6Al–4V alloy via SLM. Process parameters have great influence on the relative density, microstructure, and mechanical properties of SLM-printed Ti–6Al–4V alloy. For instance, Sun et al. [12] explored the influence of laser power and scanning speed on the relative density of SLM-printed Ti–6Al–4V parts. With an increase in the laser power and a decrease in the scanning speed, the relative density of the printed parts could increase to more than 99%. Yang et al. [13] found that the microstructure of SLM-printed Ti–6Al–4V samples was composed of a typical hierarchical martensite structure with a high density of dislocations and twins, including primary, secondary, tertiary, and quaternary α' martensite in β columnar grains. The process parameters can affect the temperature and cooling rate of the melt pools, thus affecting the microstructure and mechanical performance of the printed Ti–6Al–4V parts. Wang et al. [14] established the relationship among process parameters, microstructure evolution, and mechanical properties. With an increase in the scanning speed to 1150 mm/s, the elongation could reach the maximum value of 7.8%. The synthetic effects of the grain refinement of α (α') martensite and the nano-β particle resulted in the improvement of the elongation. The SLM-printed Ti–6Al–4V samples usually possess higher strength and microhardness than cast or forged counterparts [15–18]. If nonoptimized process parameters are applied, manufacturing defects such as balling, cracks, and porosity are prone to appear, which are detrimental to the mechanical performance of the parts [19,20].

Post-treatments by annealing and hot isostatic pressing (HIP) are commonly applied to SLM-printed Ti–6Al–4V alloy to improve its elongation by transforming the α' martensite phase into a mixture of α and β phases [21–27]. Wang et al. [23] conducted annealing on the printed samples at 840 °C (below β phase transition temperature) and found that the maximum elongation increased from 5.79% to 10.28%, while the fracture type changed from quasi-cleavage to ductile fracture. Jamshidi et al. [27] performed HIP for the printed samples along the horizontal and vertical orientations at 930 °C and 100 MPa for 4 h. The results showed that the ductility was improved 2.1- and 2.9-fold in the vertical and horizontal orientations, respectively. After the post-treatments, the SLM-printed Ti–6Al–4V alloy could obtain improved plasticity but decreased mechanical strength.

Limited research has been systematically performed on the effects of process optimization and heat treatment on the microstructure and properties of SLM-printed Ti–6Al–4V alloy. Additionally, most research focused on the investigation of the microstructure and property evolution of the SLM-printed Ti–6Al–4V alloy heat-treated below the β phase transition temperature. This study aimed to determine the densification, tailored microstructure, and mechanical properties of the SLM-printed Ti–6Al–4V alloy through annealing heat treatment. In particular, the evolution mechanisms of the microstructure and mechanical properties of the printed alloy with respect to the annealing temperature near the β phase transition temperature were investigated. The process parameter optimization of SLM was conducted to obtain a high degree of densification for the printed alloy. The influences of scanning speed and annealing temperature on the microstructure and mechanical properties of the alloy were investigated and analyzed.

2. Materials and Methods

2.1. Materials

An atomized Ti–6Al–4V alloy powder (AP&C company, Boisbriand, QC, Canada) with an average particle size of 33 μm was used. The material composition is shown in Table 1. The loose density of the powder was 2.45 g/cm^3. Figure 1 shows the morphology of the powder and its particle size distribution. The powder particles were almost entirely spherical, and the particle size distribution was 15–45 μm.

Table 1. Chemical composition of the Ti–6Al–4V alloy powder.

Element	Ti	C	O	Ni	H	Fe	Al	V	Others
Ratio (%)	Balance	0.02	0.11	0.02	0.034	0.19	6.5	3.9	<0.1

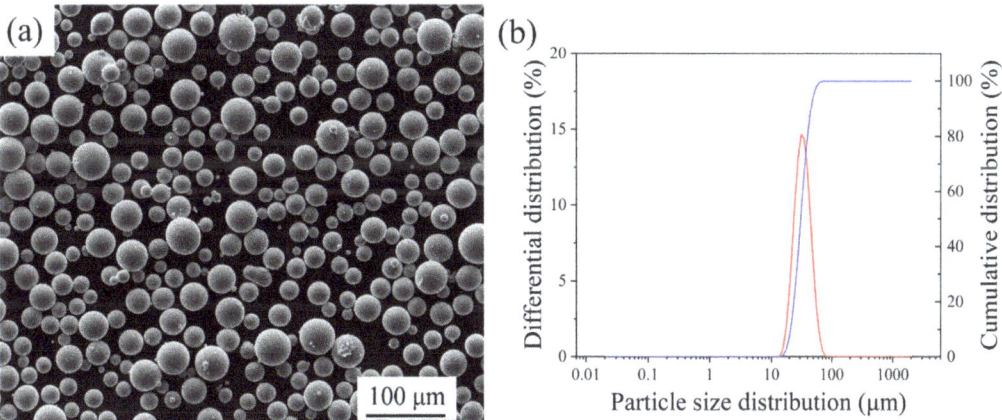

Figure 1. Characteristics of the Ti–6Al–4V alloy powder: (**a**) morphology; (**b**) particle size distribution.

2.2. SLM Process and Heat Treatment

A Dimetal-100 SLM equipment (Laseradd Technology Co., Ltd., Guangzhou, Guangdong, China) was utilized to print the Ti–6Al–4V alloy powder. The process optimization was conducted, and the parameter variables are shown in Table 2. The scanning strategy was bidirectional orthogonal scanning with a scanning starting angle of 145°. The cubic samples with dimensions of 8 mm × 8 mm × 8 mm and tensile samples were vertically printed, respectively, using various laser power, scanning speed, and hatch space values. All the samples were printed at the same position, i.e., central area of the substrate.

Annealing heat treatment was applied to the Ti–6Al–4V samples printed using optimized process parameters. It was reported that the annealing heat treatment for the SLM-printed Ti–6Al–4V alloy was mostly conducted at the temperature of 600–750 °C for 2 h, and the favorable temperature range was 800–900 °C for other heat treatments (except annealing) in the same timespan of 2 h [28]. In addition, the Ti–6Al–4V alloy annealed near to the β phase transition temperature (995 °C) could obtain the most improved mechanical properties [29]. Therefore, the annealing temperatures were set at 750 °C, 850 °C, 950 °C, and 1050 °C respectively. The samples were then held in the furnace for 2 h and then cooled. The procedure of the annealing process is shown in Figure 2.

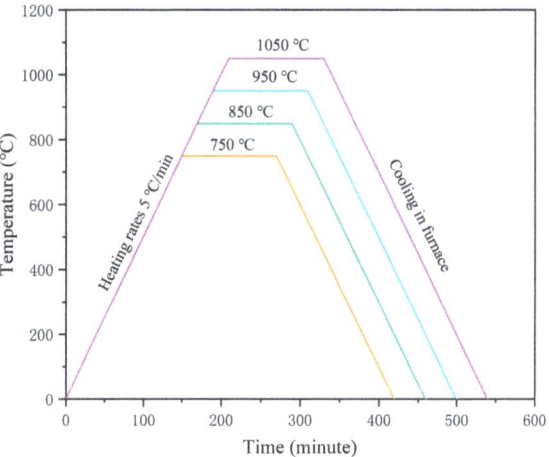

Figure 2. Annealing heat treatment procedures for the SLM-printed Ti–6Al–4V alloy.

Table 2. Process parameters of selective laser melting (SLM) for printing the Ti–6Al–4V alloy powder.

Parameter	Value
Laser power (W)	140, 150, 160, 170, 180
Scanning speed (mm/s)	700, 800, 900, 1000, 1100, 1200, 1300, 1400, 1500
Hatch space (mm)	0.06, 0.07, 0.08, 0.09, 0.1
Layer thicknesses (mm)	0.03

2.3. Characterizations

After printing, the relative density of the samples was measured to determine the optimal process parameters. The relative density of the printed Ti–6Al–4V samples was measured by an OHAUS PX124ZH electronic analytical balance (OHAUS Corporation, Parsippany, NJ, USA) according to the Archimedes drainage method. The samples were ground, polished, and etched with Kroll's agent ($H_2O/HF/HNO_3$ = 1:3:50 mL) for 20 s for microstructure characterization. The microstructure of the samples was observed with a Leica inverted optical microscope (Leica Microsystems GmbH, Wetzlar, Germany). The phase identification of the printed samples before and after annealing was conducted by a D8 ADVANCE X-ray diffractometer with copper target X-ray (Bruker AXS GmbH, Karlsruhe, Germany), with a scanning speed of 4 °/min, scan angle of 20–80°, and voltage value of 40 kV. The fracture morphology of the samples was observed by a Quanta 250 scanning electron microscope (FEI Company, Hillsboro, Oregon, USA).

2.4. Mechanical Testing

Tensile samples were printed according to ASTM-E8 (65 mm height, 25 mm gauge length, 2 mm thickness, and 5 mm width). Three samples were tested for each scanning speed and annealing temperature. The tensile tests at room temperature were carried out on a CMT5504 electronic universal testing machine (Zhuhai SUST Electrical Equipment Co., Ltd., Zhuhai, Guangdong, China) with an NCS electronic extensometer (NCS Testing Technology Co., Ltd., Beijing, China) and a speed of 0.5 mm/min.

3. Results and Discussion

3.1. Relative Density

Figure 3 shows the relative density of the SLM-printed samples with various scanning speed and laser power values when the hatch space was 0.07 mm, and the layer thickness was 0.03 mm. Under a low laser power (140–150 W), the relative density increased from 98.3% to 99.04% with the scanning speed increasing from 700 mm/s to 900 mm/s, and then decreased to 97.4% when the scanning speed was larger than 900 mm/s (Figure 3a). A low laser power is prone to producing lack of fusion pores [30]. At low scanning speeds, the laser beam can continuously heat the melt pools, resulting in greater laser energy to the Ti–6Al–4V powder particles and more unstable molten pool flow [31]. It is easy to trap gas into the melt pools to form micropores during their solidification [32]. Meanwhile, serious sputtering occurs, and splashed metal particles fall back to the surface of powder bed to form metal spheres. The reason for the decrease in the relative density is that the accumulation of the spheres leads to the generation of inclusions and pores [33].

Comparatively, under a relatively high laser power (160–180 W) (Figure 3b), the relative density could reach the highest values at a scanning speed of 1300 mm/s. A high laser power can result in a large depth of the powder layer penetrated by the laser beam, which improves the fluidity of the melt pool [34]. At high scanning speeds, a large solidification shrinkage of the melt pools tends to occur, resulting in a poor multi-track overlap and large gap between the tracks. The increase in the gap leads to the increase in layer thickness in the track gap after powder spreading. Therefore, the effective energy density is reduced, which promotes the formation of pores and reduces the relative density of the printed samples. The highest relative density of 99.51% could be achieved with a laser power of 170 W and a scanning speed of 1300 mm/s for the SLM-printed Ti–6Al–4V alloy.

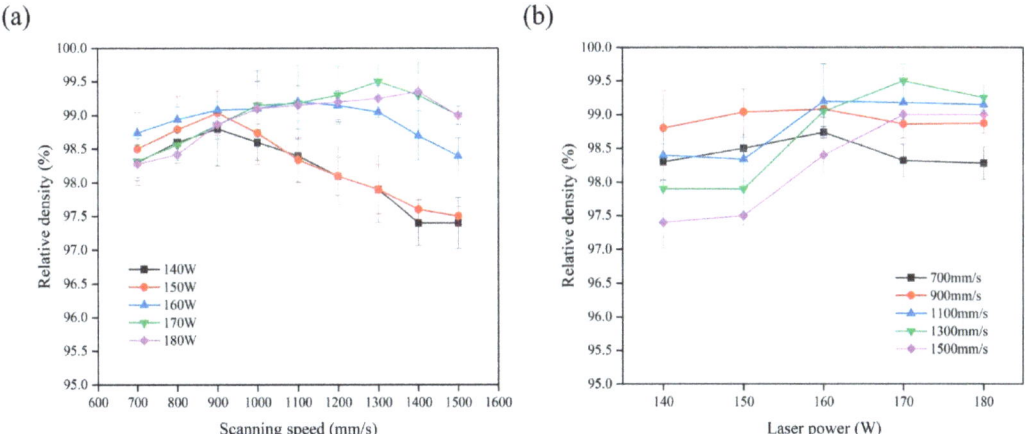

Figure 3. Variation trend of relative density of Ti–6Al–4V samples with various parameters: (**a**) scanning speed; (**b**) laser power.

Figure 4 shows the variation of the relative density of the SLM-printed samples with hatch space when the laser power was 170 W, the scanning speed was 1300 mm/s, and the layer thickness was 0.03 mm. When the hatch space increased from 0.06 mm to 0.07 mm and from 0.07 mm to 0.1 mm, the relative density of the sample increased from 99.21% to 99.5% and then gradually decreased to 98.5%, respectively. When the laser power and scanning speed were kept constant, the laser input energy was constant, and the melt pool width remained stable. A large hatch space resulted in a quite small overlap rate between adjacent melt pools, which is conducive to the formation of porosity. Comparatively, the decrease in the hatch space increased the overlap rate and reduced the heating time interval between the adjacent melt pools, resulting in a sufficient metal flow within the melt pools and resultant high relative density [35].

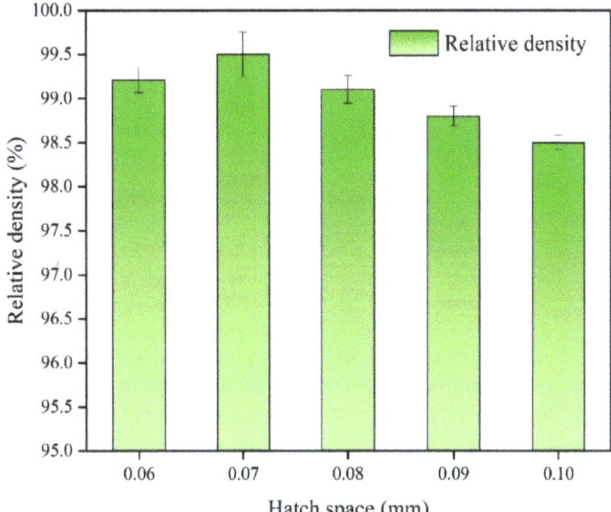

Figure 4. Histogram showing the influence of the hatch space on the relative density of the SLM-printed Ti–6Al–4V samples.

3.2. Microstructure

Figure 5 exhibits the microstructure of the SLM-printed Ti–6Al–4V samples manufactured at different scanning speeds under a laser power of 170 W, hatch space of 0.07 mm, and layer thickness of 0.03 mm. The microstructure of the samples with different scanning speeds was mainly composed of coarse epitaxial columnar grains that grew along the building direction. In the SLM process, the melt pool temperature is generally higher than that of the β phase generation. The ultrahigh cooling rates (up to 10^6 K/s) suppresses the transformation from the β phase into the α phase, and martensitic transformation occurs to form fine acicular α′ grains [36]. The primary β grains were filled with fine acicular α′ martensite that grew toward 45° upward with the building direction. When the scanning speed ranged from 900 mm/s to 1100 mm/s, the average width of the primary β grain was about 200 μm. However, the further increase in the scanning speed from 1200 mm/s to 1300 mm/s resulted in a decrease in the average width of the primary β grain to 150 μm. In addition, the scanning speed had a great impact on the porosity of the samples, which is consistent with the results shown in Figure 3.

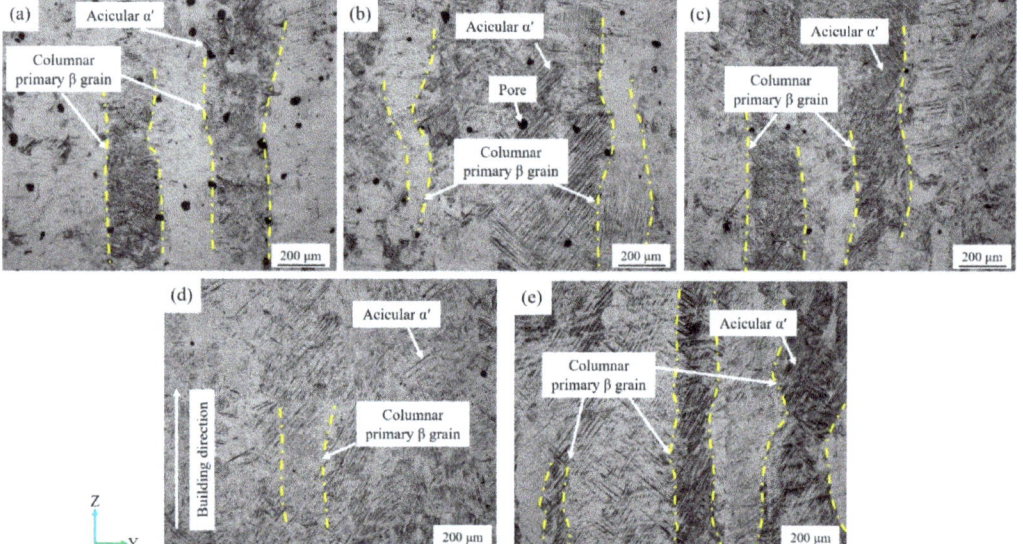

Figure 5. Microstructure of the SLM-printed Ti–6Al–4V samples at different scanning speeds: (**a**) 900 mm/s; (**b**) 1000 mm/s; (**c**) 1100 mm/s; (**d**) 1200 mm/s; (**e**) 1300 mm/s.

Figure 6 shows the X-ray Diffraction (XRD) pattern of the SLM-printed Ti–6Al–4V samples with different scanning speeds. Both α and α′ phases possessed hexagonal close-packed (hcp) structures. The diffraction angle of the strongest peak of the samples was shifted to a large value, as compared to the standard diffraction angle of 40.251°. This peak shift confirmed the formation of the martensite α′ phase. The increase in scanning speed led to the larger shift of the α′ peak, due to the increase in the cooling rate of the melt pool [37]. The cooling rate increase reduced the β precipitated phase but increased the content of V and Al in the acicular α′ phase, which decreased the lattice size of the α′ phase and the augmentation of its diffraction angle [14]. In the spectrum, the diffraction peak of the β phase was not obvious because of its low content caused by its transformation into the α′ phase during the cooling process.

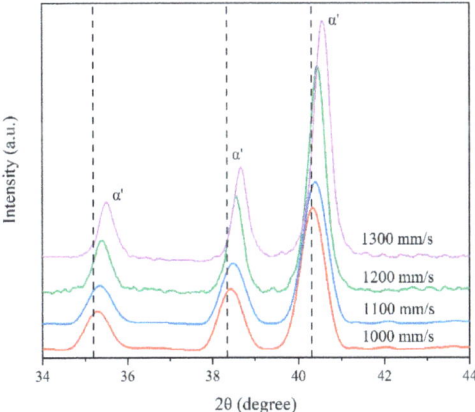

Figure 6. X-ray Diffraction (XRD) pattern of the SLM-printed Ti–6Al–4V samples at different scanning speeds from 34° to 44°.

3.3. Mechanical Properties

Figure 7 shows the mechanical properties of the SLM-printed Ti–6Al–4V samples with different scanning speeds. The scanning speed had a significant effect on the elongation but not on the tensile strength. The tensile strength of the samples was in the range of 1200–1265 MPa, and the maximum value could be obtained when the laser power was 170 W and the scanning speed was 900 mm/s. However, the elongation firstly increased from 5.5% to 7.8% and then decreased with an increase in the scanning speed. The maximum elongation could be obtained at a scanning speed of 1300 mm/s. Since the as-printed Ti–6Al–4V samples contained complete martensite structures and fine grains, their tensile strength was much larger than the minimum strength requirements for forged Ti–6Al–4V specified in the standard ASTM F1472-14, but the elongation was lower due to the brittleness of the martensite and large residual stress.

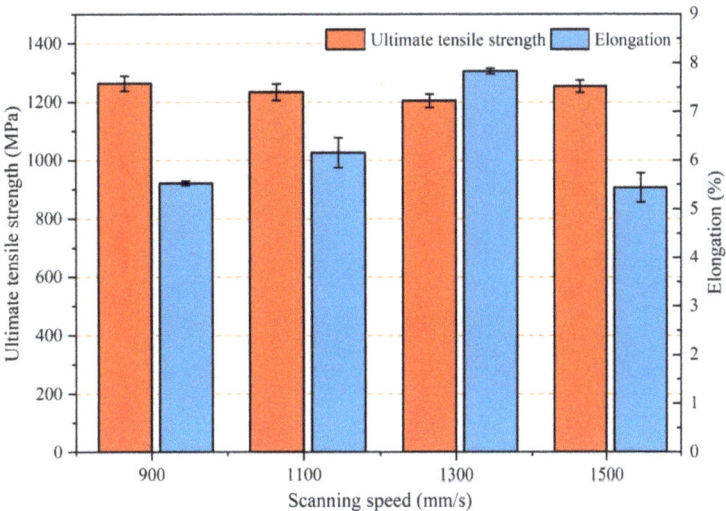

Figure 7. Effect of the scanning speed on the mechanical properties of the SLM-printed Ti–6Al–4V alloy.

The scanning speed could influence the amount of the precipitated β phase and the size of the α′ martensite in the SLM-printed samples. The β phase distributed at the boundary of the acicular α′ martensite phase had higher strength, and the decrease in elongation was mainly due to the dislocation locking by β during tensile stress [38], which hindered the movement of dislocations between the α phases. Therefore, with the increase in the scanning speed, the β phase decreased and the elongation of the samples increased. On the other hand, the reduction in the slip length of the α phase may have resulted in the increase in elongation [39]. The slip length of the α phase could be approximately equal to the width of the acicular α′ martensite. Upon increasing the scanning speed, the α′ martensite was refined, thereby reducing the slip length of the α phase and increasing the elongation.

Figure 8 presents the representative tensile fracture morphology of the SLM-printed samples under a scanning speed of 900 mm/s and 1300 mm/s. When the scanning speed was 900 mm/s, the fracture surface was mainly composed of flat cleavage steps, and also contained shallow dimples, indicating brittle fracture. The edge of the sample was exposed with more pore defects and nonmolten powder particles. These defects came from the insufficient laser energy input during the SLM process and were strongly related to the improper scanning speed. During the tensile testing, the pores led to the initiation of cracks, causing premature failure of the sample. Under a scanning speed of 1300 mm/s, the sample showed few internal defects and shearing surfaces. The higher scanning speed reduced the size of the crystal grains, which exhibited higher ductility under tension [12].

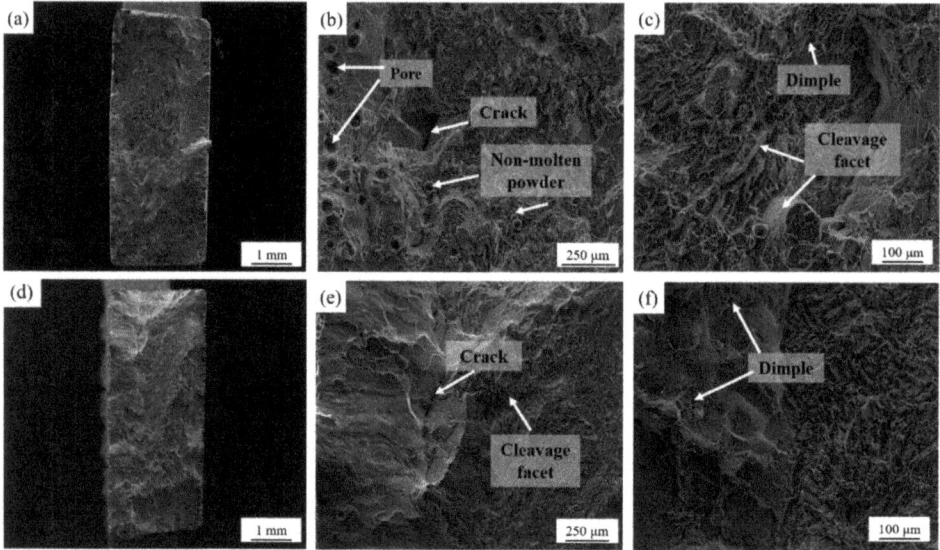

Figure 8. Fracture morphology of the SLM-printed Ti–6Al–4V samples at different scanning speeds: (**a**–**c**) 900 mm/s; (**d**–**f**) 1300 mm/s.

3.4. Effect of Annealing Temperature on Microstructure

The Ti–6Al–4V alloy samples before annealing were printed with a laser power of 170 W, a scanning speed of 1300 mm/s, and a hatch space of 0.07 mm. Figure 9 presents the XRD pattern of SLM-printed Ti–6Al–4V alloy samples after annealing at different temperatures. The results showed similar α and β diffraction peaks of the samples at different annealing temperatures. However, the diffraction peak of the β phase was weak, indicating the low volume fraction of the β phase. The full width at half maximum (FWHM) of the α/α′ peak (2θ = 40.2°) is listed in Table 3. Compared to the SLM-printed sample, the

FWHM of the heat-treated samples significantly decreased, elucidating that the residual stress within the printed samples was significantly eliminated [40].

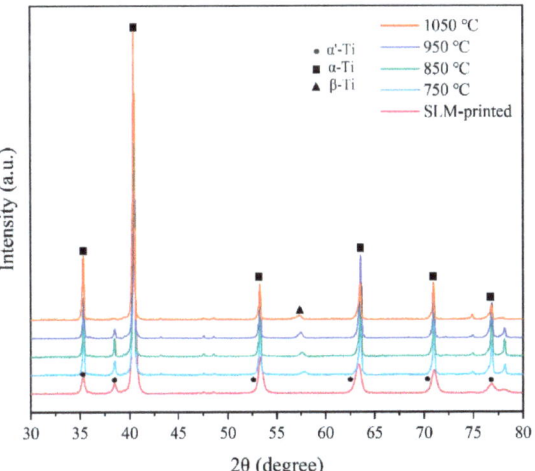

Figure 9. XRD pattern of the SLM-printed samples before and after annealing heat treatment.

Table 3. Full width at half maximum (FWHM) of the samples calculated from the XRD patterns.

Annealing Temperature	SLM-Printed	750 °C	850 °C	950 °C	1050 °C
FWHM	0.419	0.188	0.149	0.196	0.160

The microstructure of the SLM-printed samples after annealing at different temperatures is shown in Figure 10. When the annealing temperature of 750 °C was applied, the microstructure of the sample changed significantly as compared with that of the as-printed sample (Figure 10a,b). The primary β columnar grains in the annealed sample still existed, while the acicular α' martensite in the columnar grains transformed into a mixed α + β phase. The annealing temperature of 750 °C could only drive the partial decomposition of the α' martensite; thus, the α phase in the structure maintained the acicular shape [41]. At the annealing temperature of 850 °C, there was still epitaxial growth of the β columnar grains, but the β boundary became blurred and disappeared (Figure 10c,d). Compared with the annealed microstructure at 750 °C, the metastable acicular α' martensite phase in the β columnar crystals almost decomposed, and the lath-shaped α phase increased and became coarse. After annealing at 850 °C, the microstructure consisted of the α and β phase, showing a network structure. Studies have shown that the α' phase of Ti–6Al–4V alloy can completely decompose at temperatures above 800 °C [42].

After annealing at 950 °C, β columnar grains disappeared and the lamellar α phase could be observed (Figure 10e,f). The grains were further coarsened, and the distribution of α + β dual phase was more uniform. When the annealing temperature increased to 1050 °C, exceeding the β phase transition temperature, the microstructure was completely β phase in the heat preservation state, and the columnar structure in the as-printed sample could be completely eliminated (Figure 10g,h). Due to the slow cooling rate, the α phase in the β grain gathered to form a lath-shaped structure with the same orientation, and coarse Widmanstätten structures could be obtained [43].

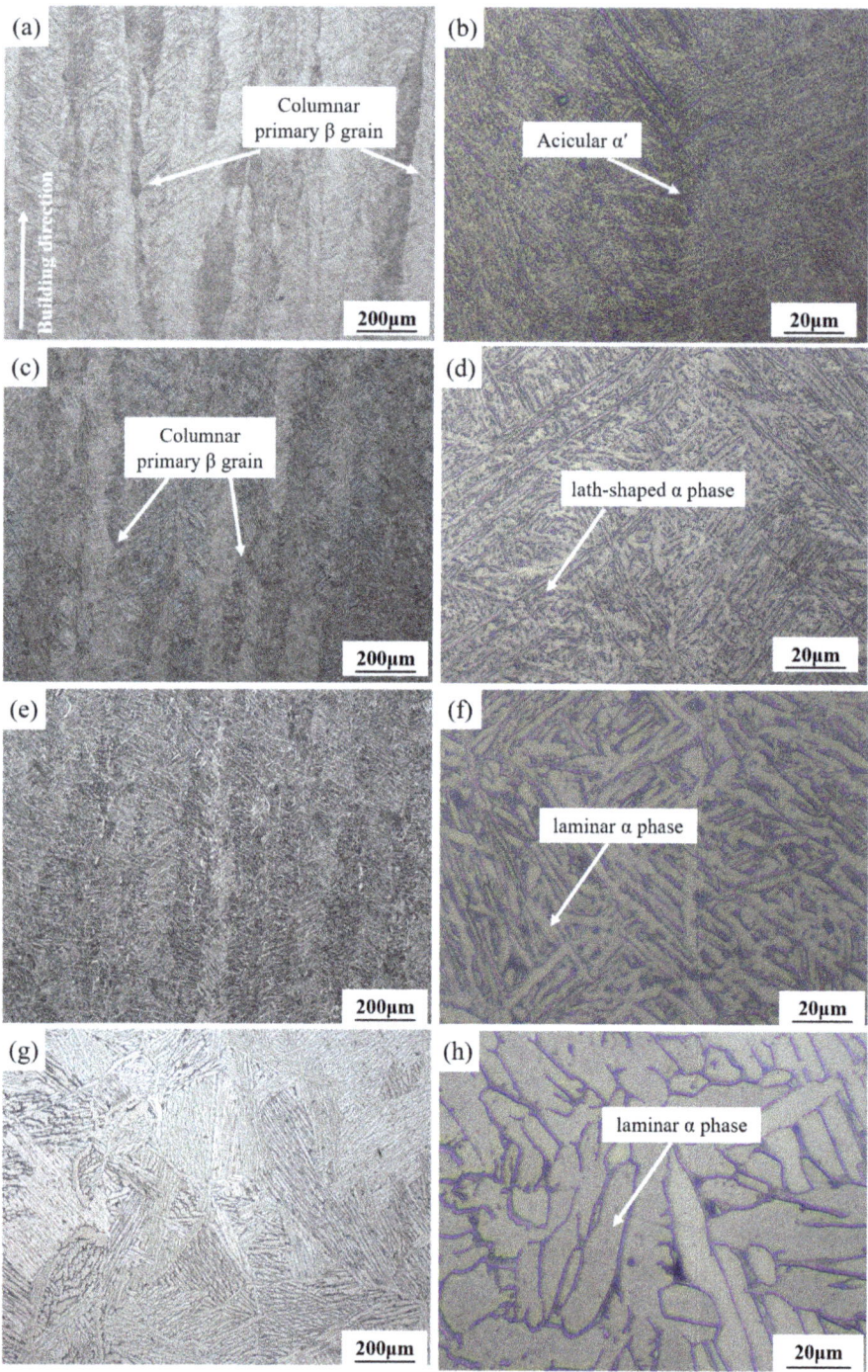

Figure 10. Microstructure of the printed Ti–6Al–4V samples along the build direction after annealing at different temperatures: (**a**,**b**) 750 °C; (**c**,**d**) 850 °C; (**e**,**f**) 950 °C; (**g**,**h**) 1050 °C.

3.5. Effect of Annealing Temperature on Mechanical Properties

Table 4 and Figure 11 show the tensile properties of the SLM-printed samples at different annealing temperatures. It can be seen that the tensile strength decreased gradually with the increase in the annealing temperature. When the annealing temperature of 750 °C was applied, the tensile strength of the sample decreased to 1094 MPa, which is 9% lower than that of the printed sample. The elongation of 7% was similar to that of the printed sample. The maximum elongation of 14% could be obtained at 950 °C, which is 79% higher than that of the as-printed sample.

Table 4. Mechanical properties of the SLM-printed Ti–6Al–4V samples.

Sample	Annealing Temperature (°C)	Ultimate Tensile Strength (MPa)	Elongation (%)
1	SLM-printed	1204 ± 32	7.8 ± 0.1
2	750	1094 ± 20	7 ± 0.5
3	850	1055 ± 1	11 ± 1.5
4	950	1007 ± 3	14 ± 0.1
5	1050	877 ± 16	11 ± 1

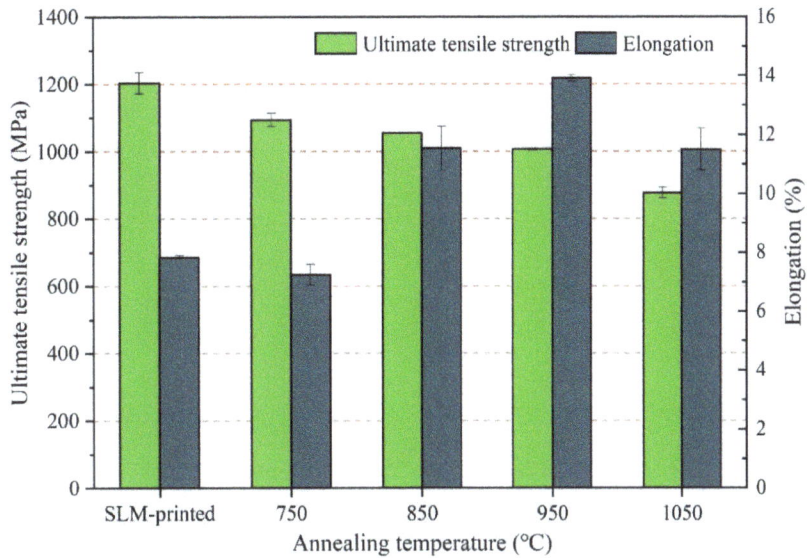

Figure 11. Effect of annealing temperature on the mechanical properties of the SLM-printed Ti–6Al–4V samples (laser power of 170 W, scanning speed of 1300 mm/s, and hatch space of 0.07 mm).

After annealing at 750 °C, some brittle and hard α' martensites decomposed into the $\alpha + \beta$ phase with relatively high ductility. However, the partial decomposition suppressed the change in the elongation of the sample but reduced its tensile strength. When the annealing temperature exceeded 800 °C, the acicular α' martensite completely decomposed into the α and β phases, which decreased the tensile strength and gradually increased the elongation. When the annealing temperature exceeded the β phase transition temperature, the formed coarse grains and the lath-shaped α phases inside them could hinder the slip of dislocations, causing stress concentration at the interface of the α and β phases and eventually reducing the ductility.

Figure 12 shows the fracture morphology of the samples at different annealing temperatures. It can be observed that the fracture morphology of the samples annealed at 750 °C and 850 °C was similar to that of the as-printed samples, including cleavage facets and dimples. However, when the annealing temperature increased to 950 °C, the cleavage facets disappeared and dense dimples with large sizes were formed, indicating a ductile fracture. When the local stress at the phase interface exceeded the interfacial bonding force in the tensile process, micropores occurred and consumed a large amount of strain energy. Micropores lead to dimples in the aggregation process, and denser dimples indicate the better ductility. After annealing at 1050 °C, the dimple size increased, and a small number of tear ridges appeared, which reduced the elongation of the sample.

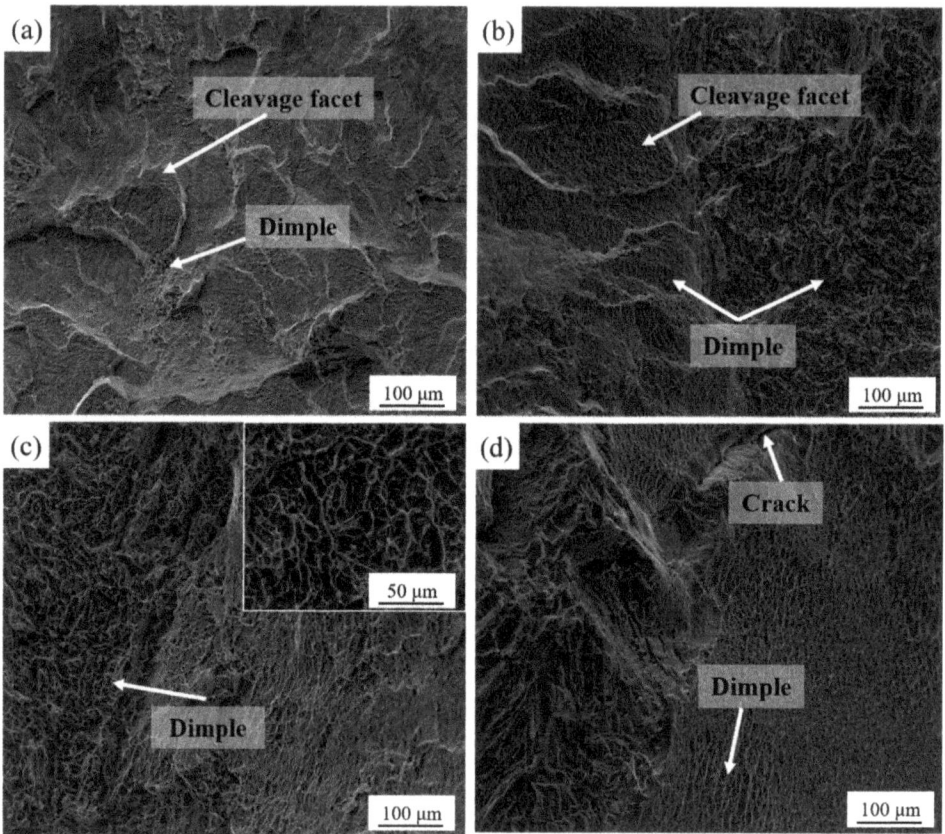

Figure 12. Fracture morphology of the printed samples after annealing at different temperatures: (**a**) 750 °C; (**b**) 850 °C; (**c**) 950 °C; (**d**) 1050 °C.

Table 5 shows the mechanical properties of Ti–6Al–4V samples after different post-treatments as compared to those reported in previous studies. It can be seen that the sample after 950 °C heat treatment exhibited a superior tensile strength and reasonable elongation. Heat treatment temperatures below 950 °C reduced the β phase [23], resulting in a higher tensile strength of the printed alloy. The samples with higher elongations were treated by HIP [26,27], which was more conducive to tailoring the microstructure and reducing the pore defects in the sample.

Table 5. Mechanical properties of the SLM-printed Ti–6Al–4V samples after different post treatments.

Sample Condition	Ultimate Tensile Strength (MPa)	Elongation (%)	Source
950 °C for 2 h	1007 ± 3	14 ± 0.1	This work
850 °C for 2 h	1004 ± 6	12.84 ± 1.36	[21]
890 °C for 2 h	998 ± 14	6 ± 2	[22]
840 °C for 2 h + furnace cooling to 450 °C + air cooling	1068.3 ± 26.7	10.28 ± 0.20	[23]
HIP at 920 °C and 100 MPa for 2 h	1088.5 ± 26.3	13.8 ± 1.3	[25]
HIP at 900 °C and 120 MPa for 2 h	941	19	[26]
HIP at 930 °C and 100 MPa for 4 h + wet polishing	936 ± 3.6	21.7 ± 2.3	[27]

4. Conclusions

This work investigated the microstructure and mechanical properties of the SLM-printed Ti–6Al–4V alloy post-treated by annealing. The effects of process parameters on the relative density, microstructure, and mechanical properties of SLM-printed samples were studied. The effects of annealing temperature on microstructure and mechanical properties of the printed samples were further studied. The main findings are presented below.

The relative density of the SLM-printed sample was significantly affected by the scanning speed. In particular, the SLM-printed sample could obtain the highest density of 99.51% with a laser power of 170 W, a scanning speed of 1300 mm/s, a layer thickness of 0.03 mm, and a hatch space of 0.07 mm.

The microstructure of the printed sample was composed of β columnar crystals, which contained a large number of acicular α′ martensite, resulting in higher strength and lower plasticity of the sample. The width of the β columnar crystals decreased with the increase in the scanning speed, as determined by the decrease in the energy density. The maximum tensile strength of 1265 MPa was achieved at a scanning speed of 900 mm/s, while its elongation could reach the highest value of 7.8% at a scanning speed of 1300 mm/s.

The annealing temperature had a significant effect on the microstructure of the sample. After annealing, the acicular α′ martensite was decomposed into the α + β dual phase. With the increase in the annealing temperature, the tensile strength gradually decreased, while the elongation increased first and then decreased. Annealing at 950 °C could result in the highest elongation of 14%, which is 79% higher than that of the as-printed sample, without a significant reduction in the tensile strength.

Author Contributions: Conceptualization, D.W. and C.H.; funding acquisition, D.W.; investigation, H.W.; methodology, X.C.; project administration, D.W.; resources, H.W. and X.C.; supervision, Y.L. and C.H.; validation, X.C.; visualization, H.W. and C.H.; writing—original draft, D.W. and H.W.; writing—review and editing, Y.L., D.L., X.L. and C.H. All authors have read and agreed to the published version of the manuscript.

Funding: This work was supported in part by the Guangdong Provincial Basic and Applied Basic Research Fund under Grant 2019B1515120094, 2021A1515110527, and in part by the open project founded by the State Key Laboratory of Vanadium and Titanium Resources Comprehensive Utilization under Grant 2021P4FZG11A.

Conflicts of Interest: The authors declare no conflict of interest.

References

1. Shipley, H.; McDonnell, D.; Culleton, M.R.; Coull, R.; Lupoi, R.; O'Donnell, G.; Trimble, D. Optimisation of process parameters to address fundamental challenges during selective laser melting of Ti-6Al-4V: A review. *Int. J. Mach. Tools Manuf.* **2018**, *128*, 1–20. [CrossRef]
2. Sallica-Leva, E.; Caram, R.; Jardini, A.L.; Fogagnolo, J.B. Ductility improvement due to martensite α′ decomposition in porous Ti–6Al–4V parts produced by selective laser melting for orthopedic implants. *J. Mech. Behav. Biomed. Mater.* **2016**, *54*, 149–158. [CrossRef] [PubMed]
3. Liu, S.; Shin, Y.C. Additive manufacturing of Ti6Al4V alloy: A review. *Mater. Des.* **2019**, *164*, 107552. [CrossRef]

4. Wang, D.; Liu, L.; Deng, G.; Deng, C.; Bai, Y.; Yand, Y.; Wu, W.; Chen, C.; Liu, Y.; Wang, Y.; et al. Recent progress on additive manufacturing of multi-material structures with laser powder bed fusion. *Virtual Phys. Prototyp.* **2022**, *2022*, 2028343. [CrossRef]
5. Li, B.; Han, C.; Lim, C.W.J.; Zhou, K. Interface formation and deformation behaviors of an additively manufactured nickel-aluminum-bronze/15-5 PH multimaterial via laser-powder directed energy deposition. *Mater. Sci. Eng. A* **2022**, *829*, 142101. [CrossRef]
6. Han, C.; Fang, Q.; Shi, Y.; Tor, S.B.; Chua, C.K.; Zhou, K. Recent advances on high-entropy alloys for 3D printing. *Adv. Mater.* **2020**, *32*, 1903855. [CrossRef]
7. Han, C.; Li, Y.; Wang, Q.; Wen, S.; Wei, Q.; Yan, C.; Hao, L.; Liu, J.; Shi, Y. Continuous functionally graded porous titanium scaffolds manufactured by selective laser melting for bone implants. *J. Mech. Behav. Biomed. Mater.* **2018**, *80*, 119–127. [CrossRef]
8. Seabra, M.; Azevedo, J.; Araújo, A.; Reis, L.; Pinto, E.; Alves, N.; Santos, R.; Mortágua, J.P. Selective laser melting (SLM) and topology optimization for lighter aerospace components. *Procedia Struct. Integr.* **2016**, *1*, 289–296. [CrossRef]
9. Spetzger, U.; Frasca, M.; König, S. A Surgical planning, manufacturing and implantation of an individualized cervical fusion titanium cage using patient-specific data. *Eur. Spine J.* **2016**, *25*, 2239–2246. [CrossRef]
10. Yan, C.; Hao, L.; Hussein, A.; Young, P. Ti–6Al–4V triply periodic minimal surface structures for bone implants fabricated via selective laser melting. *J. Mech. Behav. Biomed. Mater.* **2015**, *51*, 61–73. [CrossRef]
11. Xie, W.; Zheng, M.; Wang, J.; Li, X. The effect of build orientation on the microstructure and properties of selective laser melting Ti-6Al-4V for removable partial denture clasps. *J. Prosthet. Dent.* **2020**, *123*, 163–172. [CrossRef] [PubMed]
12. Sun, D.; Gu, D.; Lin, K.; Ma, J.; Chen, W.; Huang, J.; Sun, X.; Chu, M. Selective laser melting of titanium parts: Influence of laser process parameters on macro-and microstructures and tensile property. *Powder Technol.* **2019**, *342*, 371–379. [CrossRef]
13. Yang, J.; Yu, H.; Yin, J.; Gao, M.; Wang, Z.; Zeng, X. Formation and control of martensite in Ti-6Al-4V alloy produced by selective laser melting. *Mater. Des.* **2016**, *108*, 308–318. [CrossRef]
14. Wang, Z.; Xiao, Z.; Tse, Y.; Huang, C.; Zhang, W. Optimization of processing parameters and establishment of a relationship between microstructure and mechanical properties of SLM titanium alloy. *Opt. Laser Technol.* **2019**, *112*, 159–167. [CrossRef]
15. Murr, L.E.; Quinones, S.A.; Gaytan, S.M.; Lopez, M.I.; Rodela, A.; Martinez, E.Y.; Hernandez, D.H.; Martinez, E.; Medina, F.; Wicker, R.B. Microstructure and mechanical behavior of Ti–6Al–4V produced by rapid-layer manufacturing, for biomedical applications. *J. Mech. Behav. Biomed. Mater.* **2009**, *2*, 20–32. [CrossRef]
16. Shunmugavel, M.; Polishetty, A.; Littlefair, G. Microstructure and mechanical properties of wrought and additive manufactured Ti-6Al-4 V cylindrical bars. *Procedia Technol.* **2015**, *20*, 231–236. [CrossRef]
17. Jiao, Z.H.; Xu, R.D.; Yu, H.C.; Wu, X.R. Evaluation on tensile and fatigue crack growth performances of Ti6Al4V alloy produced by selective laser melting. *Procedia Struct. Integr.* **2017**, *7*, 124–132. [CrossRef]
18. Bartolomeu, F.; Buciumeanu, M.; Pinto, E.; Alves, N.; Silva, F.S.; Carvalho, O.; Miranda, G. Wear behavior of Ti6Al4V biomedical alloys processed by selective laser melting, hot pressing and conventional casting. *Trans. Nonferr. Met. Soc. China* **2017**, *27*, 829–838. [CrossRef]
19. Do, D.K.; Li, P. The effect of laser energy input on the microstructure, physical and mechanical properties of Ti-6Al-4V alloys by selective laser melting. *Virtual Phys. Prototyp.* **2016**, *11*, 41–47. [CrossRef]
20. Pal, S.; Gubeljak, L.; Hudák, R.; Lojen GRajťúková, V.; Brajlih, T.; Drstvenšek, I. Evolution of metallurgical properties of Ti-6Al-4V alloy fabricated in different energy densities in the Selective Laser Melting technique. *J. Manuf. Processes* **2018**, *35*, 538–546. [CrossRef]
21. Vrancken, B.; Thijs, L.; Kruth, J.P.; Van Humbeeck, J. Heat treatment of Ti6Al4V produced by Selective Laser Melting: Microstructure and mechanical properties. *J. Alloys Compd.* **2012**, *541*, 177–185. [CrossRef]
22. Cain, V.; Thijs, L.; Van Humbeeck, J.; Van Hooreweder, B.; Knutsen, R. Crack propagation and fracture toughness of Ti6Al4V alloy produced by selective laser melting. *Addit. Manuf.* **2015**, *5*, 68–76. [CrossRef]
23. Wang, D.; Dou, W.; Yang, Y. Research on selective laser melting of Ti6Al4V: Surface morphologies, optimized processing zone, and ductility improvement mechanism. *Metals* **2018**, *8*, 471. [CrossRef]
24. Liu, Y.; Xu, H.; Peng, B.; Wang, X.; Li, X.; Wang, Q.; Li, Z.; Wang, Y. Effect of heating treatment on the microstructural evolution and dynamic tensile properties of Ti-6Al-4V alloy produced by selective laser melting. *J. Manuf. Processes* **2022**, *74*, 244–255. [CrossRef]
25. Leuders, S.; Lieneke, T.; Lammers, S.; Tröster, T.; Niendorf, T. On the fatigue properties of metals manufactured by selective laser melting–The role of ductility. *J. Mater. Res.* **2014**, *29*, 1911–1919. [CrossRef]
26. Yan, X.; Yin, S.; Chen, C.; Huang, C.; Bolot, R.; Lupoi, R.; Kuang, M.; Ma, W.; Coddet, C.; Liao, H.; et al. Effect of heat treatment on the phase transformation and mechanical properties of Ti6Al4V fabricated by selective laser melting. *J. Alloys Compd.* **2018**, *764*, 1056–1071. [CrossRef]
27. Jamshidi, P.; Aristizabal, M.; Kong, W.; Villapun, V.; Cox, S.C.; Grover, L.M.; Attallah, M.M. Selective laser melting of Ti-6Al-4V: The impact of post-processing on the tensile, fatigue and biological properties for medical implant applications. *Materials* **2020**, *13*, 2813. [CrossRef]
28. Singla, A.K.; Banerjee, M.; Sharma, A.; Singh, J.; Bansal, A.; Gupta, M.K.; Khanna, N.; Shahi, A.S.; Goyal, D.K. Selective laser melting of Ti6Al4V alloy: Process parameters, defects and post-treatments. *J. Manuf. Processes* **2021**, *64*, 161–187. [CrossRef]
29. Singla, A.K.; Singh, J.; Sharma, V.S. Impact of cryogenic treatment on mechanical behavior and microstructure of Ti-6Al-4V ELI biomaterial. *J. Mater. Eng. Perform.* **2019**, *28*, 5931–5945. [CrossRef]

30. Kasperovich, G.; Haubrich, J.; Gussone, J.; Requena, J. Correlation between porosity and processing parameters in TiAl6V4 produced by selective laser melting. *Mater. Des.* **2016**, *105*, 160–170. [CrossRef]
31. Thijs, L.; Verhaeghe, F.; Craeghs, T.; Van Humbeeck, J.; Kruth, J.P. A study of the microstructural evolution during selective laser melting of Ti–6Al–4V. *Acta Mater.* **2010**, *58*, 3303–3312. [CrossRef]
32. Khairallah, S.A.; Anderson, A.T.; Rubenchik, A.; King, W.E. Laser powder-bed fusion additive manufacturing: Physics of complex melt flow and formation mechanisms of pores, spatter, and denudation zones. *Acta Mater.* **2016**, *108*, 36–45. [CrossRef]
33. Gu, D.; Hagedorn, Y.C.; Meiners, W.; Meng, G.; Batista, R.J.S.; Wissenbach, K.; Poprawe, R. Densification behavior, microstructure evolution, and wear performance of selective laser melting processed commercially pure titanium. *Acta Mater.* **2012**, *60*, 3849–3860. [CrossRef]
34. Qiu, C.; Panwisawas, C.; Ward, M.; Basoalto, H.C.; Brooks, J.W.; Attallah, M.M. On the role of melt flow into the surface structure and porosity development during selective laser melting. *Acta Mater.* **2015**, *96*, 72–79. [CrossRef]
35. Khorasani, A.M.; Gibson, I.; Awan, U.S.; Ghaderi, A. The effect of SLM process parameters on density, hardness, tensile strength and surface quality of Ti-6Al-4V. *Addit. Manuf.* **2019**, *25*, 176–186. [CrossRef]
36. Yang, J.; Han, J.; Yu, H.; Yin, J.; Gao, M.; Wang, Z.; Zeng, X. Role of molten pool mode on formability, microstructure and mechanical properties of selective laser melted Ti-6Al-4V alloy. *Mater. Des.* **2016**, *110*, 558–570. [CrossRef]
37. Dai, D.; Gu, D. Tailoring surface quality through mass and momentum transfer modeling using a volume of fluid method in selective laser melting of TiC/AlSi10Mg powder. *Int. J. Mach. Tools Manuf.* **2015**, *88*, 95–107. [CrossRef]
38. Sassi, B.H. *Morphologies Structurales de L'alliage de Titane TA6V, Incidences sur les Propriétés Mécaniques, le Comportement à la Rupture et la Tenue en Fatigue*; Centre d'édition et de documentation de l'ENSTA: Paris, France, 1977; pp. 1–162.
39. Lütjering, G. Influence of processing on microstructure and mechanical properties of (α+ β) titanium alloys. *Mater. Sci. Eng. A* **1998**, *243*, 32–45. [CrossRef]
40. Tsai, M.T.; Chen, Y.W.; Chao, C.Y.; Jang, J.S.C.; Tsai, C.C.; Su, Y.L.; Kuo, C.N. Heat-treatment effects on mechanical properties and microstructure evolution of Ti-6Al-4V alloy fabricated by laser powder bed fusion. *J. Alloys Compd.* **2020**, *816*, 152615. [CrossRef]
41. Wu, S.Q.; Lu, Y.J.; Gan, Y.L.; Huang, T.T.; Zhao, C.Q.; Lin, J.J.; Guo, S.; Lin, J.X. Microstructural evolution and microhardness of a selective-laser-melted Ti–6Al–4V alloy after post heat treatments. *J. Alloys Compd.* **2016**, *672*, 643–652. [CrossRef]
42. Huang, Q.; Liu, X.; Yang, X.; Zhang, R.; Shen, Z.; Feng, Q. Specific heat treatment of selective laser melted Ti–6Al–4V for biomedical applications. *Front. Mater. Sci.* **2015**, *9*, 373–381. [CrossRef]
43. Kim, Y.K.; Park, S.H.; Yu, J.H.; AlMangour, B.; Lee, K.A. Improvement in the high-temperature creep properties via heat treatment of Ti-6Al-4V alloy manufactured by selective laser melting. *Mater. Sci. Eng. A* **2018**, *715*, 33–40. [CrossRef]

Article

A Layer-Dependent Analytical Model for Printability Assessment of Additive Manufacturing Copper/Steel Multi-Material Components by Directed Energy Deposition

Wenqi Zhang [1], Baopeng Zhang [1], Haifeng Xiao [1], Huanqing Yang [2], Yun Wang [2] and Haihong Zhu [1,*]

[1] Wuhan National Laboratory for Optoelectronics, Huazhong University of Science and Technology, Wuhan 430074, China; vinkyz@hust.edu.cn (W.Z.); zhangbp@hust.edu.cn (B.Z.); xiaohaif@hust.edu.cn (H.X.)
[2] XI'AN Space Engine Company Limited, Xi'an 710100, China; 13991882146@163.com (H.Y.); wyun7103@163.com (Y.W.)
* Correspondence: zhuhh@mail.hust.edu.cn; Tel.: +86-27-87544774

Citation: Zhang, W.; Zhang, B.; Xiao, H.; Yang, H.; Wang, Y.; Zhu, H. A Layer-Dependent Analytical Model for Printability Assessment of Additive Manufacturing Copper/Steel Multi-Material Components by Directed Energy Deposition. *Micromachines* **2021**, *12*, 1394. https://doi.org/10.3390/mi12111394

Academic Editor: Nam-Trung Nguyen

Received: 12 October 2021
Accepted: 10 November 2021
Published: 13 November 2021

Publisher's Note: MDPI stays neutral with regard to jurisdictional claims in published maps and institutional affiliations.

Copyright: © 2021 by the authors. Licensee MDPI, Basel, Switzerland. This article is an open access article distributed under the terms and conditions of the Creative Commons Attribution (CC BY) license (https://creativecommons.org/licenses/by/4.0/).

Abstract: Copper/steel bimetal, one of the most popular and typical multi-material components (MMC), processes excellent comprehensive properties with the high strength of steel and the high thermal conductivity of copper alloy. Additive manufacturing (AM) technology is characterized by layer-wise fabrication, and thus is especially suitable for fabricating MMC. However, considering both the great difference in thermophysical properties between copper and steel and the layer-based fabrication character of the AM process, the optimal processing parameters will vary throughout the deposition process. In this paper, we propose an analytical calculation model to predict the layer-dependent processing parameters when fabricating the 07Cr15Ni5 steel on the CuCr substrate at the fixed layer thickness (0.3 mm) and hatching space (0.3 mm). Specifically, the changes in effective thermal conductivity and specific heat capacity with the layer number, as well as the absorption rate and catchment efficiency with the processing parameters are considered. The parameter maps predicted by the model have good agreement with the experimental results. The proposed analytical model provides new guidance to determine the processing windows for novel multi-material components, especially for the multi-materials whose physical properties are significantly different.

Keywords: directed energy deposition; additive manufacturing; bimetal; analytical model; printability maps

1. Introduction

Multi-material components (MMC), such as gradient materials, dissimilar joints, and sandwich structure materials, are characterized by spatial composition variation in one or more directions [1]. Due to their unique properties with progressive change in performance and function, MMC has gained notable attention and has been widely used in many fields such as electrical and aerospace over the past few decades [2]. Copper/steel bimetal, one of the most popular and typical MMC, processes excellent comprehensive properties with the high strength of steel and the high thermal conductivity of copper alloy. Because of its excellent properties, copper/steel has found its applications in the power generation, transmission, and die-casting industries [3]. Despite its attractive function and thermophysical properties, fabricating copper/steel MMC is still challenging due to its heterogeneous materials and thermophysical properties [4].

Additive manufacturing (AM) has been identified as an innovative manufacturing method that enables the build-up of components with complex geometries directly from 3D models. AM technology is characterized by the layer-wise fabrication of a part through selectively adding and melting material. It provides many advantages, including high manufacturing freedom, excellent part performance, and high production efficiency [5]. Due to the layer-wise process approach, AM is especially suitable for fabricating MMC [6,7].

With the growing requirement in industrial applications, many researchers are committed to using AM methods to fabricate MMC. The two most popular AM technologies, directed energy deposition (DED) [8,9] and powder bed fusion (PBF) [10,11], have been widely investigated to fabricate the single material including copper alloys [12–14], iron alloys [15–18], titanium alloys [19–21], aluminum alloys [22–25], and nickel alloys [26,27]. However, the AM process for heterogeneous materials is very different from that for homogeneous materials. In general, processing parameters (for example, laser power and scanning velocity) and material properties (for example, thermal conductivity, specific heat capacity, and density) influence the thermal profile and the printability of the AM process. Because effective thermal conductivity varies between layers, the processing parameters may also need to change as the deposition layer numbers increase. This phenomenon is particularly prominent for copper/steel dissimilar materials, because the thermophysical properties, such as thermal conductivity, of copper and steel differ greatly. The schematic of fabricating steel on the CuCr substrate is illustrated in Figure 1.

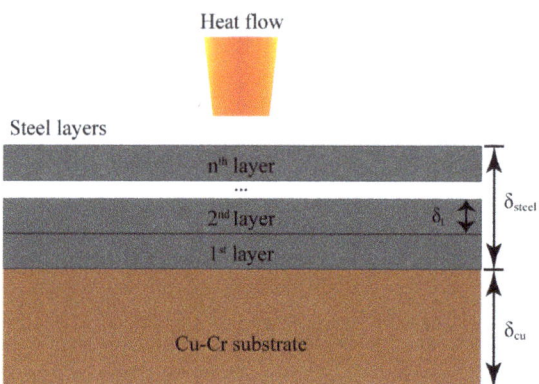

Figure 1. The schematic of fabricating the steel on the CuCr substrate layer by layer, where δ_{cu}, δ_{steel}, and δ_t represent the thickness of the CuCr substrate, the total thickness of steel coatings and the thickness of single layer steel coating, respectively.

When investigating the existing literature on AM of fabrication copper/steel or steel/copper bimetal, processing parameters optimization for every layer is absent. At present, there are mainly two methods to fabricating dissimilar materials. One common method is to use fixed parameters for steel and copper, respectively. Bai et al. [28] manufactured 316L/C52400/316L sandwich structure materials by SLM, and used the parameters for C52400 copper alloy and 316L, separately. Liu et al. [29] and Chen et al. [30] used different processing parameters for the individual alloys of steel and copper. The other common method is to utilize variable parameters, i.e., one set of parameters for the interfacial layers of copper/steel and another set of parameters for steel or copper, respectively. Chen et al. [31] optimize the interfacial layers by orthogonal experiment, and the set of parameters are fixed for steel and copper, respectively. Tan et al. [32] successfully processed steel on the copper alloy substrate by SLM, using one set of parameters for the first ten layers by remelting twice and using the optimized parameters of steel for the rest of the fabricating. In our previous work [3], the steel was built on the CuCr alloy substrate by DED. We optimize the parameters from one to four layers and the optimized parameters for steel are used for the rest of the layers. Based on the literature discussed above, there is currently no known work describing process-layer number relationships for fabricating copper/steel bimetal in the contest of AM.

Considering both the great difference in thermophysical properties between copper and steel and the layer-based fabrication character of the AM process, the optimal processing parameters will vary throughout the deposition process. Since there are numerous process variables within AM, the optimization of processing parameters for every layer is a huge amount of work [33]. Therefore, establishing a framework to implement model-based approaches to building dissimilar material parts is essential.

In this paper, we propose an analytical calculation model to predict the layer-dependent processing parameters during the fabrication of copper/steel bimetal. Specifically, the changes in effective thermal conductivity and specific heat capacity with the layer number as well as the absorption rate and catchment efficiency with the processing parameters are considered. The analytical model is established to predict molten pool temperature and thus to provide a methodology to estimate the process maps for multi-layer copper/steel bimetal dependent on the layer number. These results are compared with experimentally observed molten pool width and parameter maps for copper/steel specimens and found to have good agreement.

2. Theoretical Modeling

Figure 2 shows the framework for estimating the layer-dependent printability of the copper/steel bimetal. The general workflow starts with the calculations of layer-dependent thermophysical properties (the effective thermophysical properties [34]) and the processing parameters-dependent catchment efficiency [35]. Then the temperature fields and molten pool dimensions with different laser power and scanning velocity covering the processing space are obtained. The peak temperature (T_{max}) and dimensions of the molten pool are subsequently used to evaluate the parameter maps and verify the model. In this study, the stainless steel is fabricated on the CuCr substrate, hence regions of process space with $T_{m\_steel} < T_{max} < T_{b\_cu}$ are recognized to be the appropriate combination of processing parameters for the first layer, and $T_{m\_steel} < T_{max} < T_{b\_steel}$ for the second layer and above, where T_{m_steel}, T_{b_steel}, T_{m_cu}, and T_{m_cu} represent the melting point and boiling point of steel and Cu, respectively. The peak temperature criterion has also been applied in the selective laser melting process [36]. Details of each step in this workflow are presented in the subsequent sections.

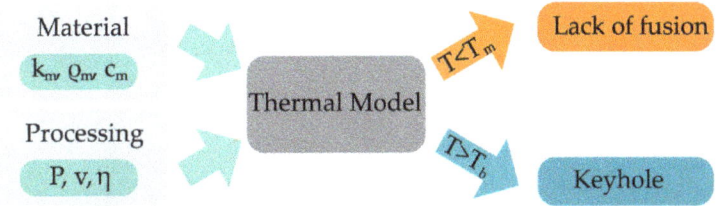

Figure 2. The framework for estimating the layer-dependent printability of the copper/steel bimetal. Processing parameters (laser power (P), scanning velocity (v), and powder catchment efficiency (η)) and layer-dependent effective material properties [35] (effective thermal conductivity (k_m), effective specific heat (c_m), and effective density (ρ_m)) are provided to the thermal model.

2.1. Thermal Model

In this framework, the temperature profile is required to calculate the temporally evolving molten pool dimensions and the temperature at the molten pool boundary (i.e., the solid-liquid interface) and the center (i.e., T_{max}). The assumption that the solidification process occurs at a constant temperature is reasonable because of the high thermal gradient and cooling rate for DED.

The Cline–Anthony model [37] is used to calculate the temperature field of the molten pool during the DED process. The Cline–Anthony model was established to analyze the thermal distribution and geometry of melting track from a moving Gaussian source on a semi-infinite substrate. Since the laser beam in this paper is small compared to the substrate, a semi-infinite geometry is a rationally good approximation. Therefore, the Cline–Anthony model is applicable in this paper. The Cline–Anthony model is beneficial to rapidly handle many calculations throughout the entire processing maps and has been widely used in the AM process [38,39]. The temperature field of the molten pool relating to the processing parameters and the thermophysical properties during DED can be calculated by the Cline–Anthony model as Equation (1):

$$T(x,y,z) = T_0 + \frac{P_e}{(2\pi^3)^{1/2} k_m r} \int_0^\infty \frac{1}{1+s^2} \exp\left\{-\frac{1}{2(1+s^2)}\left[\left(\frac{x}{r}+s^2 ar\right)^2 + \left(\frac{y}{r}\right)^2\right] - \frac{1}{2s^2}\left(\frac{z}{r}\right)^2\right\} ds \quad (1)$$

where T_0 is to the initial temperature of the substrate, P_e is the total laser power absorbed by the powder and the substrate, r is the radius of the laser spot, k_m is the effective thermal conductivity of the substrate, a is defined as $a = \rho_m c_m v/(2 k_m)$, and s is defined as $s^2 = 2 k_m t/(\rho_m c_m r)$ (ρ_m, c_m, v, and t refer to the density, specific heat capacity of the MMC, laser scanning speed, and laser scanning time, respectively).

The molten pool dimension is figured by the zone enclosed by the isosurface of the solidification temperature or melting point (T_m) as illustrated in Figure 3. Figure 3a shows the 3D diagram, where SD, TD, and BD refer to scanning, transverse, and build directions, respectively. Figure 3b shows the SD-TD cross-section. Then the width (W) and length (L) of the molten pool can be obtained, where W equals the length of CD, and L equals the length of AB.

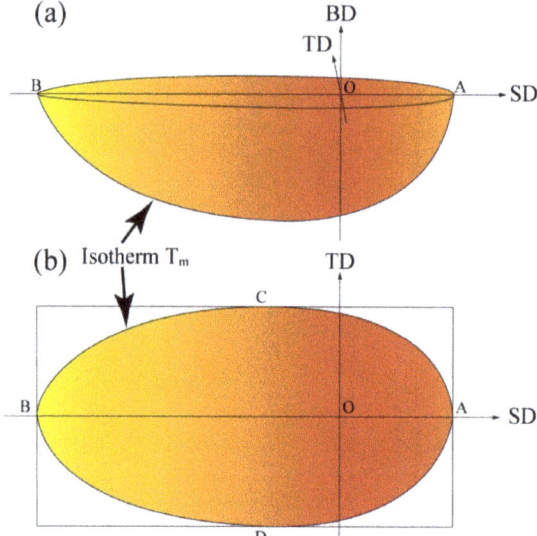

Figure 3. Schematic of the molten pool (SD, TD, BD refer to scanning, transverse, and build directions, respectively): (**a**) 3D diagram of the molten pool; (**b**) SD-TD cross-section of the molten pool.

2.2. The Laser–Materials Interaction during the DED Process

The typical shadow model is used for the laser–materials interaction during the DED process [38,40]. Figure 4 describes the laser–materials interaction during the DED process of fabricating steel on the copper substrate. The laser beam, metal powder, and protective gas are output from the nozzle coaxially. When traversing the powder stream, part of the laser beam interacts with the powder. Therefore, a portion of laser energy is attenuated (absorbed or scattered) by the powder, and the rest reaches the substrate surface [38]. However, due to the high reflectivity of the copper alloy substrate, a large portion of energy reaching the substrate is reflected and can be absorbed by the powder again. Finally, the energy carried by the heated powder particles partially falls into the molten pool for further melting. As the laser leaves, the molten pool cools and solidifies rapidly on the substrate surface to form the deposition track. The following assumptions are made for the usage of the proposed model during the DED process [40]:

(1) The laser energy attenuation is proportional to the projected area of the powder particles in the laser beam. Since the powder concentration is much smaller compared with the gas flow volume, it is reasonable to neglect the shadow between particles.
(2) The powder particles are considered homogeneous and spherical. The average diameter is used to represent the particle size. The argon gas atomized powder is used in this work, and the morphology of the powders is almost spherical from the scanning electron microscope (SEM) observation [3].
(3) The thermophysical properties of materials are regarded as constant and invariable with the temperature.
(4) The laser beam reaching the substrate is perfectly reflected upwards in the same shape as the initial beam.

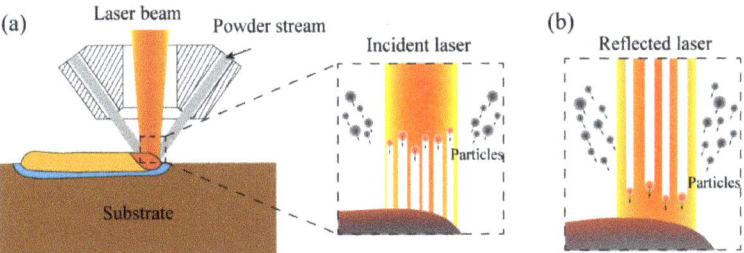

Figure 4. Schematic of the laser–material interaction: (**a**) the powder interacts with the incident laser; (**b**) the powder interacts with the reflected laser.

Based on the above assumptions, the attenuation rate of the laser energy by the powder particles can be obtained, which equals the ratio between the projected area of the powder particles and the laser beam. First, the laser beam travels through the gas–powder stream and is attenuated (absorbed or scattered) before reaching the substrate. The attenuate ratio β_{att} can be expressed as Equation (2), which is the proportion of the projected area of the powder particles to the laser beam area [38].

$$\beta_{att} = \frac{S_p}{S_l} = \frac{3m_p}{2\pi \rho_p r_p r v_p \cos\theta} \quad (2)$$

where S_p is the projected area of the powder particles, S_l is the projected area of the laser beam, m_p (kg/s) is the powder feeding rate, ρ_p (kg/m^3) is the density of the powder, r_p (m) is the average diameter of the powder particles, r (m) is the laser beam radius, v_p (m/s) is the velocity of the powder-gas stream, and θ is the angle between the gas–powder and the horizontal. The attenuated laser power (P_{att}) by the powder can be obtained as Equation (3):

$$P_{att} = \beta_{att} * P \quad (3)$$

where P is the laser power. A portion of the attenuated laser energy is absorbed by the powder stream and delivered to the substrate when the particles enter the molten pool. The absorbed laser power by the powder stream delivered to the substrate can be expressed as Equation (4).

$$P_{a1} = \eta * \beta_p * P_{att} \quad (4)$$

where β_p is the laser absorption of the powder and η the powder catchment efficiency. Second, the laser beam passes through the gas–powder stream and reaches the substrate.

Then the substrate can absorb part of the laser energy (P_s) directly based on the laser absorptivity, as is given by Equation (5):

$$P_s = \beta_s(P - P_{att}) \tag{5}$$

where β_s is the laser absorption of the substrate. Finally, the laser beam reflected by the substrate is important for the high laser reflectivity of CuCr alloy substrate. Therefore, part of the reflected laser energy (P_r) is attenuated and absorbed by the powder stream again. The powder stream transfers the absorbed energy back to the substrate (P_{a2}), as is given by Equations (6) and (7):

$$P_{a2} = \eta * \beta_{att} * \beta_p * P_r \tag{6}$$

$$P_r = (1 - \beta_s)(P - P_{att}) \tag{7}$$

Thus, the effective laser power (P_e), namely the total amount of laser power delivered into the molten pool by the powder stream and absorbed by the substrate, can be obtained by Equation (8):

$$P_e = P_{a1} + P_s + P_{a2} \tag{8}$$

2.3. The Catchment Efficiency

From Equations (4), (6), and (8), the powder catchment efficiency (η) is essential for the calculation of the temperature field. According to reference [35], the defined parameter Q, which is related to the processing parameters and material properties, can be used to evaluate the catchment efficiency as expressed by Equation (9):

$$Q = \frac{(P/v)^{2/3}}{(c\Delta T + H)^{2/3}} \tag{9}$$

where ΔT, H, and c refer to the difference between the solidus temperature and ambient temperature, the latent heat of fusion, and the specific heat capacity of the alloy, respectively. For the convenience of comparing different processing parameters, the normalized value Q^*, which is expressed as Q divided by Q_{max} (the maximum value for all the data used), is used to calculate the value of η as Equation (10) [35]:

$$\eta = -1.5(Q^*)^2 + 2.8(Q^*) - 0.3 \tag{10}$$

2.4. Evolution of the Substrate Laser Absorption

The laser absorption of the substrate varies with the track and layer numbers for the heterogeneous materials. For the first track of the first layer, the laser interacts with the copper alloy substrate, and the absorption of the laser by the substrate (β_s) equals the copper substrate $\beta_s = \beta_{cu}$, where β_{cu} represents the laser absorption of the CuCr alloy. For the second track and above the first layer, part of the laser interacts with the previous steel track, and the rest with the substrate directly as shown in Figure 5. It is considered that the absorption of the substrate is the linear sum of the laser absorption rate of copper and steel and expressed as Equation (11):

$$\beta_s = (1-l) * \beta_{cu} + l * \beta_{steel} \tag{11}$$

where β_{steel} represents the laser absorption of the steel, l is related with the laser radius (r), the width of the previous track (W), and the hatching space (HS) and can be expressed as Equation (12):

$$l = \frac{r + \left(\frac{W}{2} - HS\right)}{2r} \tag{12}$$

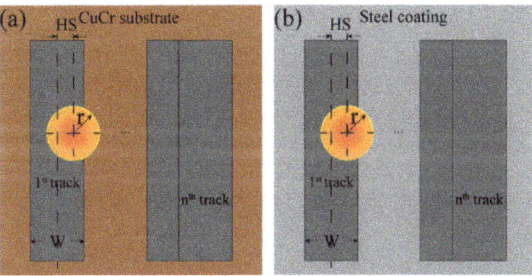

Figure 5. Evolution of the substrate laser absorption: (**a**) the first layer except for the first track; (**b**) the second layer and above.

For the second layer and above, the laser interacts with the steel coating directly, and the absorption of the laser by the substrate can be expressed as $\beta_s = \beta_{steel}$ as shown in Figure 5b.

2.5. Evolution of the Effective Thermophysical Properties

The thickness of copper and steel, which is a function of the deposition layer number, determines the thermophysical properties of the copper/steel bimetal as a single component. Equation (13) gives the thermal resistance of a single material, where R, δ, k, and A indicate the thermal resistance, the thickness of the material, the thermal conductivity, and the area of the cross-section that is perpendicular to the heat flow direction, respectively. The total thermal resistance (R_m) of the MMC, as seen in Figure 1, can be represented in Equation (14). The subscript m, steel, and cu denote the copper/steel MMC, the stainless steel, and CuCr alloy, respectively. If the copper/steel bimetal is treated as a single component, the theoretical effective thermal conductivity [34] (k_m), density (ρ_m), and specific heat capacity (c_m) of the MMC are obtained in Equations (15)–(17) [41,42]. Where $V'_{steel} = \frac{V_{steel}}{V_{steel}+V_{cu}}$, $V'_{cu} = \frac{V_{cu}}{V_{steel}+V_{cu}}$, $\rho'_{steel} = \frac{\rho_{steel}}{\rho_m}$, $\rho'_{cu} = \frac{\rho_{cu}}{\rho_m}$, $\delta_{steel} = (n-1)*\delta_t$, n represents the layer number and δ_t is the layer thickness.

$$R = \frac{\delta}{kA} \qquad (13)$$

$$R_m = \frac{\delta_{steel}}{k_{steel}A} + \frac{\delta_{cu}}{k_{cu}A} \qquad (14)$$

$$k_m = \frac{\delta_{Steel}+\delta_{Cu}}{\frac{\delta_{Steel}}{k_{Steel}}+\frac{\delta_{Cu}}{k_{Cu}}} \qquad (15)$$

$$\rho_m = \rho_{steel}V'_{steel} + \rho_{cu}V'_{cu} \qquad (16)$$

$$c_m = \rho'_{steel}V'_{steel}c_{steel} + \rho'_{cu}V'_{cu}c_{cu} \qquad (17)$$

Table 1 provides the values of the single metal thermophysical properties of CuCr and the self-developed martensite stainless steel (07Cr15Ni5), respectively. Figure 6 shows the theoretically calculated layer-dependent effective thermophysical properties (effective thermal conductivity, density, and specific heat capacity) of the copper/steel bimetal according to Equations (14)–(16).

Table 1. Thermophysical properties of CuCr and steel for the calculation [43–47].

Parameters	CuCr	Steel
Density, ρ (kg m^{-3})	8.90	7.78
Melting point, T_m (K)	1358	1654
Boiling point, T_b (K)	2835	3086
Thermal conductivity, k (W m^{-1} K^{-1})	180	80
Specific heat capacity, c (J kg^{-1} K^{-2})	385	450
Laser beam absorptivity, β	0.2	0.5

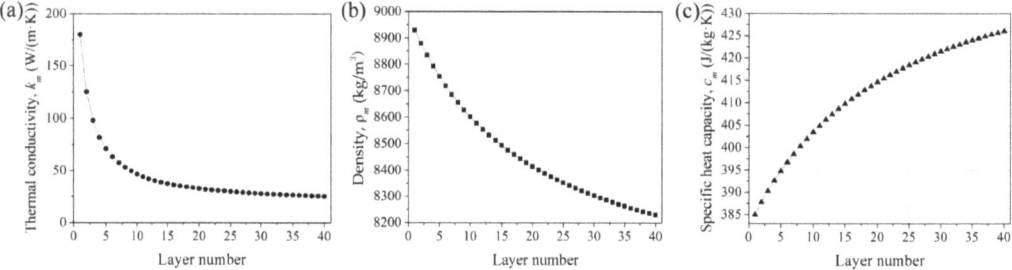

Figure 6. The calculated layer-dependent effective physical properties of the copper/steel bimetal: (a) effective thermal conductivity; (b) effective density; (c) effective specific heat capacity.

3. Printability Predictions

3.1. Printability Maps for the First Layer

During the DED process, the processing parameters such as laser power and scanning velocity determine the heat flow of laser energy to the powder and substrate and ultimately affect the temperature field. According to Equations (1) and (8), increasing laser power or decreasing scanning velocity transmits more energy to the powder and substrate. By increasing the heat input, fusion commences when the temperature of the material rises from the ambient temperature to the melting point (T_m). Further increasing the laser energy will continue to raise the temperature to the boiling point (T_b). According to reference [48], when the peak temperature (T_{max}) of the molten pool reaches T_b, the recoil pressure caused by the evaporation drives the molten pool to the keyhole mode. Therefore, the condition $T_{max} = T_b$ is identified as the keyhole mode threshold.

The printability map is determined by analyzing the temperature characteristics of the molten pool under various processing parameters. Table 2 shows the processing parameters calculated for the first layer. Figure 7 shows the peak temperature for different parameters combined with laser power and scanning velocity for the first layer. The isotherms of the melting and boiling points of copper and steel (T_{s_cu} = 1337 K, T_{s_steel} = 1654 K, T_{b_cu} = 2835 K, T_{b_steel} = 3086 K) are marked in Figure 7, respectively. According to the peak temperature at the center of the molten pool, three situations can be obtained: (1) $T_{max} < T_{s\_steel}$; (2) $T_{s\_steel} < T_{max} < T_{b\_cu}$; (3) $T_{max} > T_{b\_cu}$. For situation (1), the temperature of the molten pool cannot reach the melting point of the steel (1654 K). Therefore, no fusion will occur because of the insufficient input energy. The steel coating cannot be fabricated on the copper substrate and form a metallurgical bond. For situation (2), the molten pool temperature is in the range of T_{s_steel} and T_{b_cu}. Therefore, the fusion track can be obtained in the conduction mode. For situation (3), the molten pool temperature is higher than the T_{b_cu}, and the fusion track can be obtained in the keyhole mode. Firstly, the parameters for situation (1) are excluded because no fusion occurred between the copper substrate and steel powder. Secondly, the keyhole mode can lead to a porosity void with vapor entrapping at the bottom of the molten pool [49,50] as well as a high dilution of the molten pool [51], which is not expected to produce good mechanical properties. Hence, the parameters for situation (3) are excluded. From the results of calculation and analysis above, situation (2) is evaluated as appropriate process maps before starting the experiments. Although the criterion of $T_{s\_steel} < T_{max} < T_{b\_cu}$ may not be a precise condition, it can narrow the range of experimental parameters, and further parameters optimization can be conducted from the experiment based on the results. The peak temperature criterion has also been applied in the selective laser melting process [36,52].

Table 2. Processing parameters for the first layer.

Parameters	Values
Laser power (P, W)	500~5000 (500 increment)
Scanning velocity (V, mm/min)	400~2000 (200 increment)
Hatching space (HS, mm)	0.3
Powder feeding rate (g/min)	12
Z-axis increment (mm)	0.3
Laser spot diameter (mm)	1

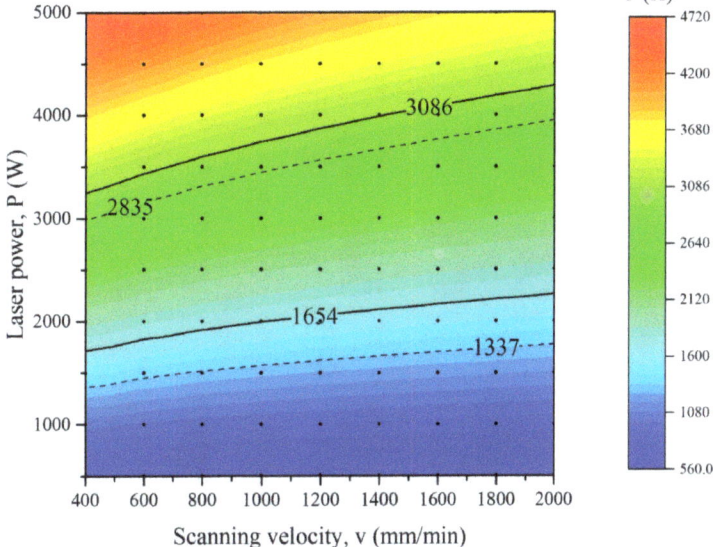

Figure 7. The peak temperature of different parameters for the first layer (T_{s_cu} = 1337 K, T_{s_steel} = 1654 K, T_{b_cu} = 2835 K, T_{b_steel} = 3086 K).

3.2. Printability Maps for Multi-Layer

Since a layer of steel has been deposited on the copper substrate, the effective thermal conductivity, laser absorption, etc., will change significantly from the second layer. Therefore, the processing parameters will vary with the layer number. Figure 8 shows the calculated peak temperature distribution for the second to seventh layers. Similarly, the isotherms of the melting and boiling points of copper and steel are marked and different regions are labeled. As described in Figure 7, the peak temperature in the range of $T_{s\_steel} < T_{max} < T_{b\_cu}$ is selected for the first layer. However, the deposited layer will be in direct contact with the steel from the second layer. Hence, the criteria for the upper limit of the peak temperature can change to the boiling point of the steel, i.e., $T_{max} < T_{b\_steel}$. Apparently, for a certain parameter combination with laser power and scanning velocity, the peak temperature increases with the layer number. Namely, the process maps move towards the low energy region and the printability maps become narrow. This phenomenon is more pronounced for the 10th to 40th layers, as is shown in Figure 9.

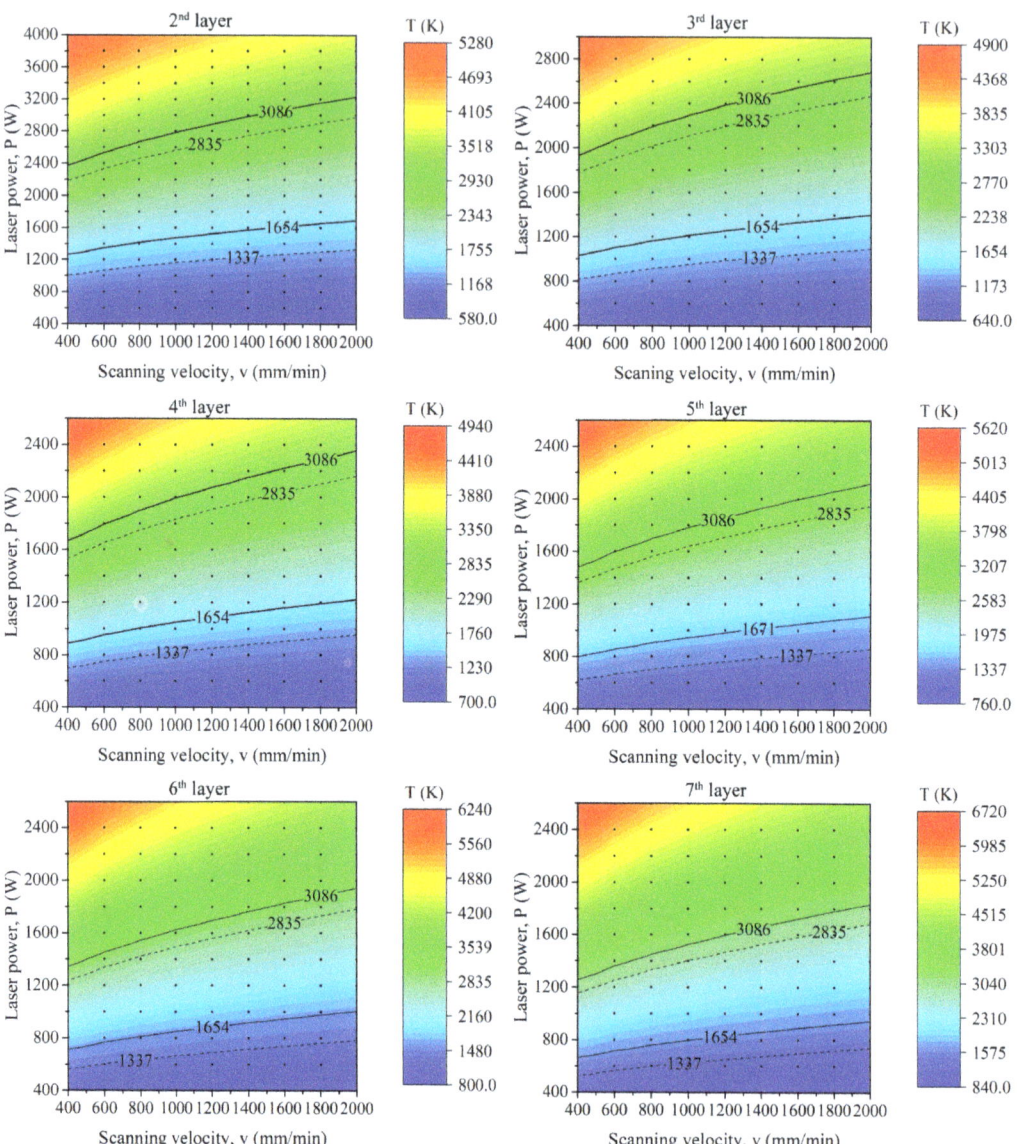

Figure 8. The peak temperature of different parameters for the 2nd to 7th layers.

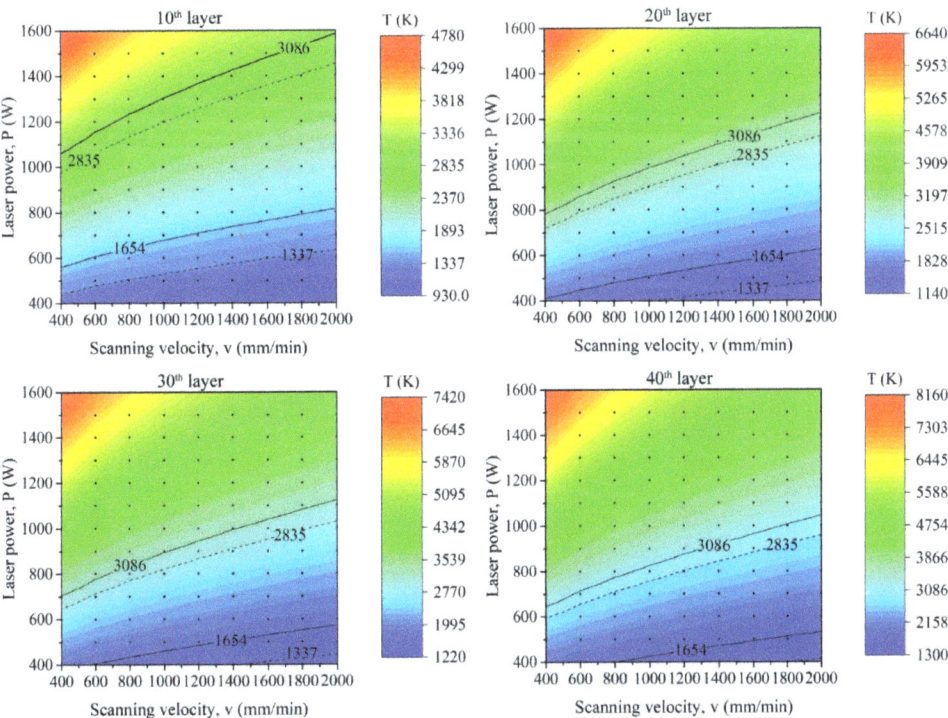

Figure 9. The peak temperature of different parameters for the 10th to 40th layers.

Figure 10 shows the optimal laser power maps dependent on the layer number when fixing the scanning velocity and hatching space. The range of laser power (ΔP) for a certain scanning velocity is defined as $\Delta P = P_{max} - P_{min}$ to evaluate the parameter range, where P_{max} and P_{min} represent the maximum and minimum laser power for the printability range. The scanning velocities ranging from 800 to 1200 mm/min are selected for the consideration of both the fabrication efficiency and the stability of powder feeding. It can be seen that the laser power changes in a large range as the layer number increases. Taking the scanning velocity of 800 mm/min as an example, the laser power should be in the range of 1920 to 3316 W for the first layer, and 384 to 773 W for the 40th layer. The decline rate of P_{max} and P_{min} is as high as 77 and 88% from the first layer to the 40th layer. Furthermore, the P_{max} and P_{min} decrease drastically for the first ten layers with the decline rate of 63 and 66% as the layer number increases and decreases slowly from the 10th to the 40th layer with the decline rate of 37 and 40%. In addition, Figure 10 also reveals a sharp fall of ΔP for the first ten layers and then a slight decline after ten layers, which corresponds to the variation of effective thermal conductivity with the number of layers from Figure 6a. The ΔP is 1395 W for the first layer and 389 W for the 40th layer with a decline rate of 72%. Compared to the high effective thermal conductivity for the first ten layers, the peak temperature response is more sensitive to the relatively low effective thermal conductivity because the energy can be conducted away quickly for the high effective thermal conductivity.

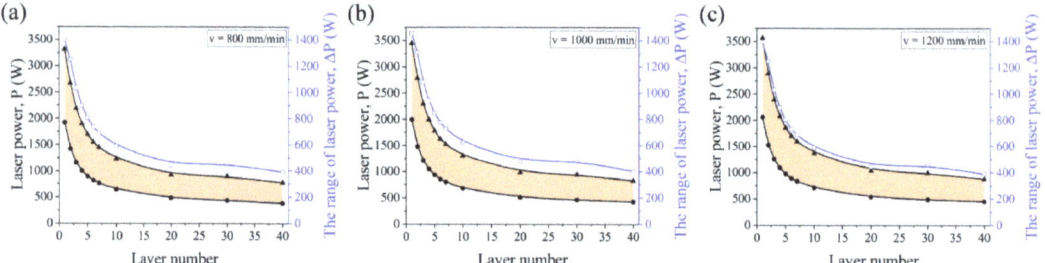

Figure 10. The variation and range of laser power with the layer number at different scanning velocities: (**a**) v = 800 mm/min; (**b**) v = 1000 mm/min, and (**c**) v = 1200 mm/min.

4. Verifications

4.1. Experiments

The initial material was self-developed martensite stainless steel (07Cr15Ni5). The argon gas atomized spherical steel powder has the average diameter of 28.9 µm. The CuCr alloy substrate with dimensions of 135 mm × 180 mm × 6 mm was rolled (900~950 °C) and annealed (400~450 °C) in the experiment. The chemical composition (wt.%) of the CuCr alloy substrate is 99.5 Cu and 0.5 Cr. The substrate surfaces were roughened, sandblasted, and cleaned with alcohol before the experiment. The DED system includes a 6 kW IPG YLR-6000 fiber laser (IPG laser GmbH, Burbach, Germany) with the beam size of 1 mm, a 6-axis robot, a powder feeder (HUST-III), and a self-made laser head. The 6-axis robot controls the movement of the laser head along the X–Y plane and/or the Z direction. The deposition area is protected from oxidation by the argon shielding gas. Details about the DED system have also been described previously [3,53]. The cross-section specimen was mechanically polished and then etched by a solution of 2 mL HF, 8 mL HNO$_3$, and 90 mL H$_2$O at room temperature to reveal the morphology of the multi-layer steel coatings.

4.2. Verification of the Single-Track Molten Pool Width

Due to the high energy input during the DED process, the temperature is difficult to measure directly. The molten pool widths are generally used for verification in thermal analysis, which is frequently adopted for the DED and SLM processes [54,55]. The molten pool width is obtained by the isothermal curves of the calculation according to the melting temperature of steel (1654 K) as depicted in Figure 3. Table 3 shows the processing parameters used for the single track.

Table 3. Processing parameters for the single track.

Parameters	Values
Laser power (P, W)	3000~3600 (200 increment)
Scanning velocity (V, mm/min)	800~1600 (200 increment)
Powder feeding rate (g/min)	12
Laser spot diameter (mm)	1

As shown in Figure 11, the calculated molten pool width matches well with the experimental measurements. Both the calculated and experiment results show that the molten pool width decreases as the scanning velocity increases. According to Equations (9) and (10), the increasing scanning velocity reduces particle catchment efficiency, hence the energy delivered by powder to the substrate is reduced. On the other hand, high scanning velocity means a shorter contact duration of the laser with the substrate, therefore, less energy is absorbed directly by the substrate. As a result, the molten pool width decreases with the scanning velocity. Since heat loss by the molten pool flow and the latent heat of fusion are not taken into account, the calculated widths are slightly higher than the experimental measurements. However, the maximum relative error is 18.59%, which demonstrates the accuracy of the thermal model.

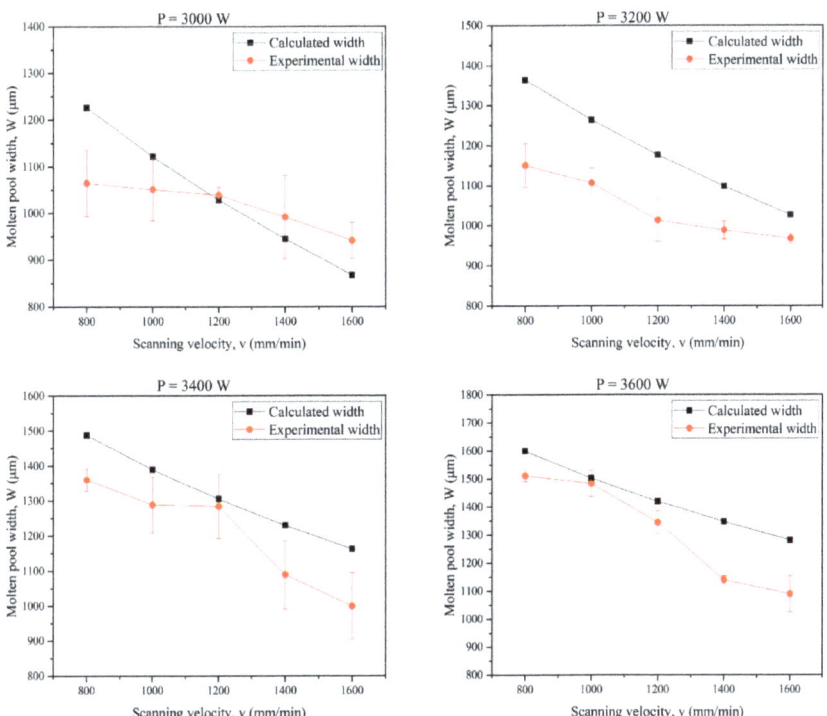

Figure 11. Comparison of experiment and calculated values of the single-track molten pool width.

4.3. Verification of Printability Maps for the First Layer

To verify the analytical model for the first layer, the laser scanning velocity was fixed at 800 mm/min and the scanning patch at 0.3 mm. As illustrated in Figure 12, the influence of laser power on the solidification behavior of the fusion track was investigated. Because of the inadequate energy, no scanning track emerged on the substrate when the laser power was 1500 W. Fusion layers can be observed when the laser power is higher than 2000 W. As shown from the cross-section of the first layer in Figure 13, the fusion tracks are formed in the conduction mode with laser power of 2500 W~3500 W, and the morphology is uniform. When increasing the laser power to 4000 W~5000 W, the fusion tracks are formed in the keyhole mode with heterogeneous morphology. According to the calculated results above, fusion tracks can be formed in the conduction mode with the laser power of 1920 W~3316 W at the fixed scanning speed of 800 mm/min. The molten pool should be the keyhole mode with the laser power of above 3316 W. However, from the cross-section morphology in Figure 13, conduction mode is observed when the laser power is 3500 W. Because heat loss by the molten pool flow and the latent heat of fusion is not taken into account, the peak temperatures may be overpredicted slightly.

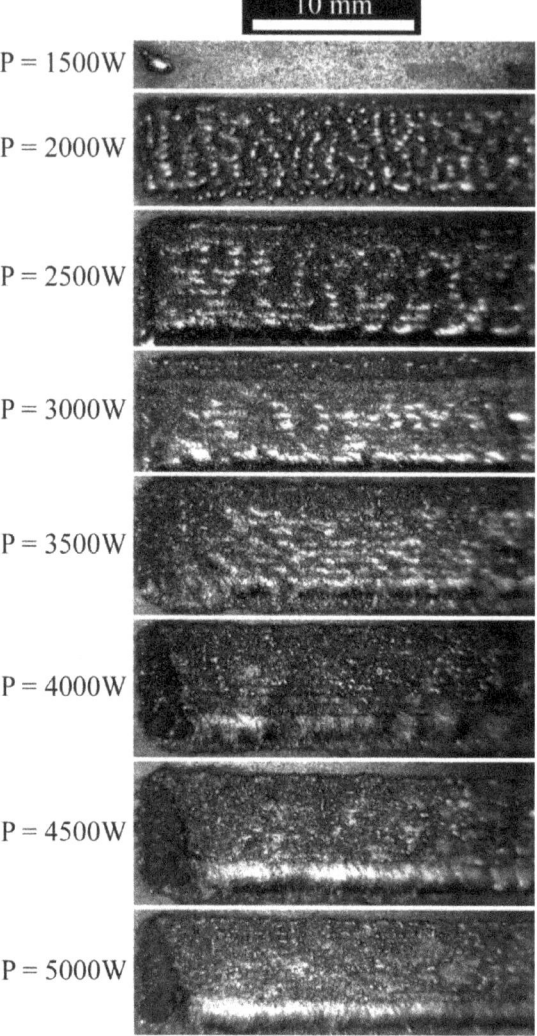

Figure 12. Surface morphologies of the single-layer specimens.

According to the experimental results shown in Figure 13, although the molten pool can be formed at 2000 to 3500 W, the forming quality is still different. When P = 2000 W, the laser energy is too low and the fused layer cannot be formed continuously. When the laser power P = 2500 W, holes can be observed near the interface with fusion in place. When the laser power is increased to P = 3000 W, the molten layer can be well formed and no cracks are produced with the flat and straight copper–steel interface, and no obvious semi-elliptical molten pool is seen, which means the melting area is small. When the laser power increases to P = 3500 W, a semi-elliptical molten pool can be observed at the interface and cracks can be observed. It should be noted that this model can only predict formability and narrow the range of experimental parameters before conducting experiments, while defects such as cracks cannot be predicted and further experimental studies are needed.

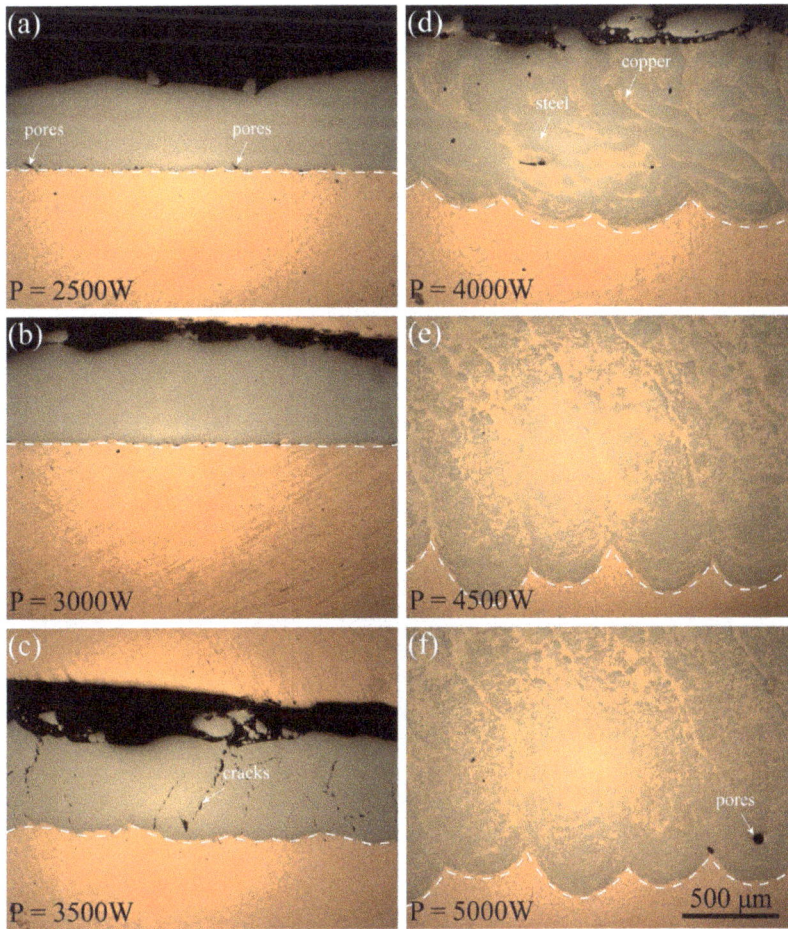

Figure 13. Cross-section morphologies of the single-layer specimens at different laser power: (**a**) P = 2500 W, (**b**) P = 3000 W, (**c**) P = 3500 W, (**d**) P = 4000 W, (**e**) P = 4500 W, and (**f**) P = 5000 W.

4.4. Verification of Printability Maps for Multi-Layer

The seven-layer specimen was fabricated based on the calculated results. The parameters used for the seven-layer specimen and bulk specimen are listed in Table 4. Figure 14a shows the cross-section morphologies of the seven-layer specimen. It should be noted that because the copper/steel interface with the mixed Cu-rich phase and Fe-rich phase is more sensitive to the etching agent, the aggressively etched areas at the interface will look like pores. In fact, the fabricated specimen is free of obvious defects, which can be confirmed compared with the unetched sample without obvious defects at the interface from Figure 13b. The interface of the copper/steel shown in Figure 14b indicates that a good copper/steel interface without cracks was obtained. Furthermore, the 40-layer bulk specimen with dimensions of about 70 × 8 × 12 mm^3 was fabricated as shown in Figure 15.

Table 4. Processing parameters used for the seven-layer and bulk specimen.

Layer Number	Laser Power (P, W)	Scanning Velocity (V, mm/min)
1	3000	800
2	2600	1000
3	2200	1000
4	1900	1000
5	1700	1000
6	1600	1000
7	1500	1000
10	1300	1000
20	950	1000
30	900	1000
40	800	1000

Figure 14. Cross-section morphologies of the seven-layer specimen: (**a**) overview of the cross-section; (**b**) magnified view of the copper/steel interface.

Figure 15. Morphology of the bulk specimen.

5. Conclusions

A layer-dependent analytical model is established to predict the layer-dependent processing parameters when fabricating the 07Cr15Ni5 steel on the CuCr substrate at the fixed layer thickness (0.3mm) and hatching space (0.3mm) and agrees well with experiments. Based on the research, the main conclusions can be drawn as follows:

(1) Considering both the great difference in thermophysical properties between copper and steel and the layer-based fabrication character of the AM process, the evolution of the effective thermophysical properties, including the effective thermal conductivity, effective density, and effective specific heat capacity with layer number are calculated. Meanwhile, the changes in absorption rate and catchment efficiency with the processing parameters are also taken into account. Then the layer-dependent effective material properties and processing parameters are provided to the Cline–Anthony thermal model. Thus, the layer-dependent analytical model is established.

(2) The model is applied to predict the printability maps as the deposition layer increases. The laser power decreases drastically for the first ten layers as the layer number increases and decreases slowly from the 10th to the 40th layer when other processing parameters are fixed. The decline rate of the maximum and minimum laser power for the printability range is as high as 77 and 88% from the first layer to the 40th layer at the scanning velocity of 800 mm/min. In addition, the ΔP is 1395 W for the first layer and 389 W for the 40th layer with a decline rate of 72%. The significant narrowing of the printability maps with the layer number is due to the decreases in effective thermal conductivity and effective specific heat capacity as well as the increase in effective density.

(3) The calculated results based on the proposed analytical model agree well with the experiments. The maximum relative error for the single-track molten pool width between the calculated and experimental results is 18.59%. Furthermore, the calculated formability of the first layer and multi-layer is in agreement with the experiment. The defects-free bulk specimen with dimensions of about $70 \times 8 \times 12$ mm^3 is successfully fabricated according to the predicted printability maps.

The proposed analytical model provides a new guidance to determine the processing windows for novel multi-material components, especially for the multi-materials whose physical properties are significantly different. Nevertheless, the criterion for the good fusion tracks is based on the peak temperature of the molten pool in the range of $T_{s\_steel} \sim T_{b\_steel}$, which may not be a precise condition to determine the final process maps. More rigorous criteria related to molten pool defects such as pores and balling formation are to be identified to further narrow the process map. Furthermore, HS and m_p also have a significant impact on printability during the fabrication of multi-material components, which will be discussed in the future.

Author Contributions: Conceptualization, methodology, and writing—original draft preparation, W.Z.; investigation, B.Z. and H.X.; writing—review and editing, project administration, H.Z.; funding acquisition, H.Y. and Y.W.; All authors have read and agreed to the published version of the manuscript.

Funding: This work is supported by the Human Spaceflight Program of China (D050302) and the Military Industry Stability Support project (2019KGW.YY4007Tm).

Acknowledgments: The authors would like to thank the Analytical and Testing Center of HUST for the SEM analysis and Jing Li for writing assistance.

Conflicts of Interest: The authors declare no conflict of interest.

References

1. Blakey-Milner, B.; Gradl, P.; Snedden, G.; Brooks, M.; Pitot, J.; Lopez, E.; Leary, M.; Berto, F.; du Plessis, A. Metal additive manufacturing in aerospace: A review. *Mater. Des.* **2021**, *209*, 110008. [CrossRef]
2. Gu, D.; Shi, X.; Poprawe, R.; Bourell, D.L.; Setchi, R.; Zhu, J. Material-structure-performance integrated laser-metal additive manufacturing. *Science* **2021**, *372*, 6545. [CrossRef]
3. Zhang, W.; Liao, H.; Hu, Z.; Zhang, S.; Chen, B.; Yang, H.; Wang, Y.; Zhu, H. Interfacial characteristics and mechanical properties of additive manufacturing martensite stainless steel on the Cu-Cr alloy substrate by directed energy deposition. *J. Mater. Sci. Technol.* **2021**, *90*, 121–132. [CrossRef]
4. Kar, J.; Roy, S.K.; Roy, G.G. Effect of beam oscillation on electron beam welding of copper with AISI-304 stainless steel. *J. Mater. Process. Technol.* **2016**, *233*, 174–185. [CrossRef]
5. Liao, H.L.; Zhu, H.H.; Xue, G.; Zeng, X.Y. Alumina loss mechanism of Al_2O_3-AlSi$_{10}$ Mg composites during selective laser melting. *J. Alloys Compd.* **2019**, *785*, 286–295. [CrossRef]
6. Zhang, X.C.; Pan, T.; Flood, A.; Chen, Y.T.; Zhang, Y.L.; Liou, F. Investigation of copper/stainless steel multi-metallic materials fabricated by laser metal deposition. *Mater. Sci. Eng. A* **2021**, *811*, 141071. [CrossRef]
7. Zhu, H.H.; Fuh, J.Y.H.; Lu, L. Microstructural evolution in direct laser sintering of Cu-based metal powder. *Rapid Prototyp. J.* **2005**, *11*, 74–81. [CrossRef]
8. Dávila, J.L.; Neto, P.I.; Noritomi, P.Y.; Coelho, R.T.; da Silva, J.V.L. Hybrid manufacturing: A review of the synergy between directed energy deposition and subtractive processes. *Int. J. Adv. Manuf. Technol.* **2020**, *110*, 3377–3390. [CrossRef]
9. Wang, X.; Lei, L.; Yu, H. A Review on Microstructural Features and Mechanical Properties of Wheels/Rails Cladded by Laser Cladding. *Micromachines* **2021**, *12*, 152. [CrossRef]
10. Xue, L.; Atli, K.C.; Picak, S.; Zhang, C.; Zhang, B.; Elwany, A.; Arroyave, R.; Karaman, I. Controlling Martensitic Transformation Characteristics in Defect-Free NiTi Shape Memory Alloys Fabricated Using Laser Powder Bed Fusion and a Process Optimization Framework. *Acta Mater.* **2021**, *5*, 117017. [CrossRef]
11. Yin, J.; Wang, D.Z.; Wei, H.L.; Yang, L.L.; Ke, L.D.; Hu, M.Y.; Xiong, W.; Wang, G.Q.; Zhu, H.H.; Zeng, X.Y. Dual-beam laser-matter interaction at overlap region during multi-laser powder bed fusion manufacturing. *Addit. Manuf.* **2021**, *46*, 102178. [CrossRef]
12. Yin, J.; Zhang, W.Q.; Ke, L.D.; Wei, H.L.; Wang, D.Z.; Yang, L.L.; Zhu, H.H.; Dong, P.; Wang, G.Q.; Zeng, X.Y. Vaporization of alloying elements and explosion behavior during laser powder bed fusion of Cu-10Zn alloy. *Int. J. Mach. Tools Manuf.* **2021**, *161*, 103686. [CrossRef]
13. Zhang, S.S.; Zhu, H.H.; Zhang, L.; Zhang, W.Q.; Yang, H.Q.; Zeng, X.Y. Microstructure and properties of high strength and high conductivity Cu-Cr alloy components fabricated by high power selective laser melting. *Mater. Lett.* **2019**, *237*, 306–309. [CrossRef]
14. Zhu, H.H.; Lu, L.; Fuh, J.Y.H.; Wu, C.C. Effect of braze flux on direct laser sintering Cu-based metal powder. *Mater. Des.* **2006**, *27*, 166–170. [CrossRef]
15. Seede, R.; Shoukr, D.; Zhang, B.; Whitt, A.; Gibbons, S.; Flater, P.; Elwany, A.; Arroyave, R.; Karaman, I. An ultra-high strength martensitic steel fabricated using selective laser melting additive manufacturing: Densification, microstructure, and mechanical properties. *Acta Mater.* **2020**, *186*, 199–214. [CrossRef]
16. Craig, O.; Bois-Brochu, A.; Plucknett, K. Geometry and surface characteristics of H13 hot-work tool steel manufactured using laser-directed energy deposition. *Int. J. Adv. Manuf. Technol.* **2021**, *116*, 699–718. [CrossRef]
17. Goll, D.; Trauter, F.; Loeffler, R.; Gross, T.; Schneider, G. Additive Manufacturing of Textured FePrCuB Permanent Magnets. *Micromachines* **2021**, *12*, 1056. [CrossRef]
18. Kozior, T.; Bochnia, J. The Influence of Printing Orientation on Surface Texture Parameters in Powder Bed Fusion Technology with 316L Steel. *Micromachines* **2020**, *11*, 639. [CrossRef]
19. Keist, J.S.; Nayir, S.; Palmer, T.A. Impact of hot isostatic pressing on the mechanical and microstructural properties of additively manufactured Ti-6Al-4V fabricated using directed energy deposition. *Mater. Sci. Eng. A* **2020**, *787*, 139454. [CrossRef]
20. Xue, A.T.; Lin, X.; Wang, L.L.; Wang, J.; Huang, W.D. Influence of trace boron addition on microstructure, tensile properties and their anisotropy of Ti6Al4V fabricated by laser directed energy deposition. *Mater. Des.* **2019**, *181*, 107943. [CrossRef]
21. Tan, H.; Guo, M.L.; Clare, A.T.; Lin, X.; Chen, J.; Huang, W.D. Microstructure and properties of Ti-6Al-4V fabricated by low-power pulsed laser directed energy deposition. *J. Mater. Sci. Technol.* **2019**, *35*, 2027–2037. [CrossRef]
22. Hu, Z.H.; Zhu, H.H.; Zhang, C.C.; Zhang, H.; Qi, T.; Zeng, X.Y. Contact angle evolution during selective laser melting. *Mater. Des.* **2018**, *139*, 304–313. [CrossRef]
23. Zhang, W.Q.; Zhu, H.H.; Hu, Z.H.; Zeng, X.Y. Study on the Selective Laser Melting of AlSi10Mg. *Acta Metall. Sin.* **2017**, *53*, 918–926.
24. Caiazzo, F.; Alfieri, V.; Bolelli, G. Residual stress in laser-based directed energy deposition of aluminum alloy 2024: Simulation and validation. *Int. J. Adv. Manuf. Technol.* **2021**, 1–15. [CrossRef]
25. Karg, M.C.H.; Ahuja, B.; Wiesenmayer, S.; Kuryntsev, S.V.; Schmidt, M. Effects of Process Conditions on the Mechanical Behavior of Aluminium Wrought Alloy EN AW-2219 (AlCu6Mn) Additively Manufactured by Laser Beam Melting in Powder Bed. *Micromachines* **2017**, *8*, 23. [CrossRef]
26. Yin, J.; Yang, L.L.; Yang, X.; Zhu, H.H.; Wang, D.Z.; Ke, L.D.; Wang, Z.M.; Wang, G.Q.; Zeng, X.Y. High-power laser-matter interaction during laser powder bed fusion. *Addit. Manuf.* **2019**, *29*, 100778. [CrossRef]

27. Gokcekaya, O.; Ishimoto, T.; Hibino, S.; Yasutomi, J.; Narushima, T.; Nakano, T. Unique crystallographic texture formation in Inconel 718 by laser powder bed fusion and its effect on mechanical anisotropy. *Acta Mater.* **2021**, *212*, 116876. [CrossRef]
28. Bai, Y.C.; Zhang, J.Y.; Zhao, C.L.; Li, C.J.; Wang, H. Dual interfacial characterization and property in multi-material selective laser melting of 316L stainless steel and C52400 copper alloy. *Mater. Charact.* **2020**, *167*, 110489. [CrossRef]
29. Liu, Z.H.; Zhang, D.Q.; Sing, S.L.; Chua, C.K.; Loh, L.E. Interfacial characterization of SLM parts in multi-material processing: Metallurgical diffusion between 316L stainless steel and C18400 copper alloy. *Mater. Charact.* **2014**, *94*, 116–125. [CrossRef]
30. Chen, J.; Yang, Y.Q.; Song, C.H.; Zhang, M.K.; Wu, S.B.; Wang, D. Interfacial microstructure and mechanical. properties of 316L/CuSn10 multi-material bimetallic structure fabricated by selective laser melting. *Mater. Sci. Eng. A* **2019**, *752*, 75–85. [CrossRef]
31. Chen, J.; Yang, Y.Q.; Song, C.H.; Wang, D.; Wu, S.B.; Zhang, M.K. Influence mechanism of process parameters on the interfacial characterization of selective laser melting 316L/CuSn10. *Mater. Sci. Eng. A* **2020**, *792*, 139316. [CrossRef]
32. Tan, C.L.; Zhou, K.S.; Ma, W.Y.; Min, L. Interfacial characteristic and mechanical performance of maraging steel-copper functional bimetal produced by selective laser melting based hybrid manufacture. *Mater. Des.* **2018**, *155*, 77–85. [CrossRef]
33. Wei, H.L.; Mukherjee, T.; Zhang, W.; Zuback, J.S.; Knapp, G.L.; De, A.; DebRoy, T. Mechanistic models for additive manufacturing of metallic components. *Prog. Mater. Sci.* **2021**, *116*, 100703. [CrossRef]
34. Zeng, K.; Pal, D.; Teng, C.; Stucker, B.E. Evaluations of effective thermal conductivity of support structures in selective laser melting. *Addit. Manuf.* **2015**, *6*, 67–73. [CrossRef]
35. Knapp, G.L.; Mukherjee, T.; Zuback, J.S.; Wei, H.L.; Palmer, T.A.; De, A.; DebRoy, T. Building blocks for a digital twin of additive manufacturing. *Acta Mater.* **2017**, *135*, 390–399. [CrossRef]
36. Tran, H.C.; Lo, Y.L. Systematic approach for determining optimal processing parameters to produce parts with high density in selective laser melting process. *Int. J. Adv. Manuf. Technol.* **2019**, *105*, 4443–4460. [CrossRef]
37. Cline, H.E.; Anthony, T.R. Heat treating and melting material with a scanning laser or electron beam. *J. Appl. Phys.* **1977**, *48*, 3895–3900. [CrossRef]
38. Picasso, M.; Marsden, C.F.; Wagniere, J.D.; Frenk, A.; Rappaz, M. A simple but realistic model for laser cladding. *Metall. Mater. Trans. B* **1994**, *25*, 281–291. [CrossRef]
39. Ahsan, M.N.; Pinkerton, A.J.; Moat, R.J.; Shackleton, J. A comparative study of laser direct metal deposition characteristics using gas and plasma-atomized Ti-6Al-4V powders. *Mater. Sci. Eng. A* **2011**, *528*, 7648–7657. [CrossRef]
40. Tabernero, I.; Lamikiz, A.; Martinez, S.; Ukar, E.; de Lacalle, L.N.L. Modelling of energy attenuation due to powder flow-laser beam interaction during laser cladding process. *J. Mater. Process. Technol.* **2012**, *212*, 516–522. [CrossRef]
41. Zhang, X.C.; Pan, T.; Chen, Y.T.; Li, L.; Zhang, Y.L.; Liou, F. Additive manufacturing of copper-stainless steel hybrid components using laser-aided directed energy deposition. *J. Mater. Sci. Technol.* **2021**, *80*, 100–116. [CrossRef]
42. Onuike, B.; Heer, B.; Bandyopadhyay, A. Additive manufacturing of Inconel 718-Copper alloy bimetallic structure using laser engineered net shaping (LENS (TM)). *Addit. Manuf.* **2018**, *21*, 133–140. [CrossRef]
43. Velu, M.; Bhat, S. Metallurgical and mechanical examinations of steel-copper joints arc welded using bronze and nickel-base superalloy filler materials. *Mater. Des.* **2013**, *47*, 793–809. [CrossRef]
44. Ready, J.F.; Farson, D.F.; Feeley, T. *LIA Handbook of Laser Materials Processing*; Springer: Berlin/Heidelberg, Germany, 2001; p. 715.
45. Jadhav, S.D.; Goossens, L.R.; Kinds, Y.; Hooreweder, B.V.; Vanmeensel, K. Laser-based powder bed fusion additive manufacturing of pure copper. *Addit. Manuf.* **2021**, *42*, 101990. [CrossRef]
46. Wei, C.; Gu, H.; Li, Q.; Sun, Z.; Chueh, Y.-H.; Liu, Z.; Li, L. Understanding of process and material behaviours in additive manufacturing of Invar36/Cu10Sn multiple material components via laser-based powder bed fusion. *Addit. Manuf.* **2020**, *37*, 101683. [CrossRef]
47. Khairallah, S.A.; Anderson, A.T.; Rubenchik, A.; King, W.E. Laser powder-bed fusion additive manufacturing: Physics of complex melt flow and formation mechanisms of pores, spatter, and denudation zones. *Acta Mater.* **2016**, *108*, 36–45. [CrossRef]
48. King, W.E.; Barth, H.D.; Castillo, V.M.; Gallegos, G.F.; Gibbs, J.W.; Hahn, D.E.; Kamath, C.; Rubenchik, A.M. Observation of keyhole-mode laser melting in laser powder-bed fusion additive manufacturing. *J. Mater. Process. Technol.* **2014**, *214*, 2915–2925. [CrossRef]
49. Guo, C.; Xu, Z.; Zhou, Y.; Shi, S.; Li, G.; Lu, H.; Zhu, Q.; Ward, R.M. Single-track investigation of IN738LC superalloy fabricated by laser powder bed fusion: Track morphology, bead characteristics and part quality. *J. Mater. Process. Technol.* **2021**, *290*, 117000. [CrossRef]
50. Yin, J.; Wang, D.Z.; Yang, L.L.; Wei, H.L.; Dong, P.; Ke, L.D.; Wang, G.Q.; Zhu, H.H.; Zeng, X.Y. Correlation between forming quality and spatter dynamics in laser powder bed fusion. *Addit. Manuf.* **2020**, *31*, 100958. [CrossRef]
51. Shao, J.; Yu, G.; He, X.; Li, S.; Li, Z.; Wang, X. Process maps and optimal processing windows based on three-dimensional morphological characteristics in laser directed energy deposition of Ni-based alloy. *Opt. Laser Technol.* **2021**, *142*, 107162. [CrossRef]
52. Tran, H.C.; Lo, Y.L.; Le, T.N.; Lau, A.K.T.; Lin, H.Y. Multi-scale simulation approach for identifying optimal parameters for fabrication of high-density Inconel 718 parts using selective laser melting. *Rapid Prototyp. J.* **2021**, ahead of print. [CrossRef]
53. Wang, D.Z.; Hu, Q.W.; Zeng, X.Y. Residual stress and cracking behaviors of Cr13Ni5Si2 based composite coatings prepared by laser-induction hybrid cladding. *Surf. Coat. Technol.* **2015**, *274*, 51–59. [CrossRef]

54. Siao, Y.-H.; Wen, C.-D. Examination of molten pool with Marangoni flow and evaporation effect by simulation and experiment in selective laser melting. *Int. Commun. Heat Mass Transf.* **2021**, *125*, 105325. [CrossRef]
55. Chen, C.P.; Yin, J.; Zhu, H.H.; Xiao, Z.X.; Zhang, L.; Zeng, X.Y. Effect of overlap rate and pattern on residual stress in selective laser melting. *Int. J. Mach. Tools Manuf.* **2019**, *145*, 103433. [CrossRef]

Article

Hybrid Dissection for Neutron Tube Shell via Continuous-Wave Laser and Ultra-Short Pulse Laser

Minqiang Kang [1,2,*], Yongfa Qiang [1], Canlin Zhu [1], Xiangjun Xiang [1], Dandan Zhou [1], Zhitao Peng [1], Xudong Xie [1,*] and Qihua Zhu [1,*]

1. Laser Fusion Research Center, China Academy of Engineering Physics, Mianyang 621900, China; yfqiang@zju.edu.cn (Y.Q.); 13141317443@163.com (C.Z.); dennis55555@163.com (X.X.); dan723@126.com (D.Z.); peng_zhitao@163.com (Z.P.)
2. Graduate School of China Academy of Engineering Physics, Beijing 100088, China
* Correspondence: kangmq@163.com (M.K.); xiexudong@caep.cn (X.X.); qihzh@163.com (Q.Z.)

Abstract: The sealed neutron tube shell dissection process utilizing the traditional lathe turning method suffers from low efficiency and high cost due to the frequency of replacement of the diamond knife. In this study, a hybrid dissection method is introduced by combining the continuous-wave (CW) laser for efficient tangential groove production with an ultra-short pulse laser for delamination scanning removal. In this method, a high-power CW laser is firstly employed to make a tapered groove on the shell's surface, and then a femtosecond pulse laser is used to micromachine the groove in order to obtain a cutting kerf. The thermal field was theoretically investigated in a finite element model. The simulation results show that the width of the area of temperature exceeding 100 °C is 1.9 mm and 0.4 mm with rotating speeds of 20 rad/s and 60 rad/s, respectively. In addition, a 2 mm deep slot in the 25 mm diameter tube was successfully produced in 1 min by a kilowatt fiber laser, and a 500-femtosecond pulse laser was employed to cut a plate with a material removal rate of 0.2 mm^3/min. By using the hybrid method, the cutting efficiency was improved about 49 times compared to the femtosecond laser cutting. According to the simulation and experimental results, this method provides a high-efficiency and non-thermal cutting technique for reclaimed metallic neutron tube shells with millimeter-level thick walls, which has the advantages of non-contact, minimal thermal diffusion, and no effect of molten slag. It is indicated that the hybrid dissection method not only offers a new solution for thick neutron tube shell cutting but also extends the application of laser cutting techniques.

Keywords: laser cutting; non-contact process; neutron tube shells; 304 stainless steel; fiber laser; femtosecond laser; thermal transmission simulation

1. Introduction

The neutron tubes based on ^{2}H(d,n)^{3}He(D-D) or ^{3}H(d,n)^{4}He(D-T) fusion reactions to generate monochromatic neutrons have been widely applied in the fields of neutron radiography [1–3], borehole logging [4], coal analysis [5], searching for water [6], etc. Typically, the neutron tube components are the ion source, acceleration electrode, hydrogen-absorbing filament, ion target, and tube shell [7,8]. The deuterium (D) and tritium (T) ions extracted from the ion source are firstly accelerated by high voltage electric field, and then focused and bombarded to the ion target, finally producing high energy neutrons. The ion target is one of the most important components, where the D and T atoms are stored, and the D-D and D-T nuclear reactions occur, and it requires a normal toleration temperature of less than 150 °C [9,10]. The neutron tube is an accelerator vacuum system with all the inside objects sealed in a tube shell made of ceramic or stainless steel [11]. Due to the massive application and limited lifetime, there is a large number of trash sealed neutron tubes which need to be reclaimed to gather back the D and T atoms. The dissection of the tube shell is the key procedure to take out the ion target. However, in order to avoid the actions of D and T ions, the safe temperature for the dissection process is required to be no more than 100 °C. During the dissection, the inner temperature is required to be lower than 50 °C and the yield molten slag is not allowed to synchronously enter the shell. In addition, it would be ideal to carry out the total dissection process in a vacuum vessel and without connection with the outside devices.

At present, the mechanical cutting method using a lathe turning process is employed to achieve the dissection of neutron tube shells. There is no coolant liquid in the diamond turning process to avoid troublesome liquid waste disposal. However, a lack of cooling makes diamond knives vulnerable to damage. Frequent replacement of diamond knives results in a low operating efficiency and high cost. Therefore, it is necessary to find a new cutting method for the effective cold cutting of a large number of waste neutron tubes. The abrasive-waterjet cutting method is a highly effective process with the absence of thermal diffusion, high machining versatility, and small machining force [12,13]. It is known that the process uses a jet of high pressure and velocity water and an abrasive slurry to cut the target material by means of erosion, and it needs a great deal of water and abrasives to finish the cutting process. Furthermore, the process is messy and characterized by huge quality variation in terms of the kerf width and striations. Obviously, it is not an appropriate approach for the dissection of neutron tube shells.

Compared with tradition methods, the laser ablation cutting process is an alternative approach to cut stainless steel plates and tubes, with the advantages of high efficiency, non-contact, low maintenance, and remote controllable [14,15]. It is known that the laser cutting of thick work-pieces provides considerable advantages over the conventional techniques owing to the short processing time and high precision operation. In addition, the flexibility of fiber lasers provides a fast and affordable way to accomplish larger work stations and more complex applications [16,17]. Nowadays, pipe sheet cutting systems equipped with a high-power fiber laser are available due to their high efficiency and reduced maintenance costs. However, in the conventional laser cutting method, the laser cutting head is usually pointed perpendicularly to the material surface and the cutting gas flows coaxially to the beam [18,19]. Moreover, the molten mass ejected from the kerf by the pressure of the cutting gas flow and the residual laser beam directed into the tube shell will both cause the temperature to rise inside the tube shell. Unlike the laser melt ablation process, ultra-short pulse laser micromachining provides a non-thermal method, utilizing high energy pulses of picosecond to femtosecond durations (e.g., <10 ps) [20–22]. With the effect of megawatt pulse peak power, the material is locally vaporized with minimal thermal diffusion and without any molten slag produced. However, ultra-short laser cutting is mainly used for machining ultra-thin materials with a thickness less than 300 μm [23,24]. When working with millimeter-level thickness sheets, it is not an appropriate approach due to the low removal efficiency and beam clipping by cut wall shadowing. Additionally, as the result of the obvious threshold effect and the shielding effect of plasma, the improvement of process efficiency is limited by increasing the single pulse energy and repetition frequency.

In a word, the high-power laser cutting process is highly efficient but with a large thermal diffusion region, and the ultra-short pulse micromachining process is a non-thermal technique but with ultra-low material removal rates: both of them are not suitable for the dissection of neutron tube shells. Therefore, a potential approach is to utilize the useable merits and to avoid the weaknesses of the two methods to finish the whole process. In this paper, by combining continuous-wave laser efficient tangential grooving with ultra-short pulse laser delamination scanning removal, a new dual laser beam asynchronous cutting method is presented for the efficient dissection of neutron tube shells with thick metallic walls. A three-dimensional finite element model (FEM) was established to validate the temperature control of the shell surface during the continuous-wave (CW) laser dissection process. Additionally, a kilowatt continuous-wave fiber laser and a femtosecond pulse laser were employed to process a tube part and plate sheet in the experiments to certify the novel method.

2. Materials and Methods

Figure 1 shows the configuration of the hybrid dissection process in this study. As shown in Figure 1a, the cutting system includes a high-power fiber laser (MFSC-1000 L, produced by MAX photonics Co., Ltd., Shenzhen, China) with a maximum output power of 1 kW at the wavelength of 1080 nm, and a homemade 7 W femtosecond pulse laser with the condition of 500 fs at 1030 nm. The CW laser beam is coupled to a laser cutting head and focused to the side direction of the tube to create a tapered groove, while the femtosecond pulse laser is passed through a beam expander, and then sent to a scanner head, and focused to the bottom of the tapered groove from the vertical direction by an F-theta lens to complete the dissection process. The tube shell is held by a rotating chuck to avoid deflections during the process, and then is fixed to a rotary unit, which maintains a controllable high turning speed movement.

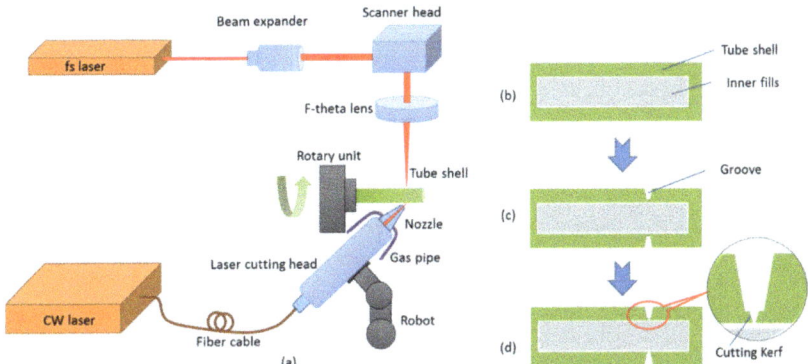

Figure 1. Configuration of the dual laser beam asynchronous dissection process: (**a**) schematic diagram of the cutting system, (**b**) cross-sectional view of the tube shell, (**c**) the tapered groove created by the fiber laser, and (**d**) the cutting kerf after the femtosecond laser completed cutting.

2.1. Cutting Method

The experiment setup is shown in Figure 2. The hybrid dissection process includes two steps. Firstly, the tapered groove operation is created by a high-power CW fiber laser using laser ablation, which is analogous to the lathe turning process, and the laser beam assumes the action of the turning knife, as shown in Figure 2a. The groove is machined by the movement of the laser beam and the synchronous rotation of the tube shell. Then, a delamination scanning removal methodology is applied to the cutting kerf operation, which is shown in Figure 2b. The cutting kerf is machined by the scanning femtosecond pulse laser and the tube shell's rotary, cooperatively. The temperature distribution is measured by a temperature monitoring device (infrared camera FLIR T620, produced by FLIR Systems Inc., Wilsonville, OR, USA). The whole cutting process holds a low temperature and does not lead to molten slag and residual laser directed at the inner substance. Table 1 indicates the main parameters of the two lasers used in this research.

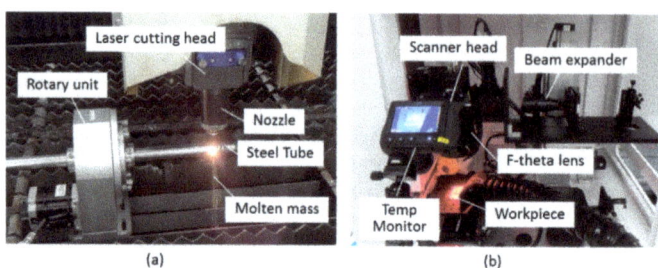

Figure 2. Experimental setup: (**a**) the first step: kilowatt fiber laser cutting; (**b**) the second step: femtosecond pulse laser cutting.

Table 1. Main parameters of the lasers.

Laser Parameters	CW Laser	Femtosecond Laser
Operation mode	Continuous wave	500 fs (Pulse)
Central wavelength	1080 nm	1030 nm
Average power	1000 W	7 W
Repletion rate	N/A	300 kHz
Focal length	150 mm	100 mm
Beam diameter	100 µm	30 µm
Beam mode	TEM_{00} Gaussion Mode	

The configuration of the high-power fiber laser fabrication is shown in Figure 3. The laser beam, with an average power of 1 kW, is fed via a process fiber (50 µm core diameter) into the laser cutting head fitted with a 75 mm collimator and a 150 mm focal lens, which results in a final theoretical spot size of 100 µm. The focused laser beam passes through the nozzle tip and irradiates the tangential direction surface of the tube shell, as shown in Figure 3b. The standoff distance between the conical nozzle and the working point is set to a fixed value to avoid the space interference. The laser cutting head contained a cylindrical tank connected to two flexible gas pipe lines. The assistive argon gas with a gauge pressure of approximately 10 bar is released into the cylindrical tank and passes through the nozzle tip toward the tube shell surface. The high-pressure gas flow is used to blow away the molten metal and accelerate fume dispersion, while cooling the shell body. The laser cutting head is mounted on a robot arm device, which is controlled via the computerized numerical control, to obtain X-Y-Z stage movement of the fiber laser beam. The tube shell is held by a rotating chuck and fixed to a rotary unit, which maintains a controllable turning movement. Usually, the kerf width of fiber laser melt cutting is only equal to the focus spot diameter [15]. With the effect of the laser beam's movement in the X-Y dimension (shown in Figure 3a), the blow down of argon gas flow, and the tube's rotary operation, the groove is created efficiently. Additionally, as a result of the cooling effect and the short time of operation, the whole tube shell is within a small thermal impact zone and holds a low temperature status during the fast-cutting procedures. This method avoids direct illumination by the high-power laser beam and the ejection of molten mass into the inner region. Additionally, the residual thin metallic wall prevents the molten metal and features from getting into the tube shell.

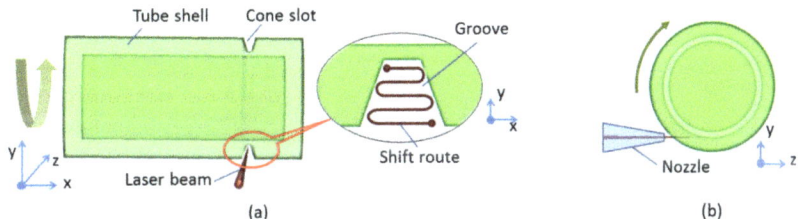

Figure 3. Configuration of the high-power fiber laser process: (**a**) schematic representation of laser beam action and drawing of the cut geometry; (**b**) cross-section view.

The schematic diagram of the femtosecond pulse laser fabrication is represented in Figure 4. The femtosecond pulse laser is passed through a beam expander to magnify the diameter of the laser spot before entering into the scanner head. Additionally, the diameter of the entrance spot is expanded beyond that of the smaller focused spot. With the use of the scanner head, the finer and faster beam with regular profile manipulation is achieved, which is reflected by two galvo-mounted mirrors. The regularly spaced profile is taken perpendicular to the cutting direction. As shown in Figure 4b, the pattern drawn is a rectangle region with the width same as the kerf width, and the length equal to a few millimeters, which is decided by the tube's radius to obtain an available reaction. In order to maintain a deep cutting kerf, the multi-line scanner is used in our work, which avoids the screen of the laser beam. The pulse overlapping value is set to about 85% according to previous experiments [25]. Then, the laser beam is focused on the bottom of the groove in the y direction by an F-theta lens (shown in Figure 4a). The F-theta lens achieves a constant beam size across the entire scanning surface region. For the cutting process, the scanner head is used to move the laser beam to the groove's bottom position, along with the rotary of the tube shell in low speed, to obtain the final dissection process.

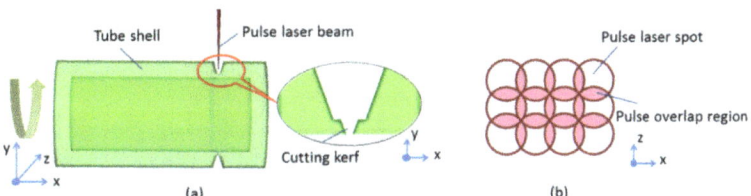

Figure 4. Configuration of the femtosecond pulse laser process: (**a**) schematic diagram of laser beam action; (**b**) schematic description of the scanning pattern drawn.

2.2. Samples

The work-piece used in our research is a non-magnetic 304 austenitic stainless steel tube to imitate the neutron tube shell described in reference [8], which has an outer diameter of 25 mm, wall thickness of 2 mm, and length of about 160 mm. Table 2 presents the chemical composition of the 304 stainless steel (data were provided by the material supplier: Shanghai Baoshao Special Steel Co., Ltd., Shanghai, China), while the physical properties are presented in Table 3. During the dissection of the neutron tube shell, the inner domination temperature is required to be less than 50 °C and the temperature range of the shell body (the region 1 mm apart from the machining point) is required to be no more than 100 °C, while the residual high-power laser and its resultant molten slag are not allowed to enter the tube shell.

Table 2. Chemical composition of 304 stainless steel for tube shell with a 25 mm diameter [26].

Type	Mass Fraction	C	Si	P	S	Mn	Ni	Cr
Tube shell	Max (%)	0.10	0.75	0.04	0.03	2	11	20
	Min (%)	0.04	-	-	-	-	8	18

Table 3. Physical properties of 304 stainless steel.

Material Properties	Symbol	Value
Density	ρ	7200 (kg·m^{-3})
Thermal conductivity (solid/liquid)	k	40/22 (W·(m·K)$^{-1}$)
Specific heat capacity (solid/liquid)	C_{ps}	720/800 (J·(kg·K)$^{-1}$)
Coefficient of thermal expansion	δ	4.95×10^{-5} (K^{-1})
Fusion latent heat	C_m	2.47×10^5 (J·kg^{-1})
Evaporation latent heat	C_v	6.34×10^6 (J·kg^{-1})
Solid temperature	T_s	1679 (K)
Liquid temperature	T_l	1727 (K)
Fusion point	T_m	1700 (K)
Boiling point	T_v	3200 (K)
Emissivity coefficient	ε_0	0.16
Convective heat transfer coefficient	h	40 (W·m^{-2}·K^{-1})
Stefan–Boltzmann constant	σ_0	5.67×10^{-8} (W·m^{-2}·K^4)

2.3. Numerical Modeling

To validate the temperature control of the shell surface during the CW laser dissection process, a three-dimensional finite element model (FEM) was established. The schematic of the dissection process using the CW laser is shown in Figure 5. The shell tube is set to be stationary due to relative motion, and the laser spot locates at the middle of the tube side surface and moves circularly along the tangential direction at a speed of 60 rad/s. The thermal effect of the light spot on the tube surface is related to the circular motion speed of the laser spot. The faster the circular motion, the shorter the time of laser action on the surface element and the slower the rate at which the surface temperature rises. The laser spot diameter is 100 µm and cuts at the outermost side of the tube, and the projection of the laser spot on the tube surface is an elliptical shape with a head and a tail as shown in Figure 5. The incident point of the laser on the tube changes with the tube rotation, as well as the local coordinate (x_1, y_1) of the laser spot, which is shown in the B-direction view in Figure 5b. The local coordinate origin of the laser spot on the tube surface can be expressed as $((R-r)\sin\omega t, (R-r)\cos\omega t)$. The projection of the laser spot on the tube surface is realized by UDF (User Defined Function), and moves around the tube with the calculation time.

When the laser is directed on the surface of the shell tube, the target element surface is heated and the energy transfer includes heat conduction, heat convection, thermal radiation and so on. Besides, with the increase in temperature, melting occurs and is even followed by vaporization if the temperature reaches the boiling point. The laser flux density reaches 10^7 W/cm^2 in the dissection process; therefore, melting and vaporization are included in the calculation.

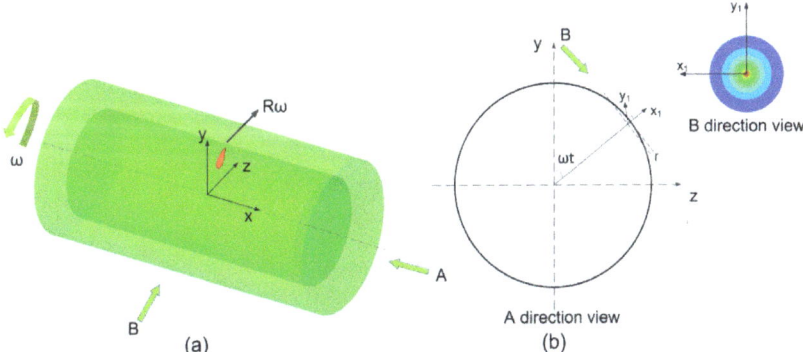

Figure 5. Schematic of the high-power fiber laser process FEM model: (**a**) axonometric view of FEM model; (**b**) direction views of FEM model.

The governing equations for the conservation of energy can be expressed in the following form,

$$\rho C_{pe}\frac{\partial T}{\partial t} - \nabla \cdot (k\nabla T) = Q - Q_{loss} \qquad (1)$$

where ρ is the density of 304 stainless steel, C_{pe} is the equivalent specific heat capacity, k is the heat conductivity, Q is the input laser energy flux density, and Q_{loss} is the energy loss due to radiation and convection, which is written in the following equation,

$$Q_{loss} = \sigma_0 \varepsilon_0 (T^4 - T_0^4) + h(T - T_0) \qquad (2)$$

where σ_0 is the Stefan–Boltzmann constant, ε_0 is the emissivity coefficient, T is the temperature of the shell tube surface, T_0 is the temperature of the surrounding environment, and h is the convective heat transfer coefficient. The flux density of laser energy follows a Gaussian distribution and can be expressed in the following form,

$$Q = \frac{3A_b P}{\pi r^2} e^{-3[\frac{(x^2+y^2)}{r^2}]} \qquad (3)$$

where A_b is the absorption coefficient of the laser by 304 stainless steel, P is the injected laser power, r is the radius of the laser spot, and finally x and y are the coordinates in the Gaussian distribution. The energy deposition profile is a moving elliptical profile after projection on the tube surface through UDF.

In the FEM calculation, melting and vaporization are also included due to a high laser energy flux density of 10^7 W/cm^2 in the dissection process, and an equivalent specific heat capacity C_{pe} is introduced, which is written in the following form,

$$C_{pe} = C_{ps} + C_m D_m + C_v D_v \qquad (4)$$

where C_{ps} is the specific heat capacity of 304 stainless steel, C_m is the latent heat of fusion, C_v is the latent heat of evaporation, and finally D_m and D_v are Gaussian functions of fusion and evaporation, respectively, which can be written in the following form,

$$D_m = \frac{\exp\left[-(T-T_m)^2/\Delta T_m^2\right]}{\Delta T_m \sqrt{\pi}}, \quad D_v = \frac{\exp\left[-(T-T_v)^2/\Delta T_v^2\right]}{\Delta T_v \sqrt{\pi}} \qquad (5)$$

The finite element mesh model for the numerical calculation is shown in Figure 6, in which a cylindrical model with a length of 40 mm and an external diameter of 25 mm is established. The center area of the cylinder is the cutting area and grid refinement is adopted.

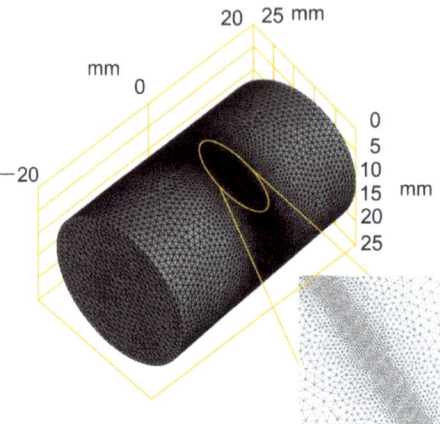

Figure 6. Component structure with mesh.

3. Results and Discussion

3.1. Numerical Modeling Results and Discussion

Figure 7 shows the contour of the temperature distribution on the tube surface and the profile of the tube wall as the calculation approaches a stable state. It can be seen from Figure 7a that the maximum temperature on the tube surface is about 3290 K. From Figure 7b, it can be seen that the temperature at the surface element irradiated by the laser is much higher and the temperature decreases rapidly from the outer wall to the inner wall. The depth of the area with a temperature higher than 100 °C is only 0.25 mm, which accounts for 1/8 of the total wall thickness, which shows that the cutting method proposed in this paper has little influence on the temperature inside the tube and meets the demand for controlling the temperature in the inner area of the tube during the dissection process. Due to the addition of jet airflow to remove residuals and for cooling the surface during dissection, a localized low temperature zone is formed around the high temperature zone, as shown in the profile view of the tube wall in Figure 7b, where the low temperature zone is located on both sides of the high temperature zone.

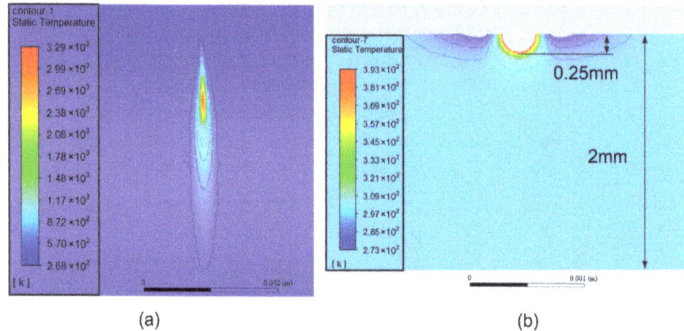

Figure 7. The contours of the temperature distribution: (**a**) temperature distribution near laser spot; (**b**) temperature profile in tube wall.

Figure 8 shows the contour of temperature distribution on the outer and inner walls of the tube with different turns of the laser moving around the tube. It can be seen that with the increase in turns, the high temperature areas on the outer and inner walls of the tube gradually expand, and after several turns, the temperature distribution approaches a stable state with a banding area of high temperature on the wall.

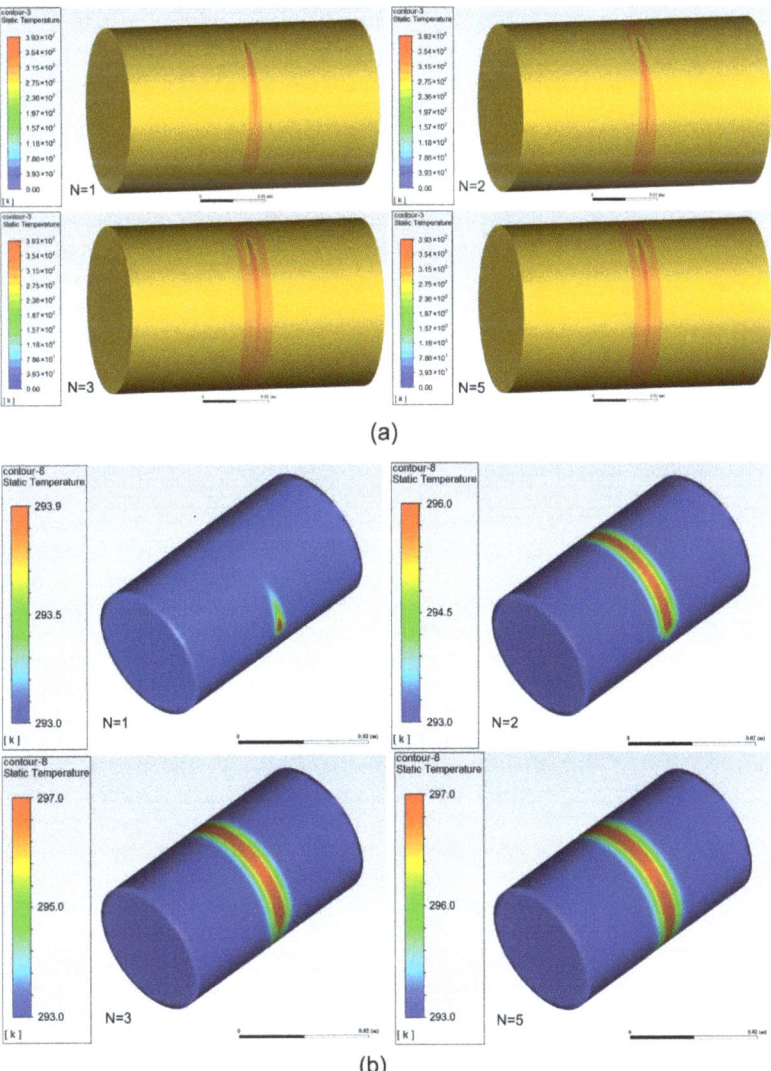

Figure 8. The contour of temperature distribution with different turns: (**a**) distribution on outer wall; (**b**) distribution on inner wall.

Figure 9 shows the area distribution of temperature beyond 100 °C at different rotating speeds of the tube. It can be seen that when the tube rotates slowly, the laser irradiates the surface element for a longer time, and thus the temperature rises higher and the area of temperature exceeding 100 °C becomes wider. The maximum width of the temperature exceeding area is about 1.9 mm and 0.4 mm when the rotating speed is 20 rad/s and 60 rad/s, respectively. This means that an appropriate rotating speed is necessary to control the whole temperature distribution during the dissection process and the rotating speed adopted in this study was 60 rad/s.

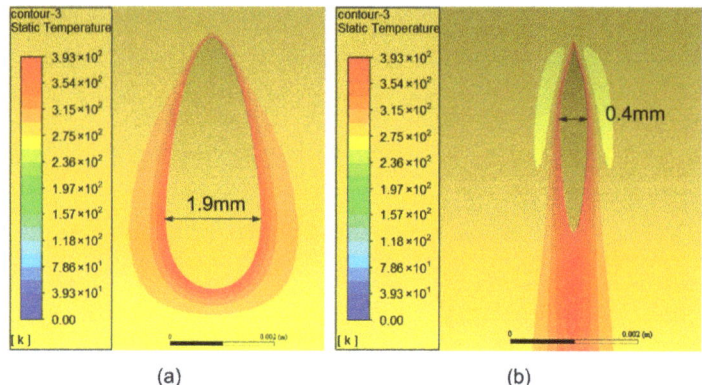

Figure 9. The area distribution of temperature beyond 100 °C: (**a**) rotating speed of 20 rad/s; (**b**) rotating speed of 60 rad/s.

3.2. Experimental Results and Discussion

In this study, the dual laser beam asynchronous dissection process was divided into two machining experiments in order to certify the practicability of this new method. Firstly, a kilowatt continuous-wave fiber laser coupled to a laser cutting head was employed to manufacture a groove on the surface of a 304 stainless steel tube. The experiment parameters of this process were set to be the same as the first-step process of the new dissection method. Then, for the second experiment, a femtosecond pulse laser was used to produce a cutting kerf on a 304 stainless steel plate part. Both of them have the same material, i.e., 304 stainless steel. Figure 10 shows the results of cutting the tube. As shown in Figure 10a, a groove was successfully produced on the surface of the tube with a 25 mm diameter and a 2.5 mm wall thickness. The size of the slot was 1 mm width and 2 mm deep. It took about 1 min to produce the slot. Additionally, some black impressions could be seen on the surface of the tube near the slot, due to the influence of melt in the process. As shown in the previous Figure 2a, due to the high-pressure assist cutting gas, the molten slag is blown to the tangential direction of the tube. Figure 10b shows the experimental results of cutting the 304 stainless steel stainless plate with a thickness of 2 mm utilizing a femtosecond pulse laser. The whole process time of the plate was 5 min, the length of the cutting kerf was 1 mm and the width was 0.4 mm. The calculated cone half-angle of the cutting kerf was about 5.7°. The cutting speed was about 0.2 mm/min and the material removal rate was 0.2 mm^3/min.

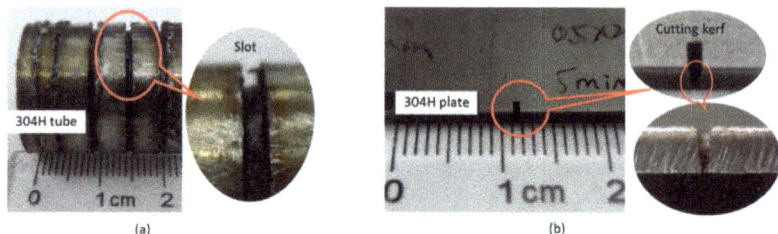

Figure 10. The results of cutting 304 stainless steel work-pieces: (**a**) the result of kilowatt fiber laser cutting of tube; (**b**) the result of 500-femtosecond pulse laser cutting of plate.

For the result of femtosecond pulse laser cutting, the side surface wall properties inside the cutting kerf were imaged and measured by a confocal microscope (SensoSCAN neox, produced by Sensofar Metrology, Terrassa, Spain), as shown in Figure 11. As shown in the side surface imaging formation (Figure 10a), it was in the region of 876 μm length and 660 µm width, and it was a periodicity smooth plane without any molten slag on the flank wall. The measured results for the x-x section indicate that the surface roughness was about 0.95 µm. In addition, within the whole femtosecond pulse laser process, operating at room temperature, i.e., 25 °C, the monitored temperature was stabilized in the range of 38 °C to 41 °C. The results indicate that the thermal effect was avoided in the process.

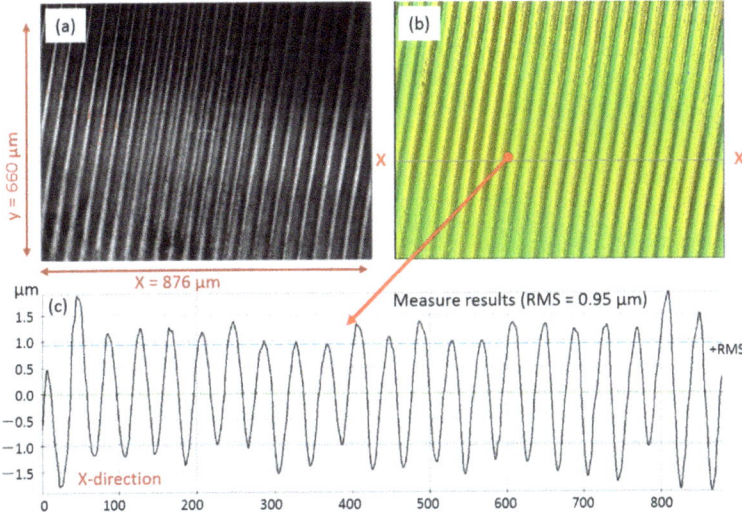

Figure 11. The side surface properties for the femtosecond pulse laser cutting kerf: (**a**) the side surface imaging formation of the cutting kerf wall; (**b**) the measured formation of the flank wall; and (**c**) the numerical value result for the x-x section.

Furthermore, with regard to the cutting of the neutron tube shell, with a size of 25 mm diameter and 2 mm wall thickness, which equals a 78.5 mm length plate, it needs 392.5 min (about 6.5 h). In addition, using the hybrid dissection method, after the first process steep of the high-power CW laser which needs 1 min, the remaining wall thickness to be cut is only 300 µm, and according to the cone half-angle of 5.7°, the cutting kerf width was less than 60 µm. This means that the material removal volume was smaller than 1.413 mm^3 (0.3 mm × 0.06 mm × 78.5 mm). Consequently, with the same material removal rate, the required cutting time is just about 7 min. Adding the first steep time of 1 min, the total dissection time is about 8 min. That is to say, by the use of the hybrid dissection method, the cutting efficiency is improved about 49 times compared to only femtosecond laser cutting.

For the actual dissection processing of the neutron tube shell, all the cutting system is located inside a closed vessel which stays at a negative pressure. However, with the use of laser processing, it just needs an optic window to put the high-power laser beam transmission on the vessel, and an air pipe to facilitate the high-pressure cutting jet-flow inside the vessel. The exhaust gas generated in the process is gathered by a purification system. Additionally, there is a way to further cool the surface of the tube during the high CW laser cutting process by employing a second air jet-flow spraying up to the other side area which will be explored in our future work.

4. Conclusions

The sealed neutron tube shell dissection process utilizing the traditional lathe turning method requires frequent replacement of the diamond knife, which makes it a low-efficiency and high-expense approach. In this work, a novel dual laser beam asynchronous cutting method was presented to dissect sealed neutron tube shells with a material of non-magnetic stainless steel. Firstly, a high-power fiber laser is employed to produce a tapered groove on the surface of the tube shell, and then a femtosecond pulse laser is used to micromachine a cutting kerf at the groove's bottom position.

A three-dimensional finite element model (FEM) was established to validate temperature control of the shell surface during the CW laser dissection process. It shows that the cutting method has little influence on the temperature inside the tube and meets the demand for controlling the temperature in the inner area of the tube during the dissection process. Due to the addition of jet airflow to remove residuals and cool the surface, a localized low temperature zone is formed around the high temperature zone. With the increase in turns, the high temperature areas on the outer and inner walls of the tube gradually expand, and after several turns, the temperature distribution approaches a stable state. When the tube rotates slowly, the laser irradiates the surface element for a longer time, and thus the temperature rise is higher and the area of temperature exceeding 100 °C becomes wider. The appropriate rotating speed is 60 rad/s to control the whole temperature distribution, with the depth of the area at a temperature higher than 100 °C equal to only 0.25 mm.

A kilowatt CW fiber laser was employed to make a groove on the surface of a 304 stainless steel tube with a 25 mm diameter and a 2.5 mm wall thickness. It took about 1 min to produce a slot of 1 mm width and 2 mm depth. Additionally, a 500-femtosecond pulse laser was used to produce a cutting kerf on a 304 stainless steel plate part with a thickness of 2 mm. It took 5 min to produce the kerf of 1 mm length and 0.4 mm width. The inner wall in the cutting kerf was smooth and without any molten slag. The monitored temperature was stabilized in the range of 38 °C to 41 °C, which indicates that the thermal effect is avoided in the process. By using the hybrid dissection method to cut a neutron tube shell of 25 mm diameter and 2.5 mm wall thickness, the cutting efficiency was improved about 49 times compared to the single femtosecond laser cutting.

According to the simulation and experiment results, this method provides a high-efficiency and non-thermal cutting technique for reclaiming neutron tube shells with advantages of non-contact, minimal thermal diffusion, and no effect of molten slag. Compared with the traditional fabrication methods, this technique provides an outstanding solution for the efficient and cold cutting of neutron tube shells, especially when the wall thickness is more than 1 mm. In summary, this work not only demonstrates an effective method for the cold cutting of steel tube shells with multi-millimeter level wall thicknesses, but also extends the application field of laser processing techniques.

Author Contributions: Conceptualization, M.K., Z.P., X.X. (Xudong Xie) and Q.Z.; methodology, M.K., X.X. (Xudong Xie) and Q.Z.; software, Y.Q.; validation, C.Z.; formal analysis, M.K. and Y.Q.; investigation, M.K., Y.Q., C.Z., Z.P., X.X. (Xiangjun Xiang), X.X. (Xudong Xie) and Q.Z.; resources, M.K., Y.Q., C.Z., X.X. (Xiangjun Xiang), X.X. (Xudong Xie) and Q.Z.; data curation, M.K. and X.X. (Xiangjun Xiang); writing—original draft preparation, M.K., Y.Q. and D.Z.; writing—review and editing, M.K. and Y.Q.; visualization, M.K. and C.Z.; supervision, X.X. (Xudong Xie) and Q.Z.; project administration, M.K., Y.Q., C.Z., Z.P., X.X. (Xudong Xie) and Q.Z.; funding acquisition, M.K., Y.Q., C.Z., X.X. (Xudong Xie) and Q.Z. All authors have read and agreed to the published version of the manuscript.

Funding: This research was funded by the Natural National Science Foundation of China (NSFC), grant number 62075201 and 12004352.

Conflicts of Interest: The authors declare no conflict of interest.

References

1. Richards, W.; Barrett, J.; Springgate, M.; Shields, K. Neutron radiography inspection of investment castings. *Appl. Radiat. Isot.* **2004**, *61*, 675–682. [CrossRef]
2. Lehmann, E.; Mannes, D.; Kaestner, A.; Grünzweig, C. Recent applications of neutron imaging methods. *Phys. Procedia* **2017**, *88*, 5–12. [CrossRef]
3. Evans, L.; Minniti, T.; Fursdon, M.; Gorley, M.; Barrett, T.; Domptail, F.; Surrey, E.; Kockelmann, W.; Müller, A.V.; Escourbiac, F.; et al. Comparison of X-ray and neutron tomographic imaging to qualify manufacturing of a fusion divertor tungsten monoblock. *Fusion Eng. Des.* **2018**, *134*, 97–108. [CrossRef]

4. Randall, R. Application of accelerator sources for pulsed neutron logging of oil and gas wells. *Nucl. Instrum. Methods Sect. B* **1985**, *10–11*, 1028–1032. [CrossRef]
5. Jing, S.-W.; Gu, D.-S.; Qiao, S.; Liu, Y.-R.; Liu, L.-M. Development of pulse neutron coal analyzer. *Rev. Sci. Instrum.* **2005**, *76*, 045110. [CrossRef]
6. Sudac, D.; Majetic, S.; Kollar, R.; Nad, K.; Obhodas, J.; Valkovic, V. Inspecting minefields and residual explosives by fast neutron activation method. *IEEE Trans. Nucl. Sci.* **2011**, *59*, 1421–1425. [CrossRef]
7. Pfutzner, H.; Groves, J.; Mahdavi, M. Performance characteristics of a compact DT neutron generator system. *Nucl. Instrum. Methods Phys. Res.* **1995**, *99*, 516–518. [CrossRef]
8. Zhou, X.; Lu, J.; Liu, Y.; Ouyang, X. A concise method to calculate the target current ion species fraction in D–D and D–T neutron tubes. *Nucl. Instrum. Methods Phys. Res. A* **2021**, *987*, 164836. [CrossRef]
9. Qiao, Y.-H. Progress in studies and applications of neutrons tube. *Nucl. Electron. Detect. Technol.* **2008**, *28*, 1134–1139.
10. Guo, W.-T.; Zhao, S.-J.; Yua, Z.-T.; Shi, G.-Y.; Jing, S.-W. Effect of target material on neutron output and sputtering yield of D-D neutron tube. *Nucl. Instrum. Methods Phys. Res. Sect. B* **2020**, *473*, 48–54. [CrossRef]
11. Zhao, S.-J.; Guo, W.-T.; Yu, Z.-T.; Li, C.; Jing, S.-W. Design of magnetic circuit and extraction system of ECR ion source for intense neutron tube. *Vacuum* **2020**, *178*, 109450. [CrossRef]
12. Chen, L.; Siores, E.; Wong, W. Kerf characteristics in abrasive waterjet cutting of ceramic materials. *Int. J. Mach. Tools Manuf.* **1996**, *36*, 1201–1206. [CrossRef]
13. Wang, J.; Wong, W. A study of abrasive waterjet cutting of metallic coated sheet steels. *Int. J. Mach. Tools Manuf.* **1999**, *39*, 855–870. [CrossRef]
14. Karatas, C.; Keles, O.; Uslan, I.; Usta, Y. Laser cutting of steel sheets: Influence of workpiece thickness and beam waist position on kerf size and striation formation. *J. Mater. Process. Technol.* **2006**, *172*, 22–29. [CrossRef]
15. Tuomas, P.A.S. A study on the effect of cutting position on performance a study on the effect of cutting position on performance. *Weld World* **2014**, *58*, 193–204.
16. Tamura, K.; Ishigami, R.; Yamagishi, R. Laser cutting of thick steel plates and simulated steel components using a 30 kW fiber laser. *J. Nucl. Sci. Technol.* **2016**, *53*, 916–920. [CrossRef]
17. Seon, S.; Shin, J.S.; Oh, S.Y.; Park, H.; Chung, C.-M.; Kim, T.-S.; Lee, L.; Lee, J. Improvement of cutting performance for thick stainless steel plates by step-like cutting speed increase in high-power fiber laser cutting. *Opt. Laser Technol.* **2018**, *103*, 311–317. [CrossRef]
18. Shin, J.S.; Oh, S.Y.; Park, H.; Chung, C.-M.; Seon, S.; Kim, T.-S.; Lee, L.; Lee, J. Cutting performance of thick steel plates up to 150 mm in thickness and large size pipes with a 10-kW fiber laser for dismantling of nuclear facilities. *Ann. Nucl. Energy* **2018**, *122*, 62–68. [CrossRef]
19. García-López, E.; Medrano-Tellez, A.G.; Ibarra-Medina, J.R.; Siller, H.R.; Rodriguez, C.A. Experimental study of back wall dross and surface roughness in fiber laser Microcutting of 316L miniature tubes. *Micromachines* **2017**, *9*, 4. [CrossRef] [PubMed]
20. Kautek, W.; Kruger, J. Femtosecond pulse laser ablation of metallic, semiconducting, ceramic and biological materials. *SPIE Proc.* **1994**, *2207*, 600–611.
21. Muthuramalingam, T.; Moiduddin, K.; Akash, R.; Krishnan, S.; Mian, S.H.; Ameen, W.; Alkhalefah, H. Influence of process parameters on dimensional accuracy of machined Titanium (Ti-6Al-4V) alloy in laser beam machining process. *Opt. Laser Technol.* **2020**, *132*, 106494. [CrossRef]
22. Wang, X.; Zheng, H.; Chu, P.; Tan, J.; Teh, K.; Liu, T.; Ang, B.C.; Tay, G. High quality femtosecond laser cutting of alumina substrates. *Opt. Lasers Eng.* **2010**, *48*, 657–663. [CrossRef]
23. Ahmmed, K.M.T.; Grambow, C.; Kietzig, A.-M. Fabrication of micro/nano structures on metals by femtosecond laser micromachining. *Micromachines* **2014**, *5*, 1219–1253. [CrossRef]
24. Lina, M.; Julius, S.; Virgilijus, V.; Romualdas, S.; Ona, B. Femtosecond laser micromachining of soda–lime glass in ambient air and under various aqueous solutions. *Micromachines* **2019**, *10*, 354.
25. Criales, L.E.; Orozco, P.F.; Medrano, A.; Rodriguez, C.; Özel, T. Effect of fluence and pulse overlapping on fabrication of microchannels in PMMA/PDMS via UV laser micromachining: Modeling and experimentation. *Mater. Manuf. Process.* **2015**, *30*, 890–901. [CrossRef]
26. Shanghai Baoshao Special Steel Co., Ltd., China. Available online: http://www.chinabshao.com (accessed on 10 November 2021).

Article

Femtosecond Laser Treatment for Improving the Corrosion Resistance of Selective Laser Melted 17-4PH Stainless Steel

Lingjian Meng [1], Jiazhao Long [1], Huan Yang [1,*], Wenjing Shen [1], Chunbo Li [1], Can Yang [1], Meng Wang [1] and Jiaming Li [2]

1. Sino-German College of Intelligent Manufacturing, Shenzhen Technology University, Shenzhen 518118, China; 2110412020@stumail.sztu.edu.cn (L.M.); 2110412018@stumail.sztu.edu.cn (J.L.); shenwenjing@sztu.edu.cn (W.S.); lichunbo@sztu.edu.cn (C.L.); yangcan@sztu.edu.cn (C.Y.); wangmeng@sztu.edu.cn (M.W.)
2. Guangdong Provincial Key Laboratory of Nanophotonic Functional Materials and Devices, School of Information and Optoelectronic Science and Engineering, South China Normal University, Guangzhou 510631, China; jmli@m.scnu.edu.cn
* Correspondence: yanghuan@sztu.edu.cn; Tel.: +86-177-2262-0530

Abstract: Currently, laser surface treatment (LST) is considered the most promising method available within the industry. It delivers precise control over surface topography, morphology, wettability, and chemistry, making the technique suitable for regulating the corrosion behavior of alloys. In this paper, femtosecond laser texturing with different parameters and atmosphere environments was adopted to clarify the effect of surface treatment on the corrosion resistance of selective laser melted (SLM-ed) 17-4PH stainless steel (SS) in a NaCl solution. The experimental results show that, after the heat treatment, the corrosion resistance of the laser-treated samples was enhanced. With the further laser treatment in an argon atmosphere, the oxidation of nanostructural surfaces was avoided. The Cr, Cu, and other alloying elements precipitated on the laser-ablated surface were beneficial to the formation of a passivation film, leading to an improved corrosion resistance performance.

Keywords: additively manufacture; 17-4PH stainless steel; femtosecond laser; corrosion resistance

1. Introduction

Precipitation hardening martensitic 17-4PH stainless steel (SS), due to its excellent mechanical properties and corrosion resistance, is widely used in various fields, such as aerospace, offshore platforms, and nuclear power plants [1]. Metal corrosion can destroy the strength of metal components, leading to the loss of reliability and safety, substantial economic losses, and even catastrophic accidents [2]. To achieve high reliability in various complex working environments, the processed parts are required to have good corrosion resistance [3].

Additive manufacturing (AM) is a layer-by-layer advanced manufacturing process that has emerged as a powerful means for producing metal parts in recent years [4]. This technology can use 3D computer-aided design (CAD) files to fabricate metal parts with complex geometries while saving time and avoiding waste [5,6]. As a typical AM technique, selective laser melting (SLM) can directly prepare high-density metal parts from micro-sized powders without post-processing [7]. AM has been proved to be a very effective and flexible technology for manufacturing high-performance metals, additionally making this process beneficial to corrosion resistance. Adrien Barroux et al. studied the corrosion resistance of 17-4PH SS produced by laser beam melting (LBM) [2] and found that, compared with the forged 17-4PH SS, the LBM-ed specimen had fewer metastable craters, a higher nucleation rate, and a longer service life [8]. Existing research is based on manufacturing processes to improve the corrosion resistance of metals. It is of great significance to develop an effective post-processing process for improving the corrosion resistance of 17-4PH SS.

With characteristics making it highly productive, contactless, and fully automatic, laser polishing is vigorously developed in order to improve surface roughness and to regulate the properties of SLM-ed metals [9]. Lan Chen et al. found that with the combination of surface roughness reduction and grain refinement, the surface properties and electrochemical corrosion behavior of 316L laser-clad SS could be improved effectively by laser polishing [10]. In addition, the laser can also prepare periodic micro-nanostructures on metals to control the surface wettability [11,12]. Superhydrophobic (SH) surfaces with low adhesiveness could be prepared by chemically modifying the laser textured surface [13,14]. Related reports showed that SH surfaces could obtain enhanced passivity, lower anodic dissolution, and corrosion current reduction [15].

Compared with conventional continuous and long-pulse lasers, the ultra-short pulse characteristics of femtosecond lasers can significantly reduce thermal effects during processing and avoid recast layers [16]. In this paper, SLM-ed 17-4PH SS was treated by a femtosecond laser with different parameters in order to regulate the surface structures and corrosion resistance behaviors. The surface characteristics of samples under different processing parameters and their effects on corrosion resistance were studied. The microstructure and phase composition of the laser-treated 17-4PH SS were investigated by optical microscopy, scanning electron microscopy (SEM), energy-dispersive X-ray spectroscopy (EDS), and X-ray diffraction (XRD). Furthermore, the corrosion behaviors were evaluated in 0.5 mol/L NaCl solution at a static temperature of 25 °C, and the passivation film components were assessed using X-ray photoelectron spectroscopy (XPS).

2. Materials and Methods

2.1. Sample Preparation

The experimental materials—gas-atomized 17-4PH SS powders—were provided by Nantong Jinyuan Intelligence Manufacturing Technology Co. Ltd (Nantong, China). The particle size range of the powders was 15–53 μm, with an average diameter of 32.85 μm, and the standard deviation was 13 μm (Figure 1a). The scanning electron microscopy (SEM) image shows that the powder (Figure 1b) is almost spherical, making it conducive to SLM forming. The chemical composition of the 17-4PH SS powder is shown in Table 1. The SLM-ed cubic 17-4PH steel with dimensions of $15 \times 15 \times 15$ mm^3 was fabricated in a nitrogen atmosphere. SLM equipment (SLM-100, Han's Laser Co. Ltd, Shenzhen, China) containing a 200 W fiber laser was used for this experiment. The laser power used for the SLM process was 180 W with a scanning speed of 800 mm/s, and the layer thickness was about 0.03 mm. During the SLM process, the scanning directions for the adjacent forming layers differed by 67°. Part of the SLM-ed samples was selected for a 0.5 h solution treatment at 1040 °C in a muffle furnace and then air-cooled to 550 °C. After that, the samples were aged for 4 h and then air-cooled to room temperature. The oxide layers on the sample surfaces were removed with 80-mesh sandpaper and then cleaned in an ethanol ultrasonic bath for 15 min. Through the use of the Archimedes method, the SLM-ed parts had a density of approximately 99.8%.

Table 1. Chemical composition of 17-4PH SS powder (wt %).

Fe	Cr	Ni	Cu	Mn	Nb	Si	O	C	P	S
balance	17.01	4.69	4.03	0.59	0.34	0.26	0.05	0.04	0.012	0.007

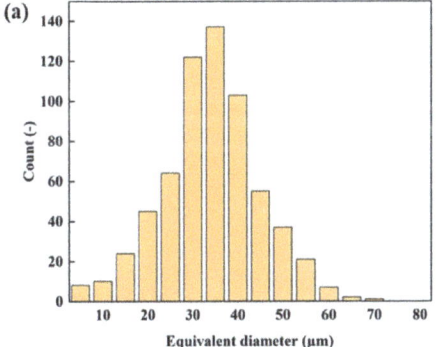 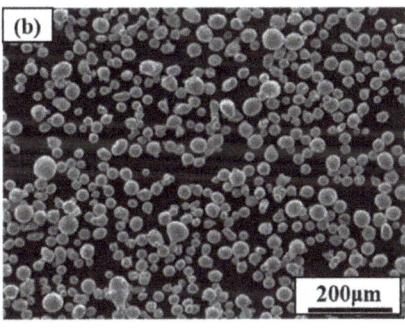

Figure 1. (a) Histogram of the powder's equivalent diameter and (b) representative powder image.

2.2. Laser Processing

After the heat treatment, the samples were textured by a femtosecond laser. The schematic diagrams for the laser treatment are shown in Figure 2. A 520 nm femtosecond laser (Spectra-Physics Spirit HE 1040-30-SHG, Boston, MA, USA) with a pulse width of 300 fs at a 250 kHz pulse repetition frequency was focused onto the sample surface by an F-theta lens with a focal spot size of 16 μm in diameter. Before laser treatment, the SLM-ed metal surfaces were pretreated with 80-mesh sandpaper. The optimized processing parameters used for laser treatment are shown in Table 2. The laser processing experiments were conducted both in air and argon atmospheres. To obtain SH surfaces, the laser textured samples were placed on a heated plate to be annealed at 110 °C for 2.5 h. Anhydrous ethanol was dripped onto the surface every 30 min during the annealing process.

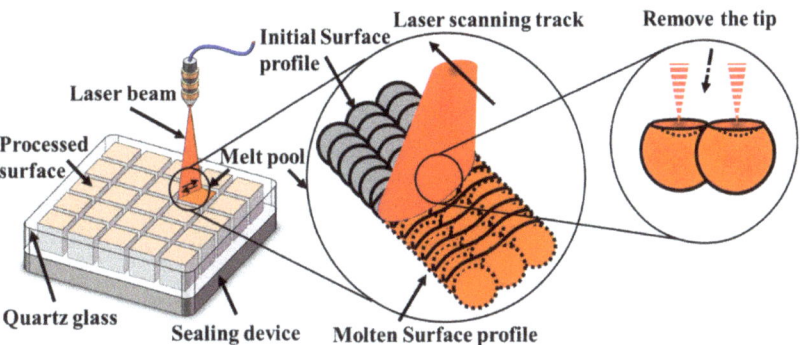

Figure 2. Schematic illustrating the laser surface treatment of the 17-4PH SS.

Table 2. Laser processing parameters used in this study.

	Air	Argon	SH
Scanning speed (mm/s)	200	200	100
Processing times (-)	1	1	5
Fluence (J/cm^2)	1.27	1.27	6.37
Scanning spacing (μm)	2	2	50

2.3. Surface Characterization and Electrochemical Analysis

The water contact angles (CAs) were measured with 3 μL distilled water using a video optic contact angle instrument (DATAPHYSICS, OCA 25, Stuttgart, Germany). The surface morphology of the laser textured SS was characterized using a 3D measuring laser microscope (Olympus, OSL 4100) and a field-emission scanning electron microscope (FESEM) (Hitachi, Su 8010, Tokyo, Japan). The sample's surface was etched with a solution consisting of nitric acid (65%), hydrochloric acid (35%), and distilled water in a ratio of 1:10:10 by volume. The phase composition was analyzed by an XRD. The chemical composition of the passivation film for the 17-4PH SS sample obtained after a 72 h immersion in the 0.5 mol/L NaCl solution was investigated by an XPS (ESCA-Lab 250 XI, Thermo-VG Scientific, Waltham, MA, USA) using Al Kα radiation (1486.6 eV).

The action potential polarization and electrochemical impedance spectroscopy (EIS) were measured in 0.5 mol/L NaCl solution using an electrochemical workstation (AUTOLAB, Herisau, Switzerland). The corrosion behaviors of the 17-4PH SS samples were investigated in 0.5 mol/L NaCl solution at a static temperature of 25 °C. In the three-electrode system, the sample's exposure area was 0.78 cm^2. In the potential polarization test, the potential scan range was from -1.5 V to $+1.5$ V (VS EOC), and the scan speed was 0.5 mV/s. The EIS test was carried out to characterize the samples soaked in 0.5 mol/L NaCl solution for 72 h. After the open circuit potential (OCP) was stable for 1800 s, the impedance test was conducted, and the impedance data were collected. The frequency range of the EIS test was $10^{-2} \sim 10^5$ Hz, and the ac signal amplitude was 10 mV. Three electrochemical tests were performed on each group of samples to ensure the repeatability of the test results.

3. Results and Discussion

3.1. Surface Morphology and Wetting Behavior

Figure 3 shows the surface morphologies and wetting behaviors of the SLM-ed SS samples treated with different parameters (Table 2). The polished original sample showed a smooth surface with only a few defects (Figure 3a), and the initial CA was 77°. Figure 3b shows the surface morphologies of the sample ablated by the femtosecond laser in air (FLAR). To describe the different samples concisely, a series of abbreviations were used (Table 3). The rough surface was covered by a large number of nanoparticles with a size of about 100 nm, and the CA was decreased to 9°. When the femtosecond laser processing was conducted in an argon (FLAN) atmosphere (Figure 3c), the surface with a CA of 10° showed an obvious periodic structure. The structure with a period of about 300 nm was perpendicular to the laser polarization direction. This periodic structure may be attributed to the interference between the plasma and the incident laser [17,18]. Moreover, argon can avoid the oxidation of materials during femtosecond laser texturing. After the laser processing in air and argon atmospheres, the surface roughness (Sa) was reduced from 1.709 μm to 1.299 μm and 1.108 μm (Figure 3e).

Table 3. Sample labels.

Full Name	Femtosecond Laser Treatment in Air	Femtosecond Laser Treatment in Argon	Superhydrophobic	Ferrite Sample	Martensite Sample
Abbreviation	FLAR	FLAN	SH	F-sample	M-sample

Preparing a periodic micro-nanostructure is usually necessary to obtain an SH surface. As shown in Figure 3d, with the increased laser fluence and scanning spacing, micro-scale grooves with a width of 10 μm were observed on the laser-ablated surface (Figure 3d). The textured surface manifested superhydrophilicity with a CA of nearly 0°. After the annealing process, the SH surface showed a high CA of 154° and a sliding angle of 3° (Figure 3d). The advancing and receding CAs were 151.03° and 146.18°, respectively, resulting in a CA hysteresis of 4.85°.

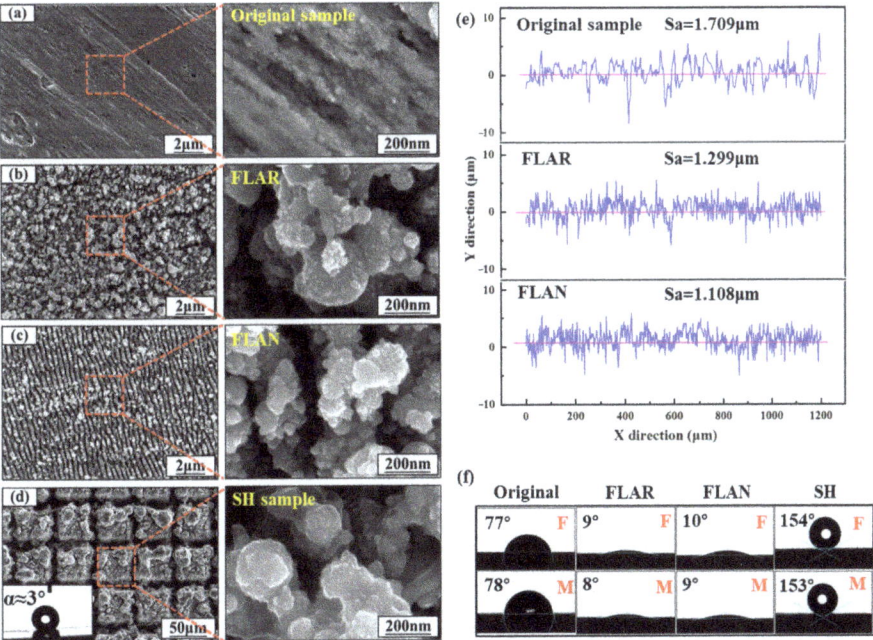

Figure 3. Surface structures of (**a**) original, (**b**) FLAR, (**c**) FLAN, and (**d**) SH samples. (**e**) Cross-section dimensions of original, FLAR, and FLAN samples. (**f**) Wetting behaviors of original, FLAR, FLAN, and SH samples.

3.2. Metallographic Analysis

Figure 4 shows the metallographic images and the XRD patterns of the SLM-ed 17-4PH SS samples before and after heat treatment. Before heat treatment, the metal sample was dominated by columnar body-centered cubic (BCC) ferrite (F) grains, composed of elongated subgrains with different growth directions. The sample without heat treatment was labeled F-sample. After heat treatment, the peak value of α' (110) was significantly increased. This was because the solution treatment refined and homogenized the microstructure. The large-grain ferrite was transformed into a fine acicular martensite (M) lath. However, more intergranular defects could be observed, and the subgrain was not apparent. The heat-treated sample was labeled M-sample. However, the high-heat treatment-induced transformation from ferrite to acicular martensite had little effect on the wetting behaviors of the SLM-ed SS samples treated with different parameters (Figure 3f).

3.3. Electrochemical Analysis
3.3.1. Potentiodynamic Polarization Studies

Figure 5 shows the polarization characteristics of each sample with different processes. The corrosion potential (E_{corr}) and current density (i_{corr}) were calculated using the Tafel extrapolation method (Table 4). Typically, a high corrosion potential indicates excellent corrosion resistance. The heat treatment caused the charge in the corrosion potential of the original samples to not be apparent. After laser treatment in argon, the F-sample showed a high corrosion potential. The laser treated M-sample showed the highest corrosion voltage of -0.3820 V. The heat treatment caused an improvement in the corrosion voltage, which can also be observed in the samples treated in air. However, the oxidized nanoparticles on the surface fell away easily, and the exposed matrix accelerated the corrosion. Therefore, the corrosion voltage was relatively low and unstable. The SH M-sample showed the lowest self-corrosion potential. The reason for this is that the polarization current destroyed the low surface energy property caused by the annealing process; therefore, the material's wetting behavior changed from SH to superhydrophilic. In general, the increased hydrophilicity can lead to an attenuation in the material's corrosion resistance.

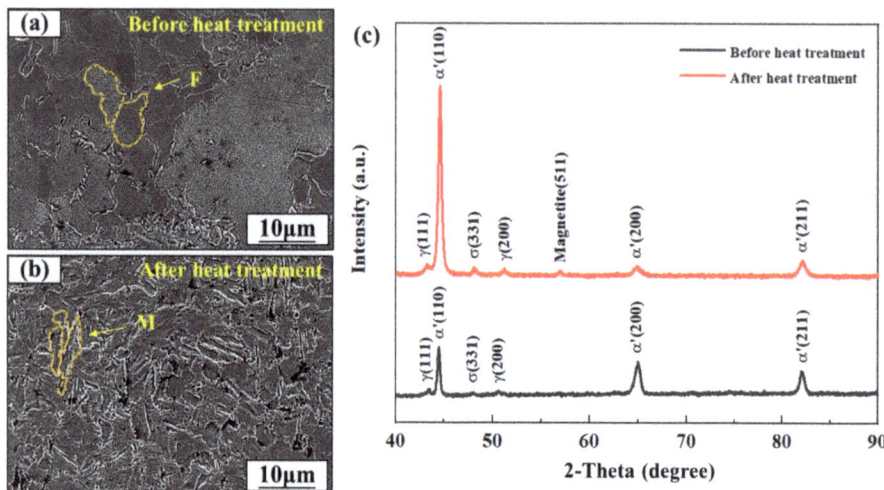

Figure 4. Microstructures of the SLM-ed 17-4PH SS (**a**) before and (**b**) after heat treatment. (**c**) XRD patterns of the SLM-ed 17-4PH SS before and after heat treatment.

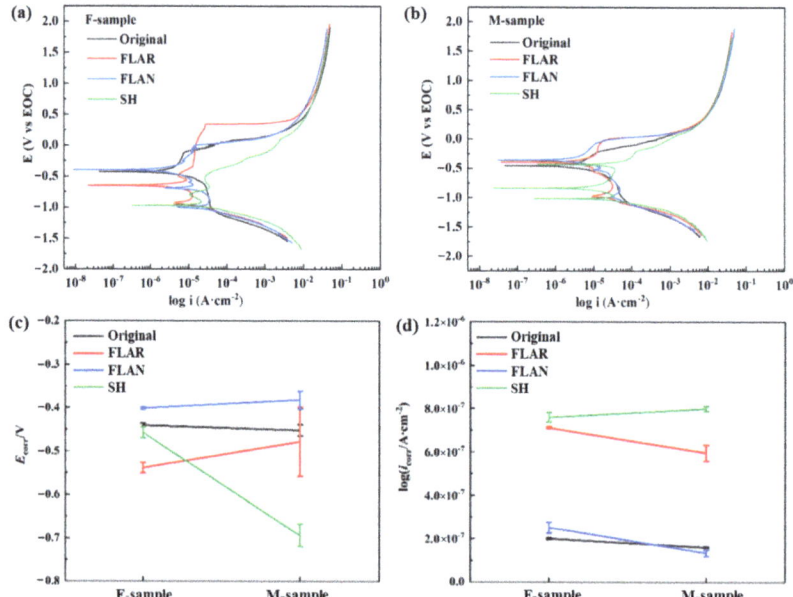

Figure 5. Potentiodynamic polarization curves of the (**a**) F-samples and (**b**) M-samples. (**c**) Corrosion voltages and (**d**) corrosion currents of SLM-ed samples with different post-treatment processes.

Table 4. Quantitative information about the potentiodynamic polarization curves of 8 samples in 0.5 mol/L NaCl solution.

Sample	Average Log i_{corr} ($\mu A \cdot cm^{-2}$)	Average E_{corr} (V)
Original F-samples	0.1999 ± 2.0%	−0.4408 ± 1.0%
Original M-samples	0.1606 ± 1.6%	−0.4521 ± 2.7%
FLAR F-samples	0.7107 ± 0.6%	−0.5389 ± 2.2%
FLAR M-samples	0.5963 ± 6.1%	−0.4786 ± 16.5%
FLAN F-samples	0.2518 ± 9.7%	−0.4017 ± 0.6%
FLAN M-samples	0.1346 ± 10.7%	−0.3820 ± 5.5%
SH F-samples	0.7583 ± 2.9%	−0.4575 ± 2.8%
SH M-samples	0.7988 ± 2.0%	−0.6940 ± 3.8%

The corrosion current density (i_{corr}) belongs to the dynamic category. The smaller the i_{corr}, the slower the corrosion rate. As shown in Figure 5d, the corrosion current densities of the original and FLAN samples were significantly lower than those of the other two groups. The FLAN M-sample showed the lowest corrosion current density of 0.1346 $\mu A \cdot cm^2$. In addition, the refined grain induced by the transformation from ferrite to acicular martensite is beneficial to the corrosion resistance [19,20].

3.3.2. Electrochemical Impedance Spectroscopic (EIS) Studies

Figure 6 shows the measured and simulated impedance characteristics of the eight samples tested in Figure 5. Capacitive arcs appeared on the Nyquist plots for all the samples, revealing that the corrosion reactions occurred at the SS/electrolyte interfaces. Without the heat treatment, the original sample's arc radius was significantly larger than that of the other three samples. The FLAR sample showed the smallest arc radius. For the heat-treated samples, the laser texturing in argon resulted in the largest capacitive arc radius, and the SH M-sample showed the smallest arc radius. The capacitive arc radius is an essential parameter for evaluating the corrosion resistance of metal materials [21,22]. The larger the capacitive arc radius, the greater the impedance value of the corrosive ions passing through the material surface, and the better the material's corrosion resistance. This indicates that laser texturing in argon significantly improved the corrosion properties of the M-sample. The high impedances and phase angles indicate that the formed passivation films were more stable for the original and FLAN samples.

To obtain the detailed characteristics of the passivation films, the two equivalent circuit models shown in Figure 7 were chosen to fit the impedance data. R_s represents solution resistance. The equivalent circuit (EC) shown in Figure 7a was named EC-1, and that shown in Figure 7b was named EC-2. The high agreement between the simulated curves and the experiment results fully verifies the circuit's validity (Figure 6). R_f and Q_f represent passivation film resistance and capacitance, respectively. R_{ct} and Q_{dl} represent charge transfer resistance and double-layer capacitance. The chi-square values (χ^2) were all less than 0.01. The correspondence between the tested samples and ECs, and the fitting results after 72 h are shown in Table 5.

The results indicate that, after 72 h of immersion, the passivation films of the original, FLAN, and SH F-samples remained intact. A similar phenomenon was also observed on the FLAN and SH M-samples. Among them, the FLAN M-sample possessed the largest R_f of 0.598 $M\Omega \cdot cm^2$, indicating its high resistance and corrosion-resistant passivation film. However, the high Q_f reveals that there may be many defects in the FLAN M-sample.

The EIS test showed that, with a small R_f and a large Q_f, the fabricated SH surfaces did not show a good anti-corrosion performance. In addition, compared with other samples, the micro-sized groove structure enabled the SH surface to process a higher Q_f value. The results also show that the passivation films of the FLAR F-sample and M-sample were damaged during the test. This is attributed to the oxide particles formed on the laser textured surface. In NaCl solution, these particles tend to fall off, resulting in the destruction of the passivation film.

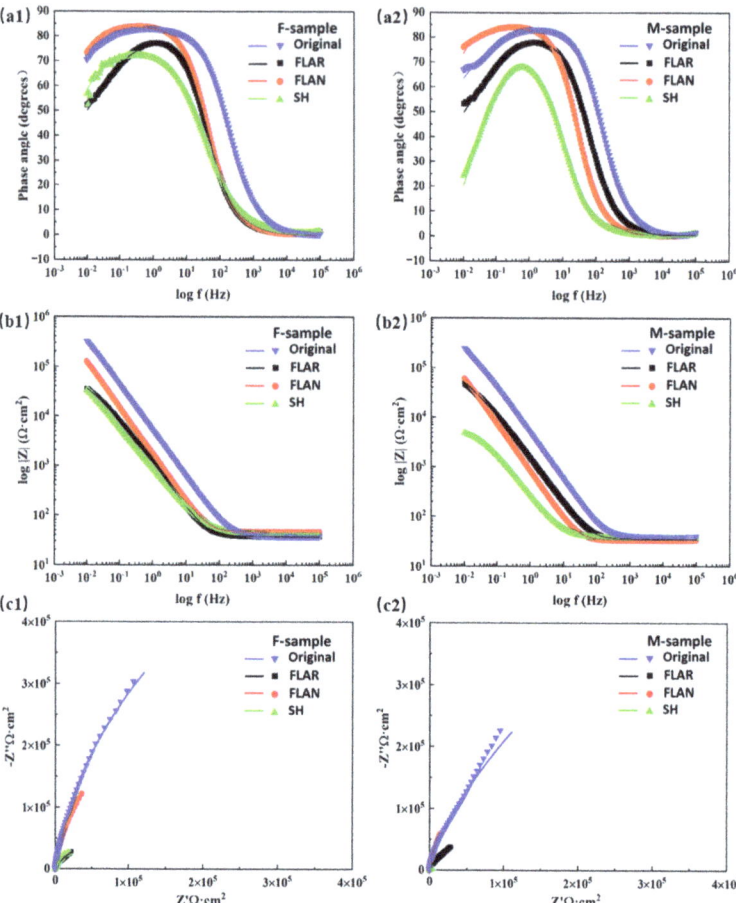

Figure 6. Measured and simulated (**a**) Bode phase angle, (**b**) Bode impedance, and (**c**) Nyquist curves of SLM-ed samples with different post-treatment processes.

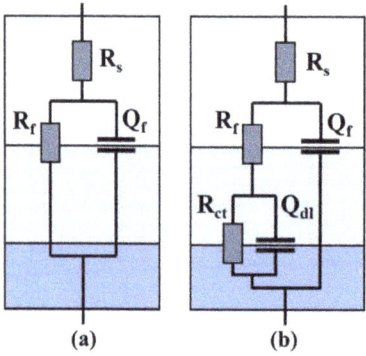

Figure 7. Electrochemical equivalent circuits for fitting the measured impedance data of (**a**) the FLAR samples and (**b**) other samples.

Table 5. Fitting results of EIS for 8 samples in 0.5 mol/L NaCl for 72 h.

Sample	R_f (MΩ·cm^2)	Q_f (µF·cm^{-2})	R_{ct} (MΩ·cm^2)	Q_{dl} (µF·cm^{-2})
Original F-samples (EC-1)	0.142 ± 16.9%	35.269 ± 5.3%	0	0
Original M-samples (EC-1)	0.173 ± 23.7%	59.290 ± 10.9%	0	0
FLAR F-samples (EC-2)	0.0000156 ± 26.3%	45.972 ± 14.9%	3.228 ± 6.7%	155.931 ± 12.4%
FLAR M-samples (EC-2)	0.0000166 ± 16.9%	33.936 ± 19.9%	8.049 ± 4.2%	115.330 ± 10.6%
FLAN F-samples (EC-1)	0.313 ± 9.3%	99.970 ± 21.8%	0	0
FLAN M-samples (EC-1)	0.598 ± 5.2%	206.392 ± 11.8%	0	0
SH F-samples (EC-1)	0.126 ± 28.6%	243.610 ± 5.1%	0	0
SH M-samples (EC-1)	0.00562 ± 6.6%	780.84 ± 5.5%	0	0

3.3.3. XPS Characterization

Figure 8 shows the XPS spectra of the passivation films on different samples. The peak spectra, such as Fe2p, Cr2p, Ni2p, Cu2p, Nb3d, and C1s, were fitted to investigate the surface components. The Fe2p spectra of the eight samples show seven peaks around 706.5 eV, 706.9 eV, 708.1 eV, 709.1 eV, 711.5 eV, 718.9 eV, and 723.7 eV, which are related to Fe, FeO/Fe$_2$O$_3$, and Fe$_3$O$_4$. The Cr2p spectrum has six peaks around 573.9 eV, 575.4 eV, 576.1 eV, 577.4 eV, 583.5 eV, and 586.4 eV, which are related to Cr, Cr$_2$O$_3$, and CrCl$_3$/Cr(OH)$_3$. The Ni2p spectrum contains three peaks around 851.8 eV, 855.0 eV, and 869.9 eV, corresponding to Ni, NiO, and Ni. Two peaks around 933.2 eV and 952.7 eV can be observed in the Cu2p spectra, corresponding to Cu/CuO and Cu$_2$O/Cu/CuO. There are three peaks in the Nb2p spectrum, namely, 202.4 eV, 207.2 eV, and 209.5 eV, which are associated with NbO, NbO/Nb$_2$O$_5$, and Nb$_2$O$_5$. The C1s spectrum has four peaks around 284.8 eV, 285.6 eV, 286.5 eV, and 588.6 eV, which are related to C.

As we all know, FeO, Fe$_2$O$_3$, Cr$_2$O$_3$, and other metal oxides are the main components of SS passivation films [23,24]. It can be seen that the PLAN samples with relatively high polarization and impedance performances showed high peak values of Fe, Cr, Cu, and Nb oxides. This is because the femtosecond laser ablation can induce the precipitation of Cr, Cu, and Nb elements to form an oxide film on the surface, resulting in increased corrosion resistance. This is consistent with the EIS results. Moreover, since the grain is refined after the heat treatment, these metal oxides tend to grow at the grain boundaries with high Gibbs free energy. The increased boundaries in the unit area are conducive to forming a dense and stable passivation film [25]. The SH surface showed the lowest peak values for all the metal oxides, which is consistent with the above impedance results. Meanwhile, this also further confirms the relatively short corrosion reaction time of the SH surface. The low peak values for the C-O bond may have contributed to the lost superhydrophobicity of the SH samples [26].

3.3.4. Electrochemical Corrosion Morphology

Figure 9 shows the surface morphologies after the impedance tests. The FLAN sample showed the fewest defects. The reason may be that the material surface possessed a dense passivation film, so the destruction speed of Cl$^-$ to the passivation film was lower than the passivation film's repair speed. However, the corrosion pits on the surface of the FLAR sample were very obvious. This can be attributed to the severely destroyed passivation film.

Interestingly, the SH surface that had relatively low polarization and impedance performances showed a good corrosion morphology. On the SH M-sample surface, the corrosion traces were very inconspicuous. Some micro-particles were distributed on the laser-fabricated micro-grooves on the SH F-sample. The EDS test showed that carbonaceous particles were derived from the surrounding air. The surface morphologies reveal that the SH surfaces had high corrosion resistance.

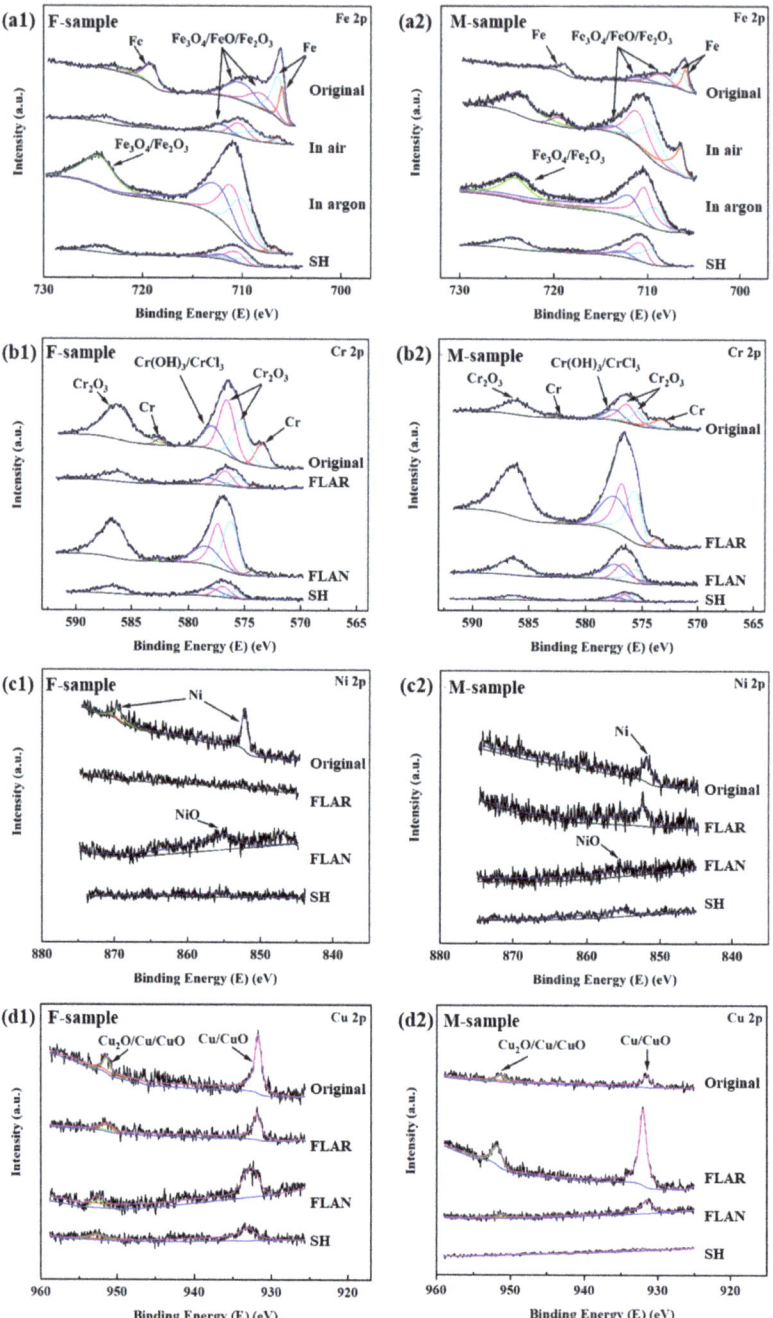

Figure 8. Cont.

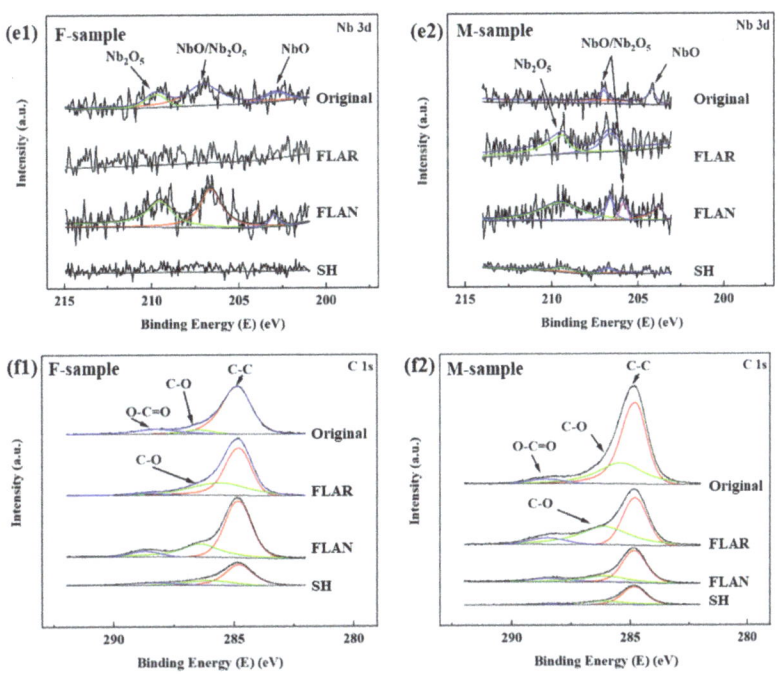

Figure 8. High-resolution XPS spectra of (**a**) Fe 2p, (**b**) Cr 2p, (**c**) Ni 2p, (**d**) Cu 2p, (**e**) Nb 3d, and (**f**) C 1s for SLM-ed samples after being immersed in 0.5 mol/L NaCl solution for 72 h.

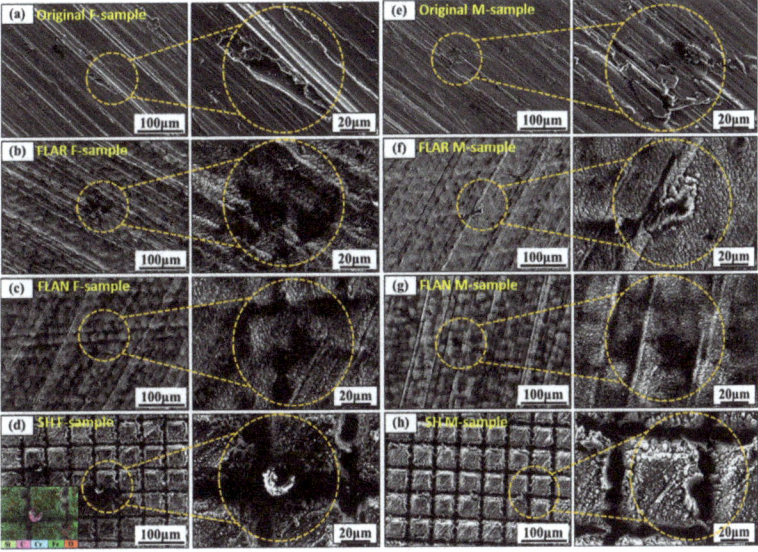

Figure 9. Surface morphologies of (**a**) original F-samples, (**b**) FLAR F-samples, (**c**) FLAN F-samples, (**d**) SH F-samples, (**e**) original M-samples, (**f**) FLAR M-samples, (**g**) FLAN M-samples, and (**h**) SH M-samples after being immersed in 0.5 mol/L NaCl solution for 72 h.

3.3.5. Influence Mechanism of Laser Polishing Treatment on Corrosion Resistance

With the little difference in the corrosion morphologies of the original, FLAR, and FLAN samples, the corresponding corrosion mechanism is explained in Figure 10a. For the SH surface with the special structure, the corrosion mechanism is shown in Figure 10b. At room temperature, the chemical reactions of the 17-4PH SS in 0.5 mol/L NaCl solution are as follows:

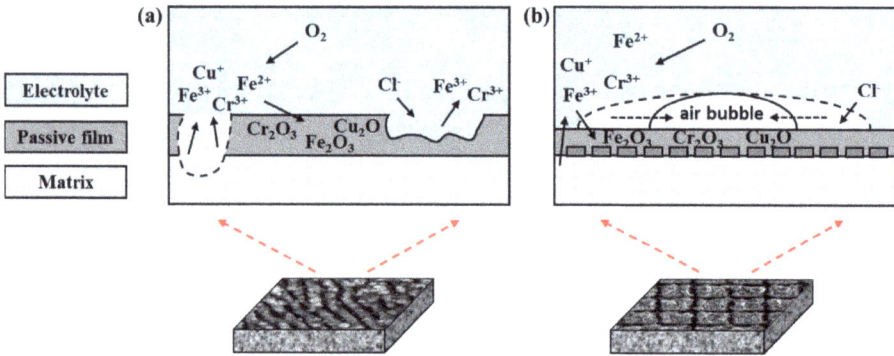

Figure 10. Schematic illustrating the corrosion mechanism of (**a**) hydrophilic and (**b**) SH 17-4PH SS.

Firstly, since the oxidation reaction occurs at the cathode, the metallic elements represented by Fe in the SLM-ed SS samples gradually dissolve into the solution after losing electrons at the anode, and the oxidation reaction occurs near the anode:

$$Cr \rightarrow Cr^{3+} + 3e^- \tag{1}$$

$$Fe \rightarrow Fe^{2+} + 2e^- \tag{2}$$

$$Fe \rightarrow Fe^{3+} + 3e^- \tag{3}$$

$$Cu \rightarrow Cu^+ + e^- \tag{4}$$

$$Nb \rightarrow Nb^{2+} + 2e^- \tag{5}$$

$$Nb \rightarrow Nb^{5+} + 5e^- \tag{6}$$

The reduction reaction occurs at the cathode:

$$O_2 + 2H_2O + 4e^- \rightarrow 4OH^- \tag{7}$$

Around the anode, the above metal ions react with the Cl^-/OH^- in the NaCl solution to form metal chloride or hydroxide, which causes the metal to continue to dissolve in the solution:

$$Cr^{3+} + 3Cl^-/OH^- \rightarrow CrCl_3/Cr(OH)_3 \tag{8}$$

$$Fe^{3+} + 3Cl^-/OH^- \rightarrow FeCl_3/Fe(OH)_3 \tag{9}$$

$$Fe^{2+} + 2Cl^-/OH^- \rightarrow FeCl_2/Fe(OH)_2 \tag{10}$$

$$Cu + Cl^- \rightarrow CuCl \tag{11}$$

$$Nb^{2+} + 2Cl^-/OH^- \rightarrow NbCl_2/Nb(OH)_2 \tag{12}$$

$$Nb^{5+} + 5Cl^-/OH^- \rightarrow NbCl_5/Nb(OH)_5 \tag{13}$$

Next, these metal compounds and residual oxygen in the water together form stable Cr_2O_3 and Cu_2O oxides on the matrix surface, as shown in the XPS results (Figure 9). The formation of these oxides causes the passivation film to be gradually repaired as follows:

$$4Cr(OH)_3 + 4Cr + 3O_2 \rightarrow 4Cr_2O_3 + 6H_2O \qquad (14)$$

$$4Fe(OH)_3 + 4Fe + 3O_2 \rightarrow 4Fe_2O_3 + 6H_2O \qquad (15)$$

$$Fe(OH)_2 + 2Fe + O_2 \rightarrow 3FeO + H_2O \qquad (16)$$

$$4Cu^+ + 4e^- + O_2 \rightarrow 2Cu_2O \qquad (17)$$

$$Nb(OH)_2 + 2Nb + O_2 \rightarrow 3NbO + H_2O \qquad (18)$$

$$2Nb(OH)_5 + 4Nb + 5O_2 \rightarrow 3Nb_2O_5 + 5H_2O \qquad (19)$$

Due to the potential difference between various elements, many micro-cells appear on the material surfaces at the initial stage. These micro-cells can facilitate the migration of Cr^{3+}, Fe^{2+}, Fe^{3+}, Cu^+, Nb^{2+}, Nb^{5+}, and Cl^- in the solution, decreasing the stability of the passivation films. The dissolved metal cations are combined with the chloride ions and then rapidly oxidize to form metal oxides such as Cr_2O_3, FeO, Fe_2O_3, Cu_2O, NbO, and Nb_2O_5 (Figure 8). The metal oxides tend to grow at the grain boundaries with a high density, which is conducive to forming passivation films. The passivation film grown along the dense grain boundaries can block the corrosion of metals, resulting in the inhibition of pitting corrosion [25]. However, if there are defects in the passivation, the Cl^- can penetrate the metal matrix, resulting in corrosion pits [27].

The Cr, Cu, and other alloying elements precipitated on the FLAN sample surface can lead to an increased thickness of the passivation film. Moreover, the surface nanostructures may hinder the diffusion of oxides, which is beneficial to the stable formation of passivation films. Therefore, the passivation films on the FLAN samples showed relatively high corrosion resistance in the impedance test (Table 5). On the FLAR samples, the loose oxide particles fell off easily during the corrosion process, resulting in substrate exposure. The passivation film was destroyed too fast to be repaired, so the pitting corrosion on the FLAR samples was very obvious (Figure 9f).

The relaively weak anti-corrosion performance of the SH samples is attributed to the unique surface structure and wetting property. Due to the surface superhydrophobicity, there was an air film between the NaCl solution and the material surface at the initial stage of the test. Therefore, the NaCl solution could not penetrate the micro-nanostructure. However, as the site in contact with the solution was corroded, the liquid gradually penetrated the rough structure (Figure 10b). After 24 h, the superhydrophobicity was wholly lost, and the material surface exhibited superhydrophilicity. As a result, the actual corrosion reaction time of the SH sample was less than that of the other samples, which may be the main reason for the relatively low resistance of the surface passivation film.

4. Conclusions

In this study, the corrosion behaviors of SLM-ed 17-4PH SS treated with different femtosecond laser parameters were investigated.

(1) The F-sample was dominated by columnar BCC ferrite grains composed of elongated subgrains with different growth directions. After the heat treatment, the large-grain ferrite was transformed into a fine acicular martensite lath, resulting in improved corrosion resistance.

(2) Femtosecond LST was used to induce periodic nanostructures on the material surfaces. The argon atmosphere effectively prevented the oxidation of the laser-ablated surface. Moreover, the Cr, Cu, and other alloying elements precipitated on the FLAN sample surface were beneficial to the formation of the passivation film, leading to excellent corrosion resistance performance.

(3) Since the wetting behavior was transformed from SH to superhydrophilic, the fabricated SH surfaces did not show a good anti-corrosion performance. However, the air film between the solution and the material surface delayed the surface corrosion, resulting in inconspicuous corrosion pits.

Author Contributions: Conceptualization, H.Y. and C.Y.; methodology, L.M.; validation, M.W., J.L. (Jiaming Li), and C.L.; formal analysis, L.M.; investigation, W.S.; writing—original draft preparation, L.M.; writing—review and editing, H.Y. and J.L. (Jiazhao Long); supervision, C.Y.; funding acquisition, H.Y. All authors have read and agreed to the published version of the manuscript.

Funding: This research was funded by the Natural Science Foundation of Top Talent of SZTU (2020103), the Project of Characteristic Innovation in Higher Education of Guangdong (2020KQNCX070), the National Natural Science Foundation of China (62005081), and the Guangdong Basic and Applied Basic Research Foundation (2021A1515011932).

Acknowledgments: The authors would like to thank the Laboratory of Advanced Additive Manufacturing, Sino-German College of Intelligent Manufacturing in SZTU, and the Analytical and Testing Centre of JNU for the XRD and XPS.

Conflicts of Interest: The authors declare no conflict of interest.

References

1. Spierings, A.B.; Schoepf, M.; Kiesel, R.; Wegener, K. Optimization of SLM productivity by aligning 17-4PH material properties on part requirements. *Rapid Prototyp. J.* **2014**, *20*, 444–448. [CrossRef]
2. Barroux, A.; Ducommun, N.; Nivet, E.; Laffont, L.; Blanc, C. Pitting corrosion of 17-4PH stainless steel manufactured by laser beam melting. *Corros. Sci.* **2020**, *169*, 108594. [CrossRef]
3. Gratton, A. Comparison of Mechanical, Metallurgical Properties of 17-4PH Stainless Steel between Direct Metal Laser Sintering (DMLS) and Traditional Manufacturing Methods. In Proceedings of the National Conference on Undergraduate Research (NCUR), Ogden, UT, USA, 29–31 March 2012.
4. Yang, H.; Xu, K.; Xu, C. Femtosecond Laser Fabricated Elastomeric Superhydrophobic Surface with Stretching-Enhanced Water Repellency. *Nanoscale Res. Lett.* **2019**, *14*, 333. [CrossRef]
5. Zhang, L.; Zhang, S.; Zhu, H.; Wang, G.; Zeng, X. Investigation on the angular accuracy of selective laser melting. *Int. J. Adv. Manuf. Technol.* **2019**, *104*, 3147–3153. [CrossRef]
6. Beaulardi, L.; Brentari, A.; Labanti, M.; Leoni, E.; Mingazzini, C.; Sangiorgi, S.; Villa, M. Microstructural and Thermo-Mechanical Characterization of Yttria Ceramic Cores for Investment Casting, with and without Particulate Reinforcement. *Adv. Sci. Technol.* **2010**, *65*, 33–38.
7. Casalino, G.; Campanelli, S.L.; Contuzzi, N.; Ludovico, A.D. Experimental investigation and statistical optimisation of the selective laser melting process of a maraging steel. *Opt. Laser Technol.* **2015**, *65*, 151–158. [CrossRef]
8. Kudzal, A.; McWilliams, B.; Hofmeister, C.; Kellogg, F.; Yu, J.; Taggart-Scarff, J.; Liang, J. Effect of scan pattern on the microstructure and mechanical properties of Powder Bed Fusion additive manufactured 17-4 stainless steel. *Mater. Des.* **2017**, *133*, 205–215. [CrossRef]
9. Bhaduri, D.; Ghara, T.; Penchev, P.; Paul, S.; Pruncu, C.I.; Dimov, S.; Morgan, D. Pulsed laser polishing of selective laser melted aluminium alloy parts. *Appl. Surf. Sci.* **2021**, *558*, 149887. [CrossRef]
10. Chen, L.; Richter, B.; Zhang, X.; Ren, X.; Pfefferkorn, F.E. Modification of surface characteristics and electrochemical corrosion behavior of laser powder bed fused stainless-steel 316L after laser polishing. *Addit. Manuf.* **2019**, *32*, 101013. [CrossRef]
11. Boinovich, L.B.; Emelyanenko, A.M.; Modestov, A.D.; Domantovsky, A.G.; Emelyanenko, K.A. Synergistic Effect of Superhydrophobicity and Oxidized Layers on Corrosion Resistance of Aluminum Alloy Surface Textured by Nanosecond Laser Treatment. *ACS Appl. Mater. Interfaces* **2015**, *7*, 19500–19508. [CrossRef]
12. Wang, Q.; Samanta, A.; Shaw, S.K.; Hu, H.; Ding, H. Nanosecond laser-based high-throughput surface nanostructuring (nHSN). *Appl. Surf. Sci.* **2019**, *507*, 145136. [CrossRef]
13. Ma, Q.; Tong, Z.; Wang, W.; Dong, G. Fabricating robust and repairable superhydrophobic surface on carbon steel by nanosecond laser texturing for corrosion protection. *Appl. Surf. Sci.* **2018**, *455*, 748–757. [CrossRef]
14. Mehran, R.; Jaffer, J.A.; Cui, C.; Duan, X.; Nasiri, A. Nanosecond Laser Fabrication of Hydrophobic Stainless Steel Surfaces: The Impact on Microstructure and Corrosion Resistance. *Materials* **2018**, *11*, 1577.
15. Trdan, U.; Hočevar, M.; Gregorčič, P. Transition from superhydrophilic to superhydrophobic state of laser textured stainless steel surface and its effect on corrosion resistance. *Corros. Sci.* **2017**, *123*, 21–26. [CrossRef]
16. Ghosh, A.; Wang, X.; Kietzig, A.-M.; Brochu, M. Layer-by-layer combination of laser powder bed fusion (LPBF) and femtosecond laser surface machining of fabricated stainless steel components. *J. Manuf. Process.* **2018**, *35*, 327–336. [CrossRef]
17. Jia, T.; Chen, H.; Huang, F.; Zhao, X.; Li, X.; Hu, S.; Sun, H.; Feng, D. Ultraviolet-infrared femtosecond laser-induced damage in fused silica and CaF$_2$ crystals. *Phys. Rev. B* **2006**, *73*, 054105. [CrossRef]

18. Levy, Y.; Derrien, T.J.-Y.; Bulgakova, N.M.; Gurevich, E.L.; Mocek, T. Relaxation dynamics of femtosecond-laser-induced temperature modulation on the surfaces of metals and semiconductors. *Appl. Surf. Sci.* **2016**, *374*, 157–164. [CrossRef]
19. Ralston, K.D.; Birbilis, N.; Davies, C.H.J. Revealing the relationship between grain size and corrosion rate of metals. *Scr. Mater.* **2010**, *63*, 1201–1204. [CrossRef]
20. Gollapudi, S. Grain size distribution effects on the corrosion behaviour of materials. *Corros. Sci.* **2012**, *62*, 90–94. [CrossRef]
21. Sulima, I.; Kowalik, R.; Hyjek, P. The corrosion and mechanical properties of spark plasma sintered composites reinforced with titanium diboride. *J. Alloys Compd.* **2016**, *688*, 1195–1205. [CrossRef]
22. Ribeiro, A.; Alves, A.; Rocha, L.; Silva, F.; Toptan, F. Synergism between corrosion and wear on CoCrMo−Al2O3 biocomposites in a physiological solution. *Tribol. Int.* **2015**, *91*, 198–205. [CrossRef]
23. Xi, T.; Shahzad, M.B.; Xu, D.; Sun, Z.; Zhao, J.; Yang, C.; Qi, M.; Yang, K. Effect of copper addition on mechanical properties, corrosion resistance and antibacterial property of 316L stainless steel. *Mater. Sci. Eng. C* **2017**, *71*, 1079–1085. [CrossRef]
24. Cheng, X.; Feng, Z.; Li, C.; Dong, C.; Li, X. Investigation of oxide film formation on 316L stainless steel in high-temperature aqueous environments. *Electrochim. Acta* **2011**, *56*, 5860–5865. [CrossRef]
25. Liu, Y.; Yang, J.; Yang, H.; Li, K.; Qiu, Y.; Zhang, W.; Zhou, S. Cu-bearing 316L stainless steel coatings produced by laser melting deposition: Microstructure and corrosion behavior in simulated body fluids. *Surf. Coat. Technol.* **2021**, *428*, 127868. [CrossRef]
26. Samanta, A.; Wang, Q.; Shaw, S.K.; Ding, H. Roles of chemistry modification for laser textured metal alloys to achieve extreme surface wetting behaviors. *Mater. Des.* **2020**, *192*, 108744. [CrossRef]
27. Prabhakaran, S.; Kulkarni, A.; Vasanth, G.; Kalainathan, S.; Shukla, P.; Vasudevan, V.K. Laser shock peening without coating induced residual stress distribution, wettability characteristics and enhanced pitting corrosion resistance of austenitic stainless steel. *Appl. Surf. Sci.* **2018**, *428*, 17–30. [CrossRef]

 *micromachines*

Review

A Review on Macroscopic and Microstructural Features of Metallic Coating Created by Pulsed Laser Material Deposition

Xinlin Wang *, Jinkun Jiang and Yongchang Tian

School of Mechanical Engineering, Dalian Jiaotong University, Dalian 116028, China; jiangjinkundybala@163.com (J.J.); s45542568@163.com (Y.T.)
* Correspondence: wxl_me@djtu.edu.cn

Abstract: Owing to the unparalleled advantages in repairing of high value-add component with big size, fabricating of functionally graded material, and cladding to enhance the surface properties of parts, the laser material deposition (LMD) is widely used. Compared to the continuous wave (CW) laser, the controllability of the laser energy would be improved and the temperature history would be different under the condition of pulse wave (PW) laser through changing the pulse parameters, such as duty cycle and pulse frequency. In this paper, the research status of temperature field simulation, surface quality, microstructural features, including microstructures, microhardness, residual stress, and cracking, as well as corrosion behavior of metallic coating created by pulsed laser material deposition have been reviewed. Furthermore, the existing knowledge and technology gaps are identified while the future research directions are also discussed.

Keywords: laser material deposition; pulse wave laser; temperature field simulation; surface quality; microstructural features; corrosion behavior

Citation: Wang, X.; Jiang, J.; Tian, Y. A Review on Macroscopic and Microstructural Features of Metallic Coating Created by Pulsed Laser Material Deposition. *Micromachines* 2022, 13, 659. https://doi.org/10.3390/mi13050659

Academic Editors: Jie Yin, Ping Zhao and Yang Liu

Received: 1 March 2022
Accepted: 19 April 2022
Published: 22 April 2022

Publisher's Note: MDPI stays neutral with regard to jurisdictional claims in published maps and institutional affiliations.

Copyright: © 2022 by the authors. Licensee MDPI, Basel, Switzerland. This article is an open access article distributed under the terms and conditions of the Creative Commons Attribution (CC BY) license (https://creativecommons.org/licenses/by/4.0/).

1. Introduction

The conventional subtractive manufacturing process presents insufficient capability in the fabrication of complex structure and high value-added parts, because of high material removal rate, time consumption, and high cost during removing material from a large stock or sheet. On the contrary, laser additive manufacturing (LAM) can cover the shortage and create the final metal part because of the incremental layer-by-layer manufacturing by adding material [1–3]. Owing to the superiorities of excellent stability, high power density, and easy controllability, LAM is widely used in direct deposition of metal materials [4–6]. LAM mainly includes powder bed fusion (PBF) mode [7,8], in which the powder is preset in powder bed, and direct laser deposition (DED) mode [9], in which the powder would be delivered by inert gas into the molten pool. Powder bed fusion LAM (Figure 1a), such as selective laser sintering/selective laser melting, is more suitable for the parts with small size and complex structure. Meanwhile, direct laser deposition LAM (Figure 1b), such as laser material deposition (LMD), presents unparalleled advantages in repairing of high value-add component with big size, fabricating of functionally graded material, and cladding to enhance the surface properties of parts [10]. In LMD process, a high-powered laser beam is acted as the heating source to melt the substrate and create a molten pool. Meanwhile, the metallic powder carried by a flowing inert gas (such as Argon) is delivered into the molten pool [11]. The molten pool would capture the delivered powder and solidify after the laser beam move out that contributes to the increased volume of molten pool. Along the scanning path guided by the computer, the laser beam and the powder nozzle would move above the substrate. After the completion of depositing the first layer on the substrate following the scanning path, the laser cladding head uparised certain height in the z-axis increment to the new position to continue deposition of the next layer. Based on the first layer, the new layer would be deposited along the scanning path and create the metallurgical bonding with first layer. Similar process, in which the latter layer is sequentially deposited on the former, would be repeated to build a three-dimensional (3D) part layer-by-layer [12–14]. Due to the advantages such as high geometry freedom, low thermal inputting, high production flexibility, LMD is a competitive additive manufacturing technology in the areas of aerospace, aviation, die and mold, etc. [15,16].

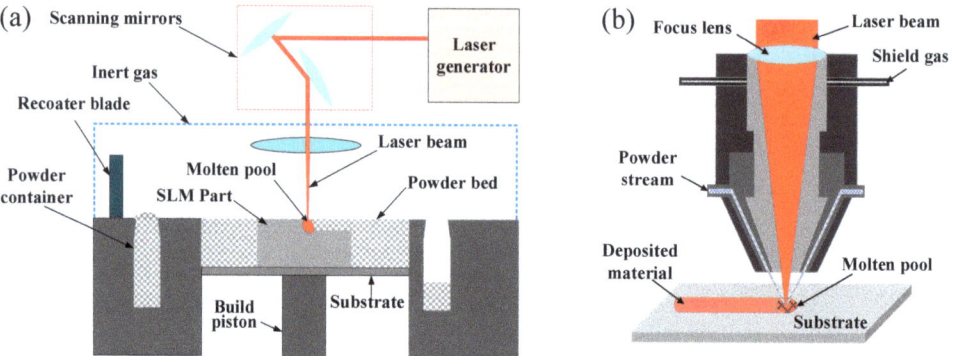

Figure 1. Schematic illustration of (**a**) powder bed fusion and (**b**) direct laser deposition LAM.

In LMD process, the continuous wave (CW) laser, which is a common laser mode used presently, inputs relatively low heat into the substrate that attributes to the obtaining of high temperature gradient and high cooling rate during laser beam scanning compared to the conventional heating source such as plasma. Owing to the process feature of rapidly melting and solidification during the moving of laser beam, the refined microstructure and superior mechanical properties are created [17]. The typical solidification structure, including the planar crystal in the interface, dendritic crystal in the central region and cellular crystal near the top surface of deposition, would be presented. The deposition layer possess high hardness, high strength but low ductility [18,19]. However, the rapid melting and solidification would result in the residual stress, especially the tensile residual stress, which is a negative factor for the mechanical properties [20–22]. Researchers has made many efforts to optimize the LMD process and eliminate the unfavorable features such as low ductility and high residual stress. Among that, the pulsed wave (PW) laser, which is acted as the heating source to melt the substrate and powder, is proved to be a suitable method to control the melting process in LMD.

The working of CW laser is fairly simple where a continuous beam of light is emitted at an average power. However, based on Q-switching, mode-locking or pulse pumping methods, a PW laser produces high energy pulses. A pulsed laser, where lasers emitting optical power in the form of pulses at constant time intervals, means a fixed amount of energy for a specified duration [23,24]. The difference between the continuous and pulsed laser can be noticed in the average power and peak power. The continuous laser is a better option during a higher average power is required. However, compared to the continuous laser, the pulsed laser can produce higher peak power which attributes to the higher melting temperature and the improvement of surface finish during the same average power is used. Another important parameter to be understand is laser energy density, which is represented by the laser energy per unit area and generally expressed as J/cm^2. The laser energy is related to the peak power and the duration time of laser. Therefore, in comparison with the continuous laser, the pulsed laser can produce higher energy density owing to the higher peak power. In addition, during the peak power is constant, the average power produced by pulsed laser is smaller compared to the CW laser mode. This is because the interval time between laser pulses enables a lower heat input into the substrate during the pulsed laser is used [24]. Furthermore, the interval time provides the blanking time for the cooling and solidification of the molten pool. So the parts with more refined microstructure and higher hardness can be produced. Through changing of the pulsed laser parameters such as pulse width, interval time and frequency etc., the controllability of the laser energy can be improved with the application of pulsed laser. The thermal history and temperature distribution can be regulated by controlling the laser energy emitted by the pulsed laser.

The capability of pulsed laser to modifying the thermal history and solidification process of molten pool has drawn researchers' much attention. Some investigations about the comparison of LMD process between the CW and PW laser as well as the effect of pulse laser parameters, such as pulse width, duty cycle, and pulse frequency, on the microstructure feature and mechanical properties have been conducted. The deposited material refers to Fe-based alloy, Co-based alloy, Ni-based alloy, titanium, ceramic and so on. Table 1 provides a summary of some common materials that have been investigated for the cladding and deposition by pulsed laser material deposition. With the changes of the deposited materials and pulsed laser parameters, the deposition would present different characteristics. This article provides an overview of the macroscopic and microstructural features of metallic coating created by pulsed laser material deposition.

Table 1. Summary of materials used in pulsed laser material deposition.

Substrate	Powder (Particle Size)		Duty Cycle	Pulse Frequency Hz	Peak Power W	Scanning Speed mm/s	Powder Feed Rate g/s	Refs.
Low carbon steel Mild steel A36 mild steel 316 L	Fe-based (54~150)	TiC-VC reinforced Fe-based powder	0.65~0.95	5, 50, 500, 4500, 5000	850, 1000	4		[25,26]
		high-nitrogen steel	0.5		1200–1800	5–11	0.167	[27]
		316 L	0.5	8.3, 12.5, 20, 25, 50, 100	450, 1000	6	0.14, 0.078	[28,29]
		Fe-20 wt.% Al	0.15~0.6	30, 60	1090, 1140, 2300	1	0.033	[30]
Ti-6Al-4V Low carbon steel	Ni-based (43~150)	Inconel 718	0.4, 0.7, 1	10, 20, 100, 1000	300, 600, 857, 1500	4.6	0.358, 0.586, 0.674	[31,32]
		K447A	0.3–0.8	100	216–720	4	0.09	[33]
ST14 plain carbon steel AISI420 Ferritic steel Inconel 713 Low carbon steel Copper alloy AISI 4135	Co-based (45~180)	Stellite 6	0.09–0.48	15, 40, 60	1110, 1330, 1530, 1660, 2220	3~9		[34–36]
		Stellite 31	0.16–0.32	16, 20		2, 4, 6		[37]
		WC–12 wt.% Co	0.09–0.63	20~90		1~10	0.05~0.13	[38]
		Co-based alloy	0.03~0.13	40, 60, 92	300, 330, 360, 390	5–11	0.22, 0.42, 0.62	[39–41]
Ti Ti-6Al-4V	Ti-based (45~75)	CoCrFeNiNbx/ CoNiTi/CoCrNiTi/ CrFeNiTi/CrNiTi	0.04–0.1	6, 20	1000	3,8		[42–45]
		Titanium	0.08–1	40, 60, 100	300–800	6.7, 10		[46,47]

2. Temperature Field Simulation and Thermal Analysis

With a view to evaluate the thermal profile during the LMD process, temperature field simulation and thermal analysis are usually conducted. The thermal profile and temperature distribution during the deposition are closely related to the process parameters which determines the laser energy inputting and the cooling rate of molten pool. The laser mode of PW, which can control the laser energy inputting as well as the interaction among the laser energy, powder, and substrate by changing the parameters of pulsed laser, has a significant effect on the thermal profile during the deposition process [48,49].

In the LMD process, the Gaussian conical heat source is recognized as the representative of laser beam. Thus a Gaussian conical heat source [50], which can be mathematically represented by Equation (1), generally was used as the thermal loading during the simulation of LMD process [51,52].

$$q(x,y,z) = \frac{9Q_0}{\pi(1-e^{-3})(z_e - z_i)(r_e^2 + r_e r_i + r_i^2)} exp\left(-\frac{x^2 + y^2}{r_0^2(z)}\right) \quad (1)$$

where, Q_0 is the heat flux and $Q_0 = \eta P$, P is the laser beam energy, η is the efficiency value, and ηP represents the heat flux. r_0 is the heat distribution coefficient and can be represented by Equation (2),

$$r_0(z) = r_i + (r_e - r_i)\frac{z - z_i}{z_e - z_i} \tag{2}$$

where r_e and r_i represent the maximum and minimum radius, respectively.

Owing to the existing of laser-on and laser-off stage in the PW laser mode, the constant about time should be defined. K. Yang et al. defined a constant about time ($\delta(t)$), as represented by Equation (3), and imported into the heat conduction equation to simulate temperature field in the LMD process with PW laser mode [27].

$$\delta(t) = \begin{cases} 0 & 0 \leq t \leq T_{pulse} \\ 1 & T_{pulse} < t \leq T_{cycle} \end{cases} \tag{3}$$

$$\delta\left(t + T_{cycle}\right) = \delta(t) \tag{4}$$

where T_{pulse} represents the pulse width and T_{cycle} represents the cycle period of one pulse. T_{pulse} and T_{cycle} can be calculated by pulse frequency f_{pulse} and the duty cycle D_u as follow:

$$T_{cycle} = 1/f_{pulse} \tag{5}$$

$$T_{pulse} = D_u T_{cycle} \tag{6}$$

With the comparison of CW laser mode, the study reported by Z. Yang et al., in which the duty cycle was 50%, showed the peak temperature variation in one pulse period with the difference of pulse frequency as shown in Figure 2a. During the LMD process of high-nitrogen steel (HNS), the peak temperature of the melt pool created by CW laser mode was about 2900 °C and remained almost stable. Under the PW laser mode regardless of the pulse frequency, the temperature of molten pool increased dramatically with a gradual decline of growth rate when the laser was ON, and then, decreased dramatically when the laser was OFF. With the increase of pulse frequency from 20 Hz to 80 Hz, the maximum temperature decreased from 2860 °C to 2690 °C and the minimum temperature increased from 1210 °C to 1480 °C. Under the pulse frequency of 60 Hz and 80 Hz, the minimum temperature of molten pool exceeded the solidus temperature (1400 °C) of HNS that meant the molten pool kept continuous during the LMD process. Thus, the fish-scale patterns were not observed on the surface morphology of single-track clad that also proofed by the experiments. The pulse frequency of 60 Hz was the critical value which decided whether the molten pool remained continuous or not [27]. Therefore, the pulse frequency should be larger than the critical value to assure the continuous molten pool.

The thermal history has significant effect on the cooling rate and then the dendrite spacing and microstructure. The influence of cooling rate on the dendrite spacing could be described by the following equation [33,53]:

$$\lambda = C \cdot \varepsilon^{-m} \tag{7}$$

where λ is the dendrite spacing, C and m represent positive constants determined by the material, ε is the cooling rate. Based on the Equation (7), it can be concluded that the dendrite spacing would increase with the increase of cooling rate. According to the simulation results about deposition of non-weldable nickel-based K447A alloy, Z. Zhang et al. pointed that the temperature history under the PW laser was more vibrational compared to that created by CW laser in the high-temperature region as shown in Figure 3. The vibrational amplitude reduced with the increase of duty cycle and the temperature curve with the duty cycle of 0.8 closed to consistent with the curve under the CW laser. The cooling rate at around 1370 °C and the dendrite spacing of depositions under different parameters were shown in Table 2. The cooling rate induced by PW laser presented higher than that by CW laser. However, the difference became smaller with the increase of duty cycle and the cooling rate under the duty cycle of 0.8 became same with that of CW laser owing to the consistent temperature curve. The dendrite spacing increased from 4.87 μm to 6.86 μm with the increase of duty cycle from 0.3 to 0.8 [33].

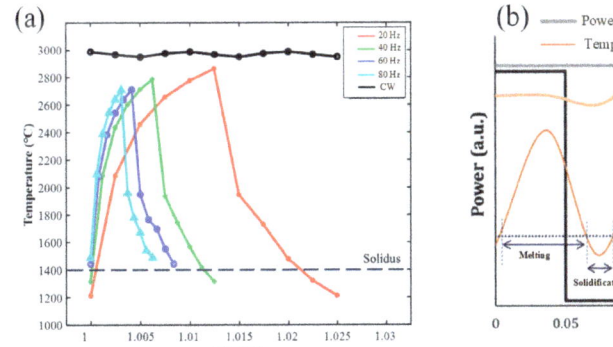
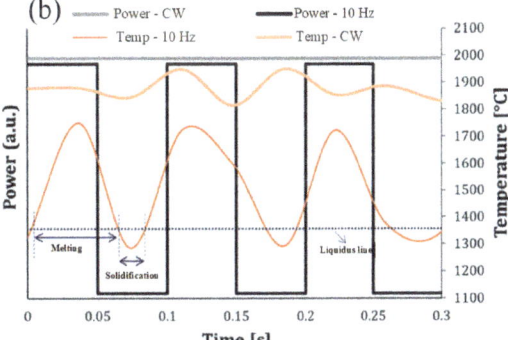

Figure 2. The temperature variation of molten pool by CW and PW temperature field based on (**a**) simulation (Reprinted from ref. [27], copyright (2021), with permission from Elsevier) and (**b**) experimental results where the peak power is constant (Reprinted from ref. [32], copyright (2020), with permission from Elsevier).

The actual thermal profile during the deposition process can be monitored by the infra-red camera and the thermocouples. With the using of CW laser mode, the measured average temperature could reach about 1900 °C during the Inconel 718 deposition was created by the laser powder of 300 W as shown in Figure 3b. However, the measured temperature presented slight oscillations owing to the powder sparks which were caused by the turbulence and expulsion of molten pool. During the PW laser with the pulse frequency of 10 Hz was used, the temperature presented periodic variation between the maximum value of 1700 °C and the minimum value of 1300 °C that attributed to the "heart-beat" behavior and thus the local thermal cycles in the molten pool. The lower temperature and local thermal cycles were obtained by using the pulsed laser within each deposited layer. It is worth noting that the temperature drop to 1300 °C, which was lower than the solidus temperature of 1350 °C during the PW laser with the pulse frequency of 10 Hz was used. Compared to the pulse frequency of 100 Hz and 1000 Hz, the measured temperature created by pulse frequency of 10 Hz led to low laser energy inputting, low heat accumulation, and high cooling rate which resulted in the measured temperature decreased under the solidus temperature. Therefore, the molten pool was subject by solidification-melting cycles. In addition, during the pulse frequency of 100 Hz was used, the temperature reached the value which was slightly higher than the solidus temperature when the laser was set on OFF. Thus, during the Inconel 718 deposition was created by the laser powder of 300 W, the pulse frequency of 100 Hz can be recognized as the transition value which determined whether the temperature down to the solidus temperature and the molten pool remains continuous or not [32].

Based on the reported results obtained by simulation and actual experiment, the temperature created by the PW laser would be lower than that by CW laser because of the existing of the interval time when the laser is set on OFF. Thus, lower heat accumulation and higher cooling rate of molten pool could be obtained. As for the PW laser mode, there is an approximate pulse frequency which can be recognized as demarcation line to distinguish the status of molten pool during the deposition process.

Table 2. Cooling rate around 1370 °C and dendrite spacing of K447A alloy deposition [33].

Duty Cycle	Cooling Rate (°C/s)		Secondary Dendrite Spacing (μm)
	PW Laser	CW Laser	
0.3	24,333	3570	4.87
0.4	5840	2393	5.94
0.5	3133	2169	6.32
0.6	2053	1709	6.68
0.8	1557	1557	6.86

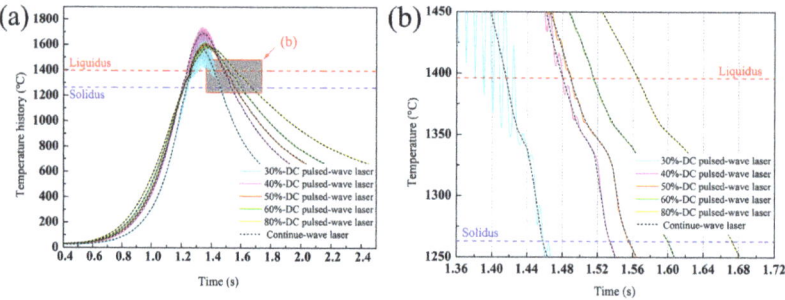

Figure 3. (**a**) Temperature history at the point that is 120 μm above the bonding interface in the depositions with different duty cycles, and (**b**) Enlargement of the shaded part in (**a**). (Reprinted from ref. [33], copyright (2020), with permission from Elsevier).

3. Surface Quality

In the LMD process, the process parameters, such as laser power, powder feed rate, z-increment, shielding gas flow and so on, govern the temperature field of molten pool then consequently affect the Marangoni flow and the surface quality [54,55]. Marangoni flow of the molten pool, which is closely related to the temperature gradient, would happen when the material absorbs the inputted laser energy and melt. Owing to the Gaussian conical heat source, the energy density near the middle of molten pool is high and it is low near the edge that contributes to the occurrence of temperature gradient and then the surface tension gradient. The Marangoni flow is mainly caused by the different of surface tension in the different region of molten pool. The high energy density corresponds to high temperature and low surface tension. Thus, the liquid metal would flow from the central region to the edge of the molten pool [56–58]. The surface quality, which is reflected by the surface roughness, is strongly linked with the intensity of Marangoni flow and vapour recoil pressure. According to the driving force of the recoil pressure and the Marangoni force, once the local momentum exceeds the pressure produced by the surface tension, the molten materials in the molten pool are ejected that is not conductive to the surface quality [59,60]. However, the strong Marangoni flow contributes to the aggravation of the surface disturbance of molten pool which is beneficial to mix and melt the powder particles within the molten pool and was found to be a vital parameter in decreasing the surface roughness [61].

Compared to the CW laser, with using of PW laser in the LMD process, the molten pool can be modified by the high surface disturbance of molten pool which improves the mixing action of powder into the molten pool and the melting efficiency of captured powder. K. Shah et al. deposited Inconel 718 powder on a Ti-6Al-4V substrate by using CW and PW laser. The experimental results presented an inverse relationship between the surface disturbance of the melt pool and the surface roughness of the part. Using the PW laser increased the mean surface disturbance by experiencing high peak power [61]. A. Pinkerton and L. Li pointed out that increasing the duty cycle, up to the boundary case of a CW laser, increased the surface roughness through conducting the multiple-layer 316 L steel deposition with the CW and PW laser [62]. In addition, as reported by S. Imbrogno et al., the width of molten pool created by different wave of laser had significant effects on the side surface quality of thin-wall deposition. As shown in Figure 4a, owing to the high laser energy inputted by CW laser, the width of molten pool was bigger than that created by PW laser with the pulse frequency of 10 Hz shown in Figure 4b. The temperature in region ABC was lower than the central region of molten pool because this region was not directly irradiated by laser beam that resulted in most of heat in this region could be conducted by the melted material. During the PW laser was used, the area of region A′B′C′ was smaller that contributed to the improvement of side surface quality. Moreover, compared to the CW laser, the less U-shaped molten pool, which was beneficial for reducing the waviness effect on the side surface, was presented when the PW laser was used because of the less energy per time unit and the lower temperature of molten pool [32]. Thus, the side surface quality of thin-wall was improved by using the PW laser. M. Gharbi et al. deposited the Titanium powder on the Ti-6Al-4V alloy with different laser wave and investigated the influence of different PW laser parameters on the surface quality. They pointed out that the surface finish of depositions, in which the average rough values Ra was 3 μm, was obtained using the PW laser owing to the decreased thermal gradients and Marangoni flow of the molten pool. However, compared to the PW Gaussian laser, the using of top-hat laser not presented further improvement to the surface quality [46].

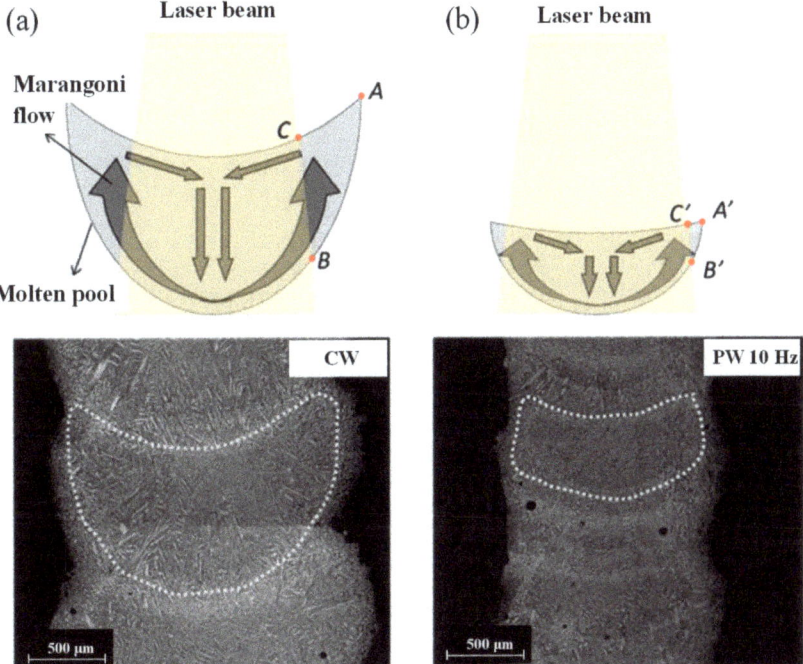

Figure 4. Marangoni flow within the cross-section of the IN718 thin walls when (**a**) CW and (**b**) 10 Hz PW laser are used. (Reprinted from ref. [32], copyright (2020), with permission from Elsevier).

4. Microstructures

The inputted laser energy has significant effect on the thermal history which can dominate the microstructure features of depositions such as morphology and grain size as shown in Figure 5a. The solidification rate of molten pool, the temperature gradient at the solid liquid interface (G), and the ratio of cooling rate to thermal gradient (R) determine the solidified microstructure. The G/R would affect the solid-liquid interface shape and the G × R, which means the cooling rate, would affect the grain size [4,63]. The G/R is large during the early stage of LMD process that contributes to the occurrence of columnar crystal. Owing to the decrease of G/R with the continuous process of LAM, the dendrite crystal tends to form [64–66].

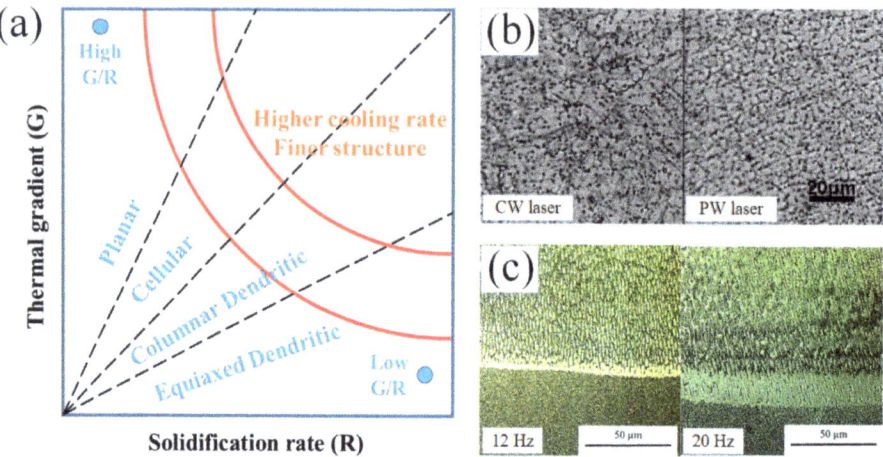

Figure 5. (**a**) Solidification map established by solidification parameters (Reprinted from ref. [27], copyright (2021), with permission from Elsevier); (**b**) Optical micrographs of CW and PW laser cladding Fe-based depositions (Reprinted from ref. [25], copyright (2014), with permission from Elsevier); (**c**) OM images of the intermixing zone of Inconel 713 deposition with a laser frequency of 12 Hz and 20 Hz (Reprinted from ref. [37], copyright (2021), with permission from Elsevier).

Although the nominal laser power is same, the PW laser heat input would be lower than that created by the CW laser owing to the existing of duty cycle. The increase of the heat input reduces the cooling rate and increases the grain size [67]. The PW laser with higher cooling rate in molten pool can result in the bigger supercooling degree which contributes to the increase of nucleation rate and nuclei number. In addition, owing to the existing of intermittent, the PW laser would urge the occurrence of thermal shock effect on the molten pool that prevents the reuniting of nuclei after separating out from the molten pool. Also, the intermittent provides the blanking time for the molten pool to solidify that results in the shorter solidification time of the molten pool by PW laser than that created by CW laser. Furthermore, compared to the CW laser, the PW laser would reduce the heat accumulation and increase cooling rate that results in the grains have not enough time to grow along with the orientation driven by the heat source. The lower pulse frequency means bigger interval time during the laser is set on OFF that contributes to longer solidification time for the melted material and higher cooling rate. This has been verified by the experiment results about Inconel 718 deposition fabricated by pulse laser with the frequency of 10 Hz through comparing to the microstructures created by the pulse frequency of 100 Hz and 1000 Hz [32]. Therefore, the microstructure in PW deposition layer was refined significantly. During the coagulation process in the molten pool, the deposition underwent nucleation. The decrease of gain size contributed to the increase of nucleation base number and nucleation rate. Therefore, the PW laser produced the refined microstructure that was verified by researchers based on the experimental results [32]. For example, the research reported by H. Zhang et al. showed the significant refined microstructure in Fe-based coating on low carbon steel though using the PW laser that concluded by the CW and PW laser deposition's average grain size of 11.68 μm and 6.86 μm, respectively as shown in Figure 5b [25]. Decreasing the heat input though reducing laser frequency could prompt the transformation of from columnar grains to the equiaxed grains with smaller size as shown in the Stellite 31 coating on the Inconel 713 substrate as shown in Figure 5c [37]. For the Ti-based coating, K. Xiang et al. fabricated the CoNiTi medium-entropy alloy depositions on the pure Ti substrate and obtained the good metallurgical bonding between the depositions and substrate using the pulsed LMD. The deposition consisted of dendritic-interdendritic microstructure. The bonding zone, which was followed by the heat affected zone with irregular-shaped bulk grains was comprised of acicular fine grains with an average width of ~320 nm [44]. The Ti coating obtained by C. Wang et al. presented disorderly distributed β phase and fine martensitic lath [47].

In addition, the secondary dendrite arm spacing (SDAS) was usually chosen as the representation of microstructure evolution. The SDAS values decreased along the direction from bottom to top of the deposition layer due to the different cooling rate in the different region of molten pool. The SDAS would increase with the increase of laser power and reduction of scanning speed because of the increased heat inputting and decreased cooling rate of molten pool. The variation of PW laser parameters exactly affects the heat inputting and the SDAS. Z. Yang et al. pointed out that the SDAS in high-nitrogen steel deposition increased dramatically with the pulse frequency increased from 20 Hz to 60 Hz during the other parameters kept constant. However, during the pulse frequency exceeded to 60 Hz, the effect on SDAS became minimal. An interesting phenomenon was presented that a sixth dendrite arm was observed in the pulsed LAM samples as shown in Figure 6a. It turned out that that as long as there were sufficient spaces and large enough solidification rates, new branches can continuously grow from the primary dendrite arms. In addition, the SDAS value map was plotted as a function of laser power, scanning speed, and pulse frequency as shown in Figure 6b [27].

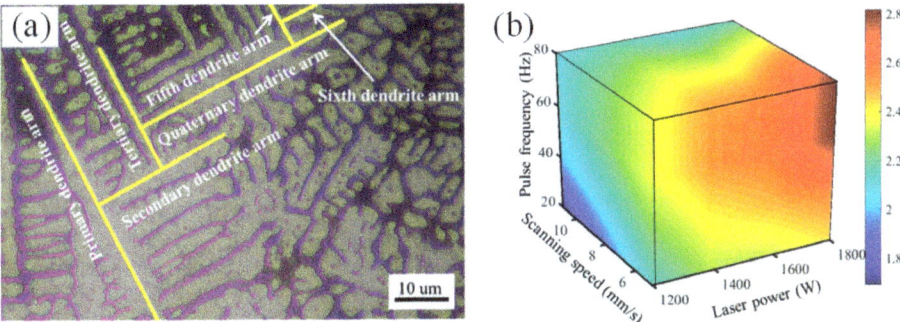

Figure 6. (a) From primary- to sixth- dendrite arm, and (b) Secondary dendrite arm spacing (SDAS) values map for single track deposition of high-nitrogen steel. (Reprinted from ref. [27], copyright (2019), with permission from Elsevier).

Furthermore, owing to the change of heat inputted and laser energy with the different wave of laser as well as the parameters of PW laser such as pulse frequency, the orientation of the grains, which is closely related to the heat source movement, would be affected. In general, during the CW laser is used, the molten pool would be stable and continuous owing the constant laser energy inputting. As shown in the investigation about Inconel 718 thin-wall deposition by LMD reported by S. Imbrogno et al., the microstructure of the longitudinal section of deposition, which was created by CW laser and PW laser with pulse frequency of 100 Hz and 1000 Hz shown in Figure 7a,b, presented epitaxial growth and the growing direction was determined by the heat source movement. When the pulse frequency of 10 Hz was used, the obvious discrete molten pool boundaries was observed and the growing of dendrites oriented to the center of each discrete molten pool as shown in Figure 7c [32]. G. Muvvala et al. indicated that the using of PW laser resulted in stacks of columnar dendrites grow along different orientation owing to the periodic remelting and solidification of irradiated volume [68]. A. Farnia et al. deposited the Stellite 6 coating on the low carbon ferritic steel with the pulse frequency ranged from 1 to 1000 Hz, pulse duration ranged from 0.2 to 20 ms, and pulse energy ranged from 0 to 40 J. The results illustrated that the pulse traces can represent the solidification fronts. The orientation of the grains showed perpendicular to the solidification front. Based on the elongation and orientation of grains near the middle region of coating, the results indicated that the shape of solidification front during the PW laser was used was similar to that produced by the CW laser with a double scanning speed. In the pulsed LMD process, the consecutive pulse resulted in the periodical repetition of grain orientations on the longitudinal section of coating [36]. Therefore, the pulse frequency plays an important role in not only the microstructure refinement but also the growing orientation of grains.

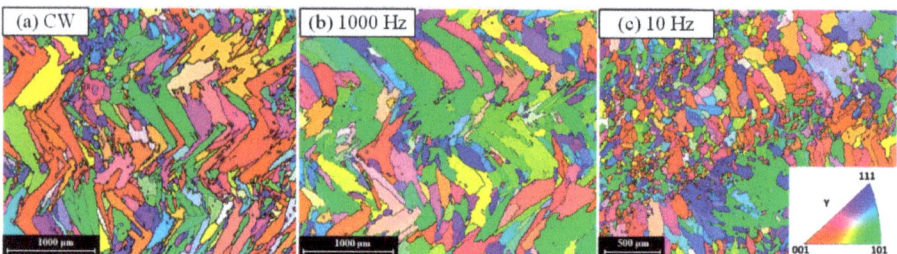

Figure 7. EBSD map of the longitudinal section zx plane during using of (**a**) CW, (**b**) PW with the frequency of 1000 Hz and (**c**) PW with the frequency of 10 Hz (Reprinted from ref. [32], copyright (2020), with permission from Elsevier).

5. Microhardness

Based on the classic Hall-Petch equation $\sigma_y = \sigma_0 + kd^{-1/2}$ [69,70], where σ_y is yield or tensile stress, σ_0 is a friction stress which is a constant stress for steel stress during dislocations move on the slip plane, k is the stress concentration factor which is reflected by the Hall-Petch slope, d is the average grain size The yield and tensile stress decrease with the increase of reciprocal root of the grain size that means the yield and tensile strength positively correlated to the grain size. H. Zhang et al. compared the microhardenss of TiC-VC reinforced Fe-based cladding created CW and PW laser. The average harnesses of CW and PW cladding layers are 950 $HV_{0.2}$ and 1160 $HV_{0.2}$, respectively. The results were consistent with grain size measurement results where the CW and PW laser cladding layers' average grain sizes are 11.68 μm and 6.86 μm, respectively. That shown the dispersion strengthening effect, in which the carbides refinement is caused by using the PW laser, and improvement of hardness [25]. The effect of grain size on strength was identical to that on the hardness [71,72]. S. Imbrogno et al. deposited Inconel 718 thin-wall by CW and PW laser, and pointed out that the hardness measured on the PW depositions was higher than that on the CW samples because of the lower heat accumulation, higher cooling rate, and smaller the primary dendrite arm spacing (PDAS). In addition, the hardness gradually decreased from the bottom to the top of depositions during the PW laser was used. However, the hardness on the sample produced by CW presented a constant distribution and a rapid drop only on the top part [32]. A. Khorram investigated the effect of parameters of PW laser on the Stellite 31 coating by LMD in which the Inconel 713 was chosen as the substrate. Based on the experimental results, they pointed out that the hardness increased almost linearly with the decrease of dilution ration which was determined by the reduction of pulse width and laser frequency [37,73]. X. Wang et al. investigated the effect of different pulsed laser shape, including continuous, rectangular, ramp up, ramp down, and hybrid ramp, on the hardness of AISI316L deposition. They pointed out that the deposition created by rectangular laser shape presented highest hardness because of the minimum value of grain size caused by the high cooling rate as shown in Figure 8 [29].

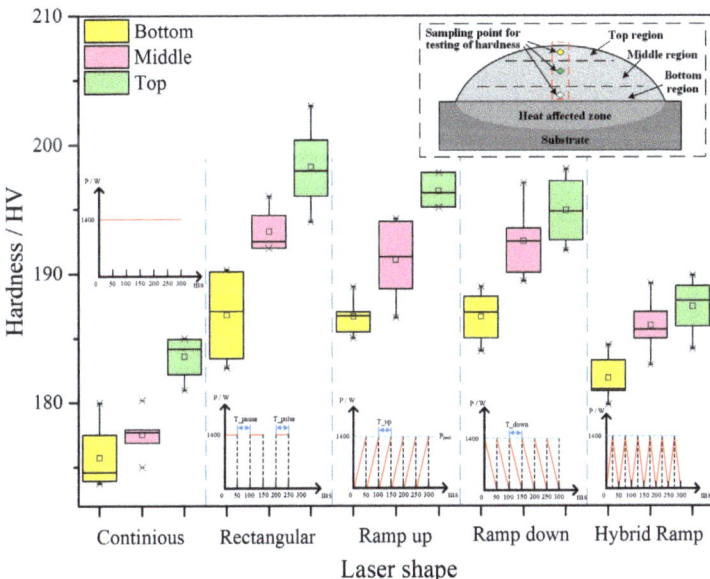

Figure 8. Microhardness at different regions of AISL316L deposition produced by different laser shape. (Reprinted from ref. [29], copyright (2020), with permission from MDPI.).

The improvement of hardness sometimes contributes to the formation of strengthening structure and hardening phase. The deposition of CoNiTi medium-entropy alloy created by PW laser reached to 571 ± 46 HV which was about five times of the hardness in the substrate (114 ± 4 HV) owing to the formation of solid-solution hardening BCC phase and second-phase hardening from the Ti_2Ni and Ti_2Co type intermetallic compounds [44]. C. Wang investigated the hardness of Ti coating created by PW laser on the Ti-6Al-4V substrate in which consisted of α grains with the grain size of 4.7 μm. The lath structure martensite with the size of 0.86 μm was presented in the coating with fine grains after pulsed laser clad that contributed to the improvement of hardness. During the pulse frequency of 16 Hz and pulse width of 13 ms were set in the deposition process, the hardness Stellite 31 coating with the value of 482 ± 10 HV, which is markedly higher than the hardness of Inconel 713 substrate with the value of 355 ± 3 HV, was presented. The improvement of hardness in the deposition contributed to the creation of Co solid solution phase and Cr_7C_3, Cr_3C_2, W_2C, and WC_x carbides [37]. During the pulse frequency was 60 Hz, H.Yan et al. deposited Co-based coating on the copper alloy substrate. The coating presented high hardness with the value about 478 HV which was about five times that of the copper substrate (92 HV) that contributed to the formation of carbide and intermetallic hard phases [41]. W. Wang et al. conducted the repairing of BT20 titanium alloy with the pulse width of 12 ms and the frequency of 4 Hz and obtained the coating without defects. The hardness of deposition presented even mean value (about 450 HV) which was slight higher than that of substrate (about 370 HV). During the LMD process, the transformation of "β→α'" martensite, in which the β stable elements, such as Mo and V, soluted into α' phase, attributed to the improvement of hardness [74]. Owing to the occurrence of dilution which was necessary for the formation of metallurgical bonding between the substrate and depositions, the hardness was affected. Through cladding the $75Cr_3C_2$ + 25 (80Ni20Cr) powder on the Inconel 718, A. Khorram et al. pointed out that with the increase of pulse frequency and pulse width as well as the decrease of laser scanning speed, the dilution ratio increased that contributed to the decrease of concentration of chromium carbides in the deposition and the formation of eutectic structure. Therefore, hardness in the deposition reduced. However, based on the optimum parameters, including laser frequency of 20 Hz, pulse width of 12.9 ms, and laser speed of 5.43 mm/s, the hardness of deposition was improved with the value of 1050 HV which was about 2.5 times that of the substrate that attributed to the abundant chromium carbide (Cr_7C_3) and refined eutectic structure ($\gamma + Cr_7C_3$) [75]. G. Muvvala et al. indicated that the increasing of cooling rate contributed to the decrease of elemental segregation, formation of Laves phase and γ matrix size that resulted in the hardness improvement of Inconel 718 deposition [68].

6. Residual Stress and Crack

The LMD process is accompanied by the repeated rapid heating and cooling of molten pool. Owing to a heterogeneous response of heat conduction and heat dissipation, the high residual stress, which is a domain drawback for the corrosion, fracture resistance, and fatigue performance, is easily to produce in the deposition itself and at the interface between clad and substrate areas as a result of the fast-cooling rates and the difference in thermal expansion coefficients [76–80]. The residual stress would be generated by the constrained thermal shrinkage, which is caused by the transient temperature gradients, the different coefficient of thermal expansion between the substrate and deposition, and the changes in specific density caused by transformation of solid phase [18,81,82]. The heat inputting affects the cooling rate which has negative correlation to the square of the melt pool length [83]. Considering the mechanical factors, the cooling rate causes the thermal strains and increases the crack initiation rate. Compared to the CW laser, the PW laser, which can result in the high cooling rate, can increase the crack resistance that verified by K. Shah [31]. In addition, the residual stress increases with the reduction of pulse length and duty cycle as shown in Figure 9a [31,84].

In the LMD process, both macro- and micro-cracks may occur in the depositions. The macro-cracks, which are known as solidification cracks, often are generated near the solidification temperature range [85]. They are caused by the insufficient supplement of liquid metal when the consolidation and contraction process suddenly release stress after the increased thermal stress discharge from the rapid cooling process [86,87]. Micro-cracks mainly include phase interface cracking, slip zone cracking, and grain boundary cracking. They are usually caused by uneven local slip and micro-cracking of metal materials. In addition, small cavities, uneven crystal grains, and impurities produced in the LMD process are also the inducements for the micro-cracks [88,89]. It can be concluded that the crack is affected by the metallurgical factors, which is mainly concern phase relationship, and mechanical factors, which are related to stress behavior [31,90]. As for the metallurgical factors, the experimental results indicated that the changing of parameters was insufficient to avoid the generation of the brittle phases in the Ni-based deposition [31]. The LMD process undergone a high cyclic heating and cooling regime and the concentrated moving heat source made the LMD process vulnerable to thermal stresses which were the source of crack formation [91,92]. The tensile stress was unfavorable for the mechanical properties of parts because it decreased the fatigue life and tensile properties of the depositions. The comprehensive residual stress was beneficial for improving the fatigue and wear resistance of parts owing the effects of forbidding the initiation and propagation of cracks [93–95]. As shown in Figure 9b, compared to the Ti-6Al-4V substrate with the residual stress of −471.75 MPa, the residual stress in Ti coating created by PW laser reached to −511.65 MPa that contributed to the decreased susceptibility to crack and high resistance for crack generation and propagation [47]. Based on the pulse shaping, M. Pleterski conducted the laser cladding with wire on the AISI D2 (EN X160CrMo12-1) tool steel. Three types of laser shape, which including rectangular (shape A), ramp-down (shape B), and ramp-up-down (shape C), were used as shown in Figure 10. When the laser peak power was set between 1 and 1.5 kW and pulse duration ranged from 30 to 60 ms, the deposition created by shape A and B presented obvious cracks but the cracks under the shape B were shallower compared to those exhibited under shape A. However, no cracks were observed in the sample deposited by the shape C that contributed to the relatively slower cooling rate which was affected by the parameter setting [96].

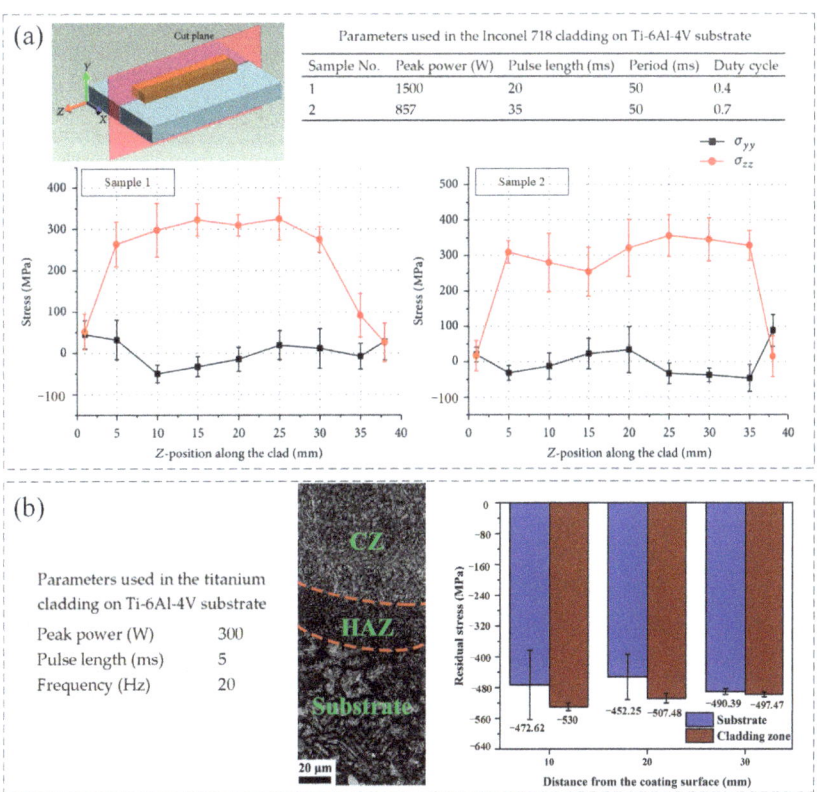

Figure 9. (**a**) Residual stress of Inconel 718 cladding creased by different parameters of PW laser (Reprinted from ref. [31], copyright (2014), with permission from Hindawi.); (**b**)Residual stress of substrate and coating zone. (Reprinted from ref. [47], copyright (2021), with permission from Elsevier).

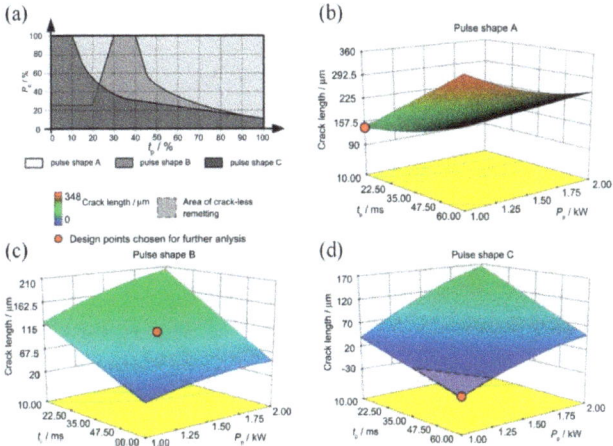

Figure 10. (**a**) Laser pulse shapes, and the 3-D RSM model plots of crack length in remelted spots during the (**b**) pulse A, (**c**) pulse B, and (**d**) pulse C was used. (Reprinted from ref. [96], copyright (2011), with permission from Elsevier).

During the cooling of molten pool in the LMD process, the shrinkage created by solidification of liquid phase and the deformation of solid structure produce the driving strain which attributes to the decrease of pressure in the liquid film and then the propagation of cracks. The driving strain increases with the increase of the dendritic spacing of the microstructure which is affected by the laser parameters. A void, which provides favorable condition for the propagation of cracks, would form when the pressure decreases below a cavitation pressure after the crack nucleates [33]. In addition, the cavitation pressure changes with the progress of solidification that can be described with the following equation [97,98]:

$$P_{crit} = \frac{2\gamma_{sl}}{h} \qquad (8)$$

where h, which has been widely recognized as the main influence factor, represents the thickness of the liquid film, γ_{sl} represents the surface tension which is closely related to the temperature. With the increase of h, the crack becomes easier to occur in the liquid film between the adjacent dendrites. According to the surface penetration test of non-weldable nickel-based superalloy deposition reported by Z. Zhang's work in which the pulse frequency was constant with the value of 10 Hz, evident cracks were observed during the duty cycle ranged from 0.6 to 0.8, and the deposition under the duty cycle ranged from 0.3 to 0.5 presented crack-free. The increase of duty cycle resulted in increasing of dendritic spacing and then bigger driving strain created by the shrinkage and deformation. The depositions with bigger dendritic spacing had weaker resistance to cracking and presented obvious cracks. In addition, the bigger duty cycle was used, the bigger width of dendrites and bigger dendrite spacing was caused by the smaller cooling rate [33]. Thus, the increase of duty cycle was not conductive to preventing from the propagation of cracks that also has been verified by the investigation of A. Odabsi et al. in which more evident and longer cracks were observed during the dendrite spacing increased from 1.06 μm to 2.3 μm [99]. Another way to preventing from the formation and propagation of cracks is preheating the substrate. J. Yin et al. obtained the defect-free Ni-Cr-Si coating on the copper substrate preheated with the preheating temperature of 700 °C [100]. With preheating of the copper substrate to 300 °C, Y. Zhang et al. deposited nickel based alloy coating which was free of cracks and composed of γ-nickel solid solution dendrites as well as some quantity of carbides and silicides [101].

7. Corrosion Behavior

The corrosion process, which is a gradual degradation process of material by chemical or electrochemical reaction with their environment, transforms a refined metal into a more chemically stable form such as sulfide, hydroxide, or oxide [102–105]. The corrosion resistance sometimes is a vital performance index to reflect the mechanical properties of the deposition by LMD when the working condition is poor such as high temperature, high pressure, corrosive atmosphere and so on. The solidified microstructure, which can be significantly affected by the laser energy inputting such as the pulse frequency in PW laser mode, determines the corrosion behavior of deposition by LMD. By using the low-pulse frequency, the refined solidified microstructure can be obtained. And the small grain size is beneficial to the corrosion behavior. The high density of grain boundaries, which is accompanied with the refinement of grains, contributes to the formation of metal passivation films followed by chloride ions [106]. The passivation films thickness is an important parameter of corrosion performance and is proportional to the corrosion resistance that have been demonstrated by many scholars [107,108].

In addition, the micro-porosity, which is related to the entrapment of gas in the molten pool, the unstable of melt flow behavior, and the local overheating of the molten pool, is also an important factor affecting the corrosion preformation of depositions [27]. However, except for the mentioned above formation mechanism, there will be other basic reason for the formation of micro-porosity during some material is used. For example, the root cause of the formation mechanism in the Inconel 718 depositions using PW laser with high frequency can be regarded as Nb segregation and the hydrogen segregation affected the micro-porosity in the aluminum alloys [109,110]. In addition, the influence factor of micro-porosity in the high-nitrogen steel depositions can be regarded as the nitrogen segregation. The high cooling rate contributed to the decrease of diffusion time of element that prevented the element segregation as proved by the macro-micro couple model developed by Xie et al. [50]. The high cooling rate of molten pool provided less time for pore growth and element diffusion that contributed to the improve of corrosion resistance. As shown in the research reported by K. Yang, the high-nitrogen steel depositions, which was created by PW laser with pulse frequency of 20 Hz, presented the most outstanding corrosion resistance with the lowest corrosion density of 1.73×10^{-6} A cm^{-2}, the highest corrosion potential of -0.2849 V and pitting potential of -0.0554 V compared to the other four samples with pulse frequency of 0 Hz, 40 Hz, 60 Hz, 80 Hz, respectively as shown in Figure 11a. In addition, two completely different corrosion morphologies, which are shallow and deep pits as shown in Figure 11b, in QCW20 sample were presented. In the areas where the pitting occurs, the different dendrite morphology and micro-porosity size in the cross-sections resulted in distinct corrosion morphology. The micro-porosity content of depositions increased with the increase of pulse frequency and decrease of the cooling rate [27].

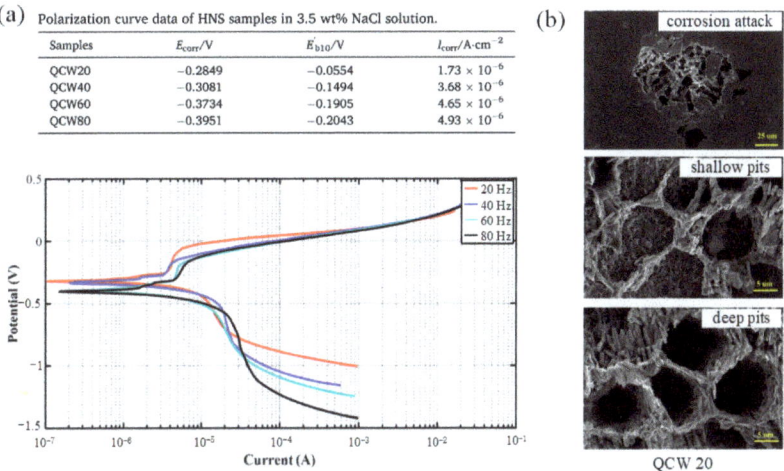

Figure 11. (a) Polarization curves of high-nitrogen steel (HNS) samples under various pulse frequencies, and (b) SEM images (2000×) of the QCW20 sample after the potentiodynamic polarization experiments. (Reprinted from ref. [27], copyright (2021), with permission from Elsevier).

8. Conclusions

This review addresses the research status of temperature simulation, surface quality, microstructures, microhardness, residual stress, cracking, and corrosion behavior of metallic coating created by pulsed laser material deposition. The main conclusions are drawn as follows:

(1) According to the simulation and experimental results about temperature field during the LMD process, the using of PW laser would contribute to the lower heat accumulation and higher cooling rate of molten pool. Under the condition of PW laser, the minimum temperature of molten pool sometimes would be lower than the solidus temperature of the material that means the molten pool can not keep continuous during the LMD process. Thus, the fish-scale patterns would be observed. For example, the pulse frequency should be larger than the critical value to assure the continuous molten pool. The thermal history has significant effect on the cooling rate and then the dendrite spacing and microstructure.

(2) With using of PW laser in the LMD process, the molten pool would be modified by the high surface disturbance of molten pool created by the strong Marangoni flow that improves the mixing action of powder into the molten pool, the melting efficiency of captured powder, and thus the surface quality.

(3) The heat input of PW laser would be lower than that created by the CW laser owing to the existing of duty cycle that contributes to the increase of cooling rate, decrease of the grain size and the secondary dendrite arm spacing (SDAS). Owing to the change of heat inputted and laser energy with the different wave of laser as well as the parameters of PW laser such as pulse frequency, the orientation of the grains, which is closely related to the heat source movement, would be affected.

(4) By optimizing the parameters of PW laser, the hardness would improve because of the decrease of grain size, the formation of strengthening structure and hardening phase.

(5) Compared to the CW laser, the PW laser, which can result in the high cooling rate, can increase the crack resistance. The increase of duty cycle is not conductive to preventing from the propagation of cracks because of the increasing of dendritic spacing and then bigger driving strain created by the shrinkage and deformation. In addition, the residual stress increases with the reduction of pulse length and duty cycle.

(6) By using the low-pulse frequency of PW laser, the refined solidified microstructure can be obtained that is beneficial to the corrosion behavior. the micro-porosity, which is also an important factor affecting the corrosion preformation of depositions, would decrease with the decrease of pulse frequency and thus the increase of cooling rate that contributes to the improvement of corrosion behavior.

9. Future Perspectives

LMD is, by far, one of the most extensively employed AM technologies. The pulsed LMD technology is expected to maintain and even increase the impact on laboratory investigation and industrial application. The reason behind the continuous impact is without any doubt related to the improved controllability of the LMD process, solidification of molten pool, and then the mechanical properties of deposition by using the PW laser, but also to the recent developments of pulsed LMD using a large variety of different materials, such as Fe-based alloy, Co-based alloy, Ni-based alloy, titanium, and so on. Despite the advancements, the pulsed LAM still have some drawbacks, such as residual stress caused by high cooling rate and high challenge of thermal distribution controlling, that require consideration. This review also addressed the relevant limitations and analyzed the solutions reported in the recent literature.

The optimization of PW laser parameters, the analysis of temperature field by PW laser, and the improvement of mechanical properties including hardness, wear resistance, corrosion, and so on, are currently the center of multiple investigations about pulsed LAM. However, the application of pulsed LAM process for metal materials which have a higher melting temperature, such as high temperature alloy, has a good foreground because the PW laser can produce a higher peak power during the average power is same compared to the CW laser. Furthermore, the PW laser is a promising method to improve the mechanical properties and surface finish of metal parts owing to the better capacity of controlling over the temperature of molten pool. In addition, a smaller molten pool can be created by using the PW laser that is ideal for the fabrication of thin-wall part. Taking into consideration all these facts, it can be concluded that a leap forward for the pulsed LAM, which is mainly investigated in a laboratory at present, could be prompted to the industrial application in the future.

Author Contributions: Investigation, X.W. and J.J.; writing—original draft preparation, X.W.; writing—review and editing, X.W., J.J. and Y.T. All authors have read and agreed to the published version of the manuscript.

Funding: This research was funded by the Natural Science Foundation of Liaoning Province, grant number 2020-BS-206, Liaoning Department of Education Scientific Research Foundation, grant number JDL2020022, and Scientific and technological research program of China National Railway Group Limited, grant number N2021T010.

Data Availability Statement: Data is contained within the article.

Conflicts of Interest: The authors declare no conflict of interest.

References

1. Herzog, D.; Seyda, V.; Wycisk, E.; Emmelmann, C. Additive manufacturing of metals. *Acta Mater.* **2016**, *117*, 371–392. [CrossRef]
2. Huang, S.H.; Liu, P.; Mokasdar, A.; Hou, L. Additive manufacturing and its societal impact: A literature review. *Int. J. Adv. Manuf. Technol.* **2012**, *67*, 1191–1203. [CrossRef]
3. Shim, D.-S.; Baek, G.-Y.; Seo, J.-S.; Shin, G.-Y.; Kim, K.-P.; Lee, K.-Y. Effect of layer thickness setting on deposition characteristics in direct energy deposition (DED) process. *Opt. Laser Technol.* **2016**, *86*, 69–78. [CrossRef]
4. Thompson, S.M.; Bian, L.; Shamsaei, N.; Yadollahi, A. An overview of Direct Laser Deposition for additive manufacturing; Part I: Transport phenomena, modeling and diagnostics. *Addit. Manuf.* **2015**, *8*, 36–62. [CrossRef]
5. Das, S.; Beama, J.J.; Wohlert, M.; Bourell, D.L. Direct laser freeform fabrication of high performance metal components. *Rapid Prototyp. J.* **1998**, *4*, 112–117. [CrossRef]
6. Wang, X.; Deng, D.; Hu, Y.; Ning, F.; Wang, H.; Cong, W.; Zhang, H. Overhang structure and accuracy in laser engineered net shaping of Fe-Cr steel. *Opt. Laser Technol.* **2018**, *106*, 357–365. [CrossRef]
7. Wang, D.; Wang, H.; Chen, X.; Liu, Y.; Lu, D.; Liu, X.; Han, C. Densification, Tailored Microstructure, and Mechanical Properties of Selective Laser Melted Ti–6Al–4V Alloy via Annealing Heat Treatment. *Micromachines* **2022**, *13*, 331. [CrossRef]
8. Song, C.; Hu, Z.; Xiao, Y.; Li, Y.; Yang, Y. Study on Interfacial Bonding Properties of NiTi/CuSn10 Dissimilar Materials by Selective Laser Melting. *Micromachines* **2022**, *13*, 494. [CrossRef]
9. Zhang, W.; Zhang, B.; Xiao, H.; Yang, H.; Wang, Y.; Zhu, H. A Layer-Dependent Analytical Model for Printability Assessment of Additive Manufacturing Copper/Steel Multi-Material Components by Directed Energy Deposition. *Micromachines* **2021**, *12*, 1394. [CrossRef]
10. Liu, Z.; Wang, X.; Kim, H.; Zhou, Y.; Cong, W.; Zhang, H. Investigations of Energy Density Effects on Forming Accuracy and Mechanical Properties of Inconel 718 Fabricated by LENS Process. *Procedia Manuf.* **2018**, *26*, 731–739. [CrossRef]
11. El Cheikh, H.; Courant, B.; Branchu, S.; Huang, X.; Hascoët, J.-Y.; Guillén, R. Direct Laser Fabrication process with coaxial powder projection of 316L steel. Geometrical characteristics and microstructure characterization of wall structures. *Opt. Lasers Eng.* **2012**, *50*, 1779–1784. [CrossRef]
12. Wang, X.; Deng, D.; Qi, M.; Zhang, H. Influences of deposition strategies and oblique angle on properties of AISI316L stainless steel oblique thin-walled part by direct laser fabrication. *Opt. Laser Technol.* **2016**, *80*, 138–144. [CrossRef]
13. Cong, W.; Ning, F. A fundamental investigation on ultrasonic vibration-assisted laser engineered net shaping of stainless steel. *Int. J. Mach. Tools Manuf.* **2017**, *121*, 61–69. [CrossRef]
14. Wang, X.; Liu, Z.; Guo, Z.; Hu, Y. A fundamental investigation on three–dimensional laser material deposition of AISI316L stainless steel. *Opt. Laser Technol.* **2020**, *126*, 106107. [CrossRef]
15. Zhang, K.; Liu, W.; Shang, X. Research on the processing experiments of laser metal deposition shaping. *Opt. Laser Technol.* **2007**, *39*, 549–557. [CrossRef]
16. D'Oliveira, A.S.C.M.; da Silva, P.S.C.P.; Vilar, R.M.C. Microstructural features of consecutive layers of Stellite 6 deposited by laser cladding. *Surf. Coat. Technol.* **2002**, *153*, 203–209. [CrossRef]
17. Farahmand, P.; Kovacevic, R. An experimental–numerical investigation of heat distribution and stress field in single- and multi-track laser cladding by a high-power direct diode laser. *Opt. Laser Technol.* **2014**, *63*, 154–168. [CrossRef]
18. Wang, X.; Lei, L.; Yu, H. A Review on Microstructural Features and Mechanical Properties of Wheels/Rails Cladded by Laser Cladding. *Micromachines* **2021**, *12*, 152. [CrossRef]
19. Calleja, A.; Tabernero, I.; Fernández, A.; Celaya, A.; Lamikiz, A.; López de Lacalle, L.N. Improvement of strategies and parameters for multi-axis laser cladding operations. *Opt. Lasers Eng.* **2014**, *56*, 113–120. [CrossRef]
20. Moat, R.J.; Pinkerton, A.J.; Li, L.; Withers, P.J.; Preuss, M. Residual stresses in laser direct metal deposited Waspaloy. *Mater. Sci. Eng. A* **2011**, *528*, 2288–2298. [CrossRef]
21. Rangaswamy, P.; Griffith, M.L.; Prime, M.B.; Holden, T.M.; Rogge, R.B.; Edwards, J.M.; Sebring, R.J. Residual stresses in LENS®components using neutron diffraction and contour method. *Mater. Sci. Eng. A* **2005**, *399*, 72–83. [CrossRef]
22. Kahlen, F.J.; Kar, A. Residual stresses in laser-deposited metal parts. *J. Laser Appl.* **2001**, *13*, 60–69. [CrossRef]
23. Kanzler, K. How Much Energy Are You Throwing Away?-A uniform, or "top hat," beam profile often is obtained at too great a cost. Laser remapping systems offer a high-throughput approach to beam shaping. *Photonics Spectra* **2006**, *40*, 74–75.
24. Pratheesh Kumar, S.; Elangovan, S.; Mohanraj, R.; Sathya Narayanan, V. Significance of continuous wave and pulsed wave laser in direct metal deposition. *Mater. Today Proc.* **2021**, *46*, 8086–8096. [CrossRef]
25. Zhang, H.; Zou, Y.; Zou, Z.; Zhao, W. Comparative study on continuous and pulsed wave fiber laser cladding in-situ titanium–vanadium carbides reinforced Fe-based composite layer. *Mater. Lett.* **2015**, *139*, 255–257. [CrossRef]
26. Zhang, H.; Chong, K.; Zhao, W.; Sun, Z. Effects of pulse parameters on in-situ Ti-V carbides size and properties of Fe-based laser cladding layers. *Surf. Coat. Technol.* **2018**, *344*, 163–169. [CrossRef]

27. Yang, K.; Wang, Z.D.; Chen, M.Z.; Lan, H.F.; Sun, G.F.; Ni, Z.H. Effect of pulse frequency on the morphology, microstructure, and corrosion resistance of high-nitrogen steel prepared by laser directed energy deposition. *Surf. Coat. Technol.* **2021**, *421*, 127450. [CrossRef]
28. Pinkerton, A.J.; Li, L. An investigation of the effect of pulse frequency in laser multiple-layer cladding of stainless steel. *Appl. Surf. Sci.* **2003**, *208*, 405–410. [CrossRef]
29. Wang, X.; Yu, H.; Jiang, J.; Xia, C.; Zhang, Z. Influences of Pulse Shaping on Single-Track Clad of AISI316L Stainless Steel by Laser Material Deposition. *Coatings* **2022**, *12*, 248. [CrossRef]
30. Corbin, S.F.; Toyserkani, E.; Khajepour, A. Cladding of an Fe-aluminide coating on mild steel using pulsed laser assisted powder deposition. *Mater. Sci. Eng. A* **2003**, *354*, 48–57. [CrossRef]
31. Shah, K.; Izhar Ul, H.; Shah, S.A.; Khan, F.U.; Khan, M.T.; Khan, S. Experimental study of direct laser deposition of Ti-6Al-4V and Inconel 718 by using pulsed parameters. *Sci. World J.* **2014**, *2014*, 841549. [CrossRef] [PubMed]
32. Imbrogno, S.; Alhuzaim, A.; Attallah, M.M. Influence of the laser source pulsing frequency on the direct laser deposited Inconel 718 thin walls. *J. Alloys Compd.* **2021**, *856*, 158095. [CrossRef]
33. Zhang, Z.; Zhao, Y.; Chen, Y.; Su, Z.; Shan, J.; Wu, A.; Sato, Y.S.; Gu, H.; Tang, X. The role of the pulsed-wave laser characteristics on restraining hot cracking in laser cladding non-weldable nickel-based superalloy. *Mater. Des.* **2021**, *198*, 109346. [CrossRef]
34. Farnia, A.; Ghaini, F.M.; Sabbaghzadeh, J. Effects of pulse duration and overlapping factor on melting ratio in preplaced pulsed Nd: YAG laser cladding. *Opt. Lasers Eng.* **2013**, *51*, 69–76. [CrossRef]
35. Sun, S.; Durandet, Y.; Brandt, M. Parametric investigation of pulsed Nd: YAG laser cladding of stellite 6 on stainless steel. *Surf. Coat. Technol.* **2005**, *194*, 225–231. [CrossRef]
36. Farnia, A.; Malek Ghaini, F.; Ocelík, V.; De Hosson, J.T.M. Microstructural characterization of Co-based coating deposited by low power pulse laser cladding. *J. Mater. Sci.* **2013**, *48*, 2714–2723. [CrossRef]
37. Khorram, A. Microstructural evolution of laser clad Stellite 31 powder on Inconel 713 LC superalloy. *Surf. Coat. Technol.* **2021**, *423*, 127633. [CrossRef]
38. Paul, C.; Alemohammad, H.; Toyserkani, E.; Khajepour, A.; Corbin, S. Cladding of WC–12 Co on low carbon steel using a pulsed Nd: YAG laser. *Mater. Sci. Eng. A* **2007**, *464*, 170–176. [CrossRef]
39. Acquaviva, S.; Caricato, A.P.; D'Anna, E.; Fernández, M.; Luches, A.; Frait, Z.; Majkova, E.; Ozvold, M.; Luby, S.; Mengucci, P. Pulsed laser deposition of Co- and Fe-based amorphous magnetic films and multilayers. *Thin Solid Films* **2003**, *433*, 252–258. [CrossRef]
40. Lee, H.-K. Effects of the cladding parameters on the deposition efficiency in pulsed Nd:YAG laser cladding. *J. Mater. Process. Technol.* **2008**, *202*, 321–327. [CrossRef]
41. Yan, H.; Wang, A.; Xu, K.; Wang, W.; Huang, Z. Microstructure and interfacial evaluation of Co-based alloy coating on copper by pulsed Nd:YAG multilayer laser cladding. *J. Alloys Compd.* **2010**, *505*, 645–653. [CrossRef]
42. Xiang, K.; Chai, L.; Zhang, C.; Guan, H.; Wang, Y.; Ma, Y.; Sun, Q.; Li, Y. Investigation of microstructure and wear resistance of laser-clad CoCrNiTi and CrFeNiTi medium-entropy alloy coatings on Ti sheet. *Opt. Laser Technol.* **2022**, *145*, 107518. [CrossRef]
43. Chai, L.; Wang, C.; Xiang, K.; Wang, Y.; Wang, T.; Ma, Y. Phase constitution, microstructure and properties of pulsed laser-clad ternary CrNiTi medium-entropy alloy coating on pure titanium. *Surf. Coat. Technol.* **2020**, *402*, 126503. [CrossRef]
44. Xiang, K.; Chai, L.; Wang, Y.; Wang, H.; Guo, N.; Ma, Y.; Murty, K.L. Microstructural characteristics and hardness of CoNiTi medium-entropy alloy coating on pure Ti substrate prepared by pulsed laser cladding. *J. Alloys Compd.* **2020**, *849*, 156704. [CrossRef]
45. Xiang, K.; Chen, L.-Y.; Chai, L.; Guo, N.; Wang, H. Microstructural characteristics and properties of CoCrFeNiNbx high-entropy alloy coatings on pure titanium substrate by pulsed laser cladding. *Appl. Surf. Sci.* **2020**, *517*, 146214. [CrossRef]
46. Gharbi, M.; Peyre, P.; Gorny, C.; Carin, M.; Morville, S.; Masson, P.L.; Carron, D.; Fabbro, R. Influence of a pulsed laser regime on surface finish induced by the direct metal deposition process on a Ti64 alloy. *J. Mater. Process. Tech.* **2014**, *214*, 485–495. [CrossRef]
47. Wang, C.; Li, J.; Wang, T.; Chai, L.; Deng, C.; Wang, Y.; Huang, Y. Microstructure and properties of pure titanium coating on Ti-6Al-4V alloy by laser cladding. *Surf. Coat. Technol.* **2021**, *416*, 127137. [CrossRef]
48. Guo, L.; Yue, T.M.; Man, H.C. A Numerical Model for Studying Laser Cladding under Pulse Mode. *Lasers Eng.* **2012**, *22*, 209–225.
49. Sun, S.; Durandet, Y.; Brandt, M. Correlation between melt pool temperature and clad formation in pulsed and continuous wave Nd:YAG laser cladding of Stellite 6. In Proceedings of the Pacific International Conference on Applications of Lasers and Optics, Melbourne, Australia, 19 April 2004; Volume 2004, p. 702.
50. Xie, H.; Yang, K.; Li, F.; Sun, C.; Yu, Z. Investigation on the Laves phase formation during laser cladding of IN718 alloy by CA-FE. *J. Manuf. Process.* **2020**, *52*, 132–144. [CrossRef]
51. Zhan, M.; Sun, G.; Wang, Z.; Shen, X.; Yan, Y.; Ni, Z. Numerical and experimental investigation on laser metal deposition as repair technology for 316L stainless steel. *Opt. Laser Technol.* **2019**, *118*, 84–92. [CrossRef]
52. Toyserkani, E.; Khajepour, A.; Corbin, S. 3-D finite element modeling of laser cladding by powder injection: Effects of laser pulse shaping on the process. *Opt. Lasers Eng.* **2004**, *41*, 849–867. [CrossRef]
53. Zeng, Y.; Li, L.; Huang, W.; Zhao, Z.; Yang, W.; Yue, Z. Effect of thermal cycles on laser direct energy deposition repair performance of nickel-based superalloy: Microstructure and tensile properties. *Int. J. Mech. Sci.* **2022**, *221*, 107173. [CrossRef]
54. Zhu, G.; Li, D.; Zhang, A.; Pi, G.; Tang, Y. The influence of laser and powder defocusing characteristics on the surface quality in laser direct metal deposition. *Opt. Laser Technol.* **2012**, *44*, 349–356. [CrossRef]

55. Bi, G.; Gasser, A.; Wissenbach, K.; Drenker, A.; Poprawe, R. Investigation on the direct laser metallic powder deposition process via temperature measurement. *Appl. Surf. Sci.* **2006**, *253*, 1411–1416. [CrossRef]
56. Lepski, D.; Brückner, F. *Laser Cladding*; Springer: Dordrecht, The Netherlands, 2009; pp. 235–279.
57. Vilar, R. Laser cladding. *J. Laser Appl.* **2001**, *11*, 64–79. [CrossRef]
58. Fuhrich, T.; Berger, P.; Hügel, H. Marangoni effect in laser deep penetration welding of steel. *J. Laser Appl.* **2001**, *13*, 178–186. [CrossRef]
59. Yin, J.; Wang, D.; Yang, L.; Wei, H.; Dong, P.; Ke, L.; Wang, G.; Zhu, H.; Zeng, X. Correlation between forming quality and spatter dynamics in laser powder bed fusion. *Addit. Manuf.* **2020**, *31*, 100958. [CrossRef]
60. Yin, J.; Zhang, W.; Ke, L.; Wei, H.; Wang, D.; Yang, L.; Zhu, H.; Dong, P.; Wang, G.; Zeng, X. Vaporization of alloying elements and explosion behavior during laser powder bed fusion of Cu–10Zn alloy. *Int. J. Mach. Tools Manuf.* **2021**, *161*, 103686. [CrossRef]
61. Shah, K.; Pinkerton, A.J.; Salman, A.; Li, L. Effects of Melt Pool Variables and Process Parameters in Laser Direct Metal Deposition of Aerospace Alloys. *Adv. Manuf. Process.* **2010**, *25*, 1372–1380. [CrossRef]
62. Pinkerton, A.J.; Li, L. The effect of laser pulse width on multiple-layer 316L steel clad microstructure and surface finish. *Appl. Surf. Sci.* **2003**, *208–209*, 411–416. [CrossRef]
63. Selcuk, C. Laser metal deposition for powder metallurgy parts. *Powder Metall.* **2011**, *54*, 94–99. [CrossRef]
64. Zhu, Y.; Yang, Y.; Mu, X.; Wang, W.; Yao, Z.; Yang, H. Study on wear and RCF performance of repaired damage railway wheels: Assessing laser cladding to repair local defects on wheels. *Wear* **2019**, *430–431*, 126–136. [CrossRef]
65. Zhou, S.; Dai, X.; Zheng, H. Microstructure and wear resistance of Fe-based WC coating by multi-track overlapping laser induction hybrid rapid cladding. *Opt. Laser Technol.* **2012**, *44*, 190–197. [CrossRef]
66. Yang, Y.L.; Zhang, D.; Yan, W.; Zheng, Y. Microstructure and wear properties of TiCN/Ti coatings on titanium alloy by laser cladding. *Opt. Lasers Eng.* **2010**, *48*, 119–124. [CrossRef]
67. Kumar, K.S. Analytical modeling of temperature distribution, peak temperature, cooling rate and thermal cycles in a solid work piece welded by laser welding process. *Procedia Mater. Sci.* **2014**, *6*, 821–834. [CrossRef]
68. Muvvala, G.; Patra Karmakar, D.; Nath, A.K. Online monitoring of thermo-cycles and its correlation with microstructure in laser cladding of nickel based super alloy. *Opt. Lasers Eng.* **2017**, *88*, 139–152. [CrossRef]
69. Yuan, W.; Panigrahi, S.; Su, J.-Q.; Mishra, R. Influence of grain size and texture on Hall–Petch relationship for a magnesium alloy. *Scr. Mater.* **2011**, *65*, 994–997. [CrossRef]
70. Wang, N.; Wang, Z.; Aust, K.T.; Erb, U. Effect of grain size on mechanical properties of nanocrystalline materials. *Acta Metall. Mater.* **1995**, *43*, 519–528. [CrossRef]
71. Morris, J.W., Jr. *The Influence of Grain Size on the Mechanical Properties of Steel*; Lawrence Berkeley National Laboratory (LBNL): Berkely, CA, USA, 2001.
72. Ning, F.; Cong, W. Microstructures and mechanical properties of Fe-Cr stainless steel parts fabricated by ultrasonic vibration-assisted laser engineered net shaping process. *Mater. Lett.* **2016**, *179*, 61–64. [CrossRef]
73. Khorram, A.; Taheri, M.; Fasahat, M. Laser cladding of Inconel 713 LC with Stellite 31 powder: Statistical modeling and optimization. *Laser Phys.* **2021**, *31*, 096001. [CrossRef]
74. Wang, W.; Wang, M.; Jie, Z.; Sun, F.; Huang, D. Research on the microstructure and wear resistance of titanium alloy structural members repaired by laser cladding. *Opt. Lasers Eng.* **2008**, *46*, 810–816. [CrossRef]
75. Khorram, A.; Davoodi Jamaloei, A.; Paidar, M.; Cao, X. Laser cladding of Inconel 718 with 75Cr$_3$C$_2$ + 25(80Ni20Cr) powder: Statistical modeling and optimization. *Surf. Coat. Technol.* **2019**, *378*, 124933. [CrossRef]
76. Wang, L.; Felicelli, S.D.; Pratt, P. Residual stresses in LENS-deposited AISI 410 stainless steel plates. *Mater. Sci. Eng. A* **2008**, *496*, 234–241. [CrossRef]
77. Chew, Y.; Pang, J.H.L.; Bi, G.; Song, B. Thermo-mechanical model for simulating laser cladding induced residual stresses with single and multiple clad beads. *J. Mater. Process. Technol.* **2015**, *224*, 89–101. [CrossRef]
78. Trojan, K.; Ocelík, V.; Čapek, J.; Čech, J.; Canelo-Yubero, D.; Ganev, N.; Kolařík, K.; De Hosson, J.T.M. Microstructure and Mechanical Properties of Laser Additive Manufactured H13 Tool Steel. *Metals* **2022**, *12*, 243. [CrossRef]
79. Köhler, H.; Partes, K.; Kornmeier, J.R.; Vollertsen, F. Residual Stresses in Steel Specimens Induced by Laser Cladding and their Effect on Fatigue Strength. *Phys. Procedia* **2012**, *39*, 354–361. [CrossRef]
80. Cottam, R.; Wang, J.; Luzin, V. Characterization of microstructure and residual stress in a 3D H13 tool steel component produced by additive manufacturing. *J. Mater. Res.* **2014**, *29*, 1978–1986. [CrossRef]
81. Roy, T.; Abrahams, R.; Paradowska, A.; Lai, Q.; Mutton, P.; Soodi, M.; Fasihi, P.; Yan, W. Evaluation of the mechanical properties of laser cladded hypereutectoid steel rails. *Wear* **2019**, *432–433*, 202930. [CrossRef]
82. Niederhauser, S.; Karlsson, B. Fatigue behaviour of Co–Cr laser cladded steel plates for railway applications. *Wear* **2005**, *258*, 1156–1164. [CrossRef]
83. Hofmeister, W.; Griffith, M. Solidification in direct metal deposition by LENS processing. *JOM* **2001**, *53*, 30–34. [CrossRef]
84. Moat, R.J.; Pinkerton, A.J.; Hughes, D.J.; Li, L.; Withers, P.J.; Preuss, M. Stress distributions in multilayer laser deposited Waspaloy parts measured using neutron diffraction. In Proceedings of the International Congress on Applications of Lasers & Electro-Optics, Orlando, FL, USA, 29 October 2007; Volume 2007, p. 101.
85. Yu, J.; Rombouts, M.; Maes, G. Cracking behavior and mechanical properties of austenitic stainless steel parts produced by laser metal deposition. *Mater. Des.* **2013**, *45*, 228–235. [CrossRef]

86. Caiazzo, F. Laser-aided Directed Metal Deposition of Ni-based superalloy powder. *Opt. Laser Technol.* **2018**, *103*, 193–198. [CrossRef]
87. Song, X.; Lei, J.; Xie, J.; Fang, Y. Microstructure and electrochemical corrosion properties of nickel-plated carbon nanotubes composite Inconel718 alloy coatings by laser melting deposition. *Opt. Laser Technol.* **2019**, *119*, 105593. [CrossRef]
88. Leo Prakash, D.G.; Walsh, M.J.; Maclachlan, D.; Korsunsky, A.M. Crack growth micro-mechanisms in the IN718 alloy under the combined influence of fatigue, creep and oxidation. *Int. J. Fatigue* **2009**, *31*, 1966–1977. [CrossRef]
89. Li, X.; Jiang, X. Effects of dislocation pile-up and nanocracks on the main crack propagation in crystalline metals under uniaxial tensile load. *Eng. Fract. Mech.* **2019**, *212*, 258–268. [CrossRef]
90. Dong, S.Y.; Ren, W.B.; Bin-Shi, X.U.; Yan, S.X.; Fang, J.X. Experiment Optimization of Impulse Laser Remanufacture Forming Process for Compresssor Thin-wall Blade. *J. Acad. Armored Force Eng.* **2015**, *43*, 6–12.
91. Alimardani, M.; Toyserkani, E.; Huissoon, J.P.; Paul, C.P. On the delamination and crack formation in a thin wall fabricated using laser solid freeform fabrication process: An experimental–numerical investigation. *Opt. Lasers Eng.* **2009**, *47*, 1160–1168. [CrossRef]
92. Zhai, Y.; Lados, D.A.; Brown, E.J.; Vigilante, G.N. Fatigue crack growth behavior and microstructural mechanisms in Ti-6Al-4V manufactured by laser engineered net shaping. *Int. J. Fatigue* **2016**, *93*, 51–63. [CrossRef]
93. Zhang, C.; Shen, X.; Wang, J.; Xu, C.; He, J.; Bai, X. Improving surface properties of Fe-based laser cladding coating deposited on a carbon steel by heat assisted ultrasonic burnishing. *J. Mater. Res. Technol.* **2021**, *12*, 100–116. [CrossRef]
94. Jun, H.-K.; Seo, J.-W.; Jeon, I.-S.; Lee, S.-H.; Chang, Y.-S. Fracture and fatigue crack growth analyses on a weld-repaired railway rail. *Eng. Fail. Anal.* **2016**, *59*, 478–492. [CrossRef]
95. Ringsberg, J.W.; Skyttebol, A.; Josefson, B.L. Investigation of the rolling contact fatigue resistance of laser cladded twin-disc specimens: FE simulation of laser cladding, grinding and a twin-disc test. *Int. J. Fatigue* **2005**, *27*, 702–714. [CrossRef]
96. Pleterski, M.; Tušek, J.; Muhic, T.; Kosec, L. Laser Cladding of Cold-Work Tool Steel by Pulse Shaping. *J. Mater. Sci. Technol.* **2011**, *27*, 707–713. [CrossRef]
97. Miller, W.; Chadwick, G. On the magnitude of the solid/liquid interfacial energy of pure metals and its relation to grain boundary melting. *Acta Metall.* **1967**, *15*, 607–614. [CrossRef]
98. Lahaie, D.; Bouchard, M. Physical modeling of the deformation mechanisms of semisolid bodies and a mechanical criterion for hot tearing. *Metall. Mater. Trans. B* **2001**, *32*, 697–705. [CrossRef]
99. Odabaşı, A.; Ünlü, N.; Göller, G.; Eruslu, M.N. A Study on Laser Beam Welding (LBW) Technique: Effect of Heat Input on the Microstructural Evolution of Superalloy Inconel 718. *Metall. Mater. Trans. A* **2010**, *41*, 2357–2365. [CrossRef]
100. Yin, J.; Wang, D.; Meng, L.; Ke, L.; Hu, Q.; Zeng, X. High-temperature slide wear of Ni-Cr-Si metal silicide based composite coatings on copper substrate by laser-induction hybrid cladding. *Surf. Coat. Technol.* **2017**, *325*, 120–126. [CrossRef]
101. Zhang, Y.-z.; Tu, Y.; Xi, M.-z.; Shi, L.-k. Characterization on laser clad nickel based alloy coating on pure copper. *Surf. Coat. Technol.* **2008**, *202*, 5924–5928. [CrossRef]
102. Menghani, J.; Vyas, A.; Patel, P.; Natu, H.; More, S. Wear, erosion and corrosion behavior of laser cladded high entropy alloy coatings–A review. *Mater. Today Proc.* **2021**, *38*, 2824–2829. [CrossRef]
103. Hsu, Y.-J.; Chiang, W.-C.; Wu, J.-K. Corrosion behavior of FeCoNiCrCux high-entropy alloys in 3.5% sodium chloride solution. *Mater. Chem. Phys.* **2005**, *92*, 112–117. [CrossRef]
104. Li, W.; Guo, W.; Zhang, H.; Xu, H.; Chen, L.; Zeng, J.; Liu, B.; Ding, Z. Influence of Mo on the Microstructure and Corrosion Behavior of Laser Cladding FeCoCrNi High-Entropy Alloy Coatings. *Entropy* **2022**, *24*, 539. [CrossRef]
105. Zheng, C.; Liu, Z.; Liu, Q.; Kong, Y.; Liu, C. Effect of Cr on Corrosion Behavior of Laser Cladding Ni-Cr-Mo Alloy Coatings in Sulfuric Acid Dew Point Corrosion Environment. *Coatings* **2022**, *12*, 421. [CrossRef]
106. Andreatta, F.; Lanzutti, A.; Vaglio, E.; Totis, G.; Sortino, M.; Fedrizzi, L. Corrosion behaviour of 316L stainless steel manufactured by selective laser melting. *Mater. Corros.* **2019**, *70*, 1633–1645. [CrossRef]
107. Feng, Z.; Cheng, X.; Dong, C.; Xu, L.; Li, X. Passivity of 316L stainless steel in borate buffer solution studied by Mott–Schottky analysis, atomic absorption spectrometry and X-ray photoelectron spectroscopy. *Corros. Sci.* **2010**, *52*, 3646–3653. [CrossRef]
108. Shang, F.; Chen, S.; Zhou, L.; Jia, W.; Cui, T.; Liang, J.; Liu, C.; Wang, M. Effect of laser energy volume density on wear resistance and corrosion resistance of 30Cr15MoY alloy steel coating prepared by laser direct metal deposition. *Surf. Coat. Technol.* **2021**, *421*, 127382. [CrossRef]
109. Zhang, Y.; Li, Z.; Nie, P.; Wu, Y. Effect of ultrarapid cooling on microstructure of laser cladding IN718 coating. *Surf. Eng.* **2013**, *29*, 414–418. [CrossRef]
110. Gu, C.; Ridgeway, C.D.; Cinkilic, E.; Lu, Y.; Luo, A.A. Predicting gas and shrinkage porosity in solidification microstructure: A coupled three-dimensional cellular automaton model. *J. Mater. Sci. Technol.* **2020**, *49*, 91–105. [CrossRef]

Article

Microstructure and Corrosion Behavior of Iron Based Biocomposites Prepared by Laser Additive Manufacturing

Yan Zhou [1], Lifeng Xu [1], Youwen Yang [2,*], Jingwen Wang [1], Dongsheng Wang [1,*] and Lida Shen [3]

[1] Key Laboratory of Construction Hydraulic Robots, Anhui Higher Education Institutes, Tongling University, Tongling 244061, China; zhouyan099@163.com (Y.Z.); abc402@163.com (L.X.); wangjingwener@126.com (J.W.)
[2] Institute of Bioadditive Manufacturing, Jiangxi University of Science and Technology, Nanchang 330013, China
[3] Jiangsu Key Laboratory of Precision and Micro-Manufacturing Technology, Nanjing University of Aeronautics and Astronautics, Nanjing 210016, China; ldshen@nuaa.edu.cn
* Correspondence: yangyouwen@jxust.edu.cn (Y.Y.); wangdongsheng@tlu.edu.cn (D.W.)

Abstract: Iron (Fe) has attracted great attention as bone repair material owing to its favorable biocompatibility and mechanical properties. However, it degrades too slowly since the corrosion product layer prohibits the contact between the Fe matrix and body fluid. In this work, zinc sulfide (ZnS) was introduced into Fe bone implant manufactured using laser additive manufacturing technique. The incorporated ZnS underwent a disproportionation reaction and formed S-containing species, which was able to change the film properties including the semiconductivity, doping concentration, and film dissolution. As a result, it promoted the collapse of the passive film and accelerated the degradation rate of Fe matrix. Immersion tests proved that the Fe matrix experienced severe pitting corrosion with heavy corrosion product. Besides, the in vitro cell testing showed that Fe/ZnS possessed acceptable cell viabilities. This work indicated that Fe/ZnS biocomposite acted as a promising candidate for bone repair material.

Keywords: iron bone implant; zinc sulfide; degradation properties; passivation film; laser powder bed fusion

1. Introduction

Metal materials have excellent comprehensive mechanical properties (high strength, toughness, fatigue resistance) and good processing and forming ability [1,2]. Thus, medical metal implants have been widely used in the field of orthopedics. Degradable metals not only have excellent comprehensive mechanical properties, but also the degradation products can be absorbed by the human body [3,4]. As a new type of medical implant, it is able to be completely degraded and absorbed in the human body after service in vivo, avoiding the pain of patients' secondary operation. Among the several representative degradable metal, iron (Fe) has gained intensive attention recently [5,6]. It is an essential nutrient element and participates in a variety of metabolic processes, which is able to maintain the normal function of bone cells. However, its degradation rate is too slow, which will hinder the growth of new bone as an implant.

Destroying the passive film is an effective way to accelerate its degradation of metallic matrix. Owing to the multivalent character of sulfur (S), sulfide is able to generate various S-containing species, which can attach on the metal surface and induce severe damaging effect on the passive film [7]. Previously, some scholars studied the S-induced corrosion of Fe-based metal material and indicated that the adsorbed S catalyzes the metal dissolution, thereby resulting in a decreased dissolution activation energy and accelerated anodic dissolution process [8]. It was also reported that the S-containing species changed the film properties including the semiconductivity, doping concentration, film dissolution rate, and film composition. Particularly, the synergistic effect of S species and chloride (Cl) was also confirmed by previous research [9].

Among S-containing species, zinc sulfide (ZnS) possessed relatively good biocompatibility and water solubility [10,11]. Basing on the above consideration, in this work, ZnS was incorporated into Fe implants aiming to accelerate the corrosion of Fe matrix. Meanwhile, Zn ion, as a trace element, could promote cell proliferation and differentiation, which was expected to improve the biocompatibility of Fe implant [12]. The Fe based implant was fabricated by the laser powder bed fusion (LPBF) technique. LPBF is a powder bed melting technology. The focused laser beam selectively melts the powder layer by layer to produce the required geometry. Since LPBF meets the requirements of high melting point, high dimensional accuracy, high performance and design flexibility, it has become the main additive manufacturing technology of metal implants [13–15]. The microstructure, corrosion behavior, and biocompatibility of Fe/ZnS composite fabricated by LPBF were investigated. Additionally, the corrosion mechanism was deeply studied.

2. Materials and Methods

2.1. Original Materials and Laser Powder Bed Fusion (LPBF) Process

Sphere Fe powder (mean particle size 35 μm) and ZnS powder (5–10 μm) were utilized in this work. Fe and ZnS (9 wt %) were mixed by a miniature planet ball mill (PULVERISETTE 6, Fritsch, Germany). The ball mill was operated at a rotation speed of 220 rpm for 2 h, with a 15 min pause every half an hour. During operation, high purity argon (99.9%) was offered to reduce the oxidation.

The mixed powder was adopted to fabricate Fe/ZnS biocomposite using LPBF system, which was consisted of a fiber laser and a computer control system. A series of pilot experiments were carried out before the LPBF experiments to obtain an optimized processing parameter and as follows: laser power 210 W, scanning rate 80 mm/s, hatching space 50 μm and layer thickness 50 μm.

2.2. Microstructural Characterization

The LPBF-processed parts were grounded and polished using SiC paper. The microstructure was characterized using a scanning electron microscopy (SEM, Zeiss, Oberkochen, Germany) equipped with an energy dispersive spectroscopy (EDS). The phase composition was determined using X-ray diffractometer (XRD, D8 Advance, Bremen, Germany) with Cu Kα radiation at 45 kV and 40 mA. The scanning range was 20–90°, and the scan rate was 8°/min.

2.3. Electrochemical Tests

An electrochemical experiment was carried out to study the corrosion behavior. The self-prepared simulated body fluid (SBF) was used at testing solution. A three-electrode system was adopted in electrochemical tests. The nominal chemical composition of SBF was listed in Table 1. The system consisted of platinum as counter electrode, saturated calomel as reference electrode and the test part as working electrode. The initial open-loop circuit (OCP) tests were firstly performed. Then, the Tafel polarization curve was recorded at a rate of 0.05 mV/s. The corrosion rate (Pi) was determined by corrosion current (I_{corr}):

$$Pi = 3.27 \times 10^{-3} \times I_{corr} \, E/\rho \tag{1}$$

Table 1. Chemical composition of SBF.

Composition	NaCl	NaHCO$_3$	KCl	K$_2$HPO$_4$·3H$_2$O	MgCl$_2$·6H$_2$O	CaCl$_2$
Weight (g/L)	8.035	0.355	0.225	0.231	0.311	0.292

E was the weight equivalent, and ρ was the material densigty. Besides, the electrochemical impedance spectroscopy (EIS) testing was carried out within the scope of 0.01 Hz to 1000 kHz. Zsimpwin software was adopted to analyze the result. Furthermore, the transient time-current curve was determined at 1 mV/s. The Mott-Schottky curve was recorded to study the semiconductor properties of the corrosion film.

2.4. Immersion Tests

SBF immersion testing was performed to further study the degradation behavior of as-built parts. The parts were immersed in SBF at an exposure ratio of 0.1 cm^2/mL. After immersion for 7, 14, and 28 days, the parts were washed with distilled water and then observed by SEM. The samples were washed using 200 g/L of CrO_3 solution to remove corrosion products. Subsequently, the surface morphology was investigated by an atomic force microscope (AFM, Veeco Instruments, Plainview, NY, USA). Meanwhile, the corrosion rate (*Cr*) was calculated by using the weight loss method after immersion tests.

2.5. Cytotoxicity Evaluation

MG-63 cells were used to evaluate the cytotoxicity of Fe-based biocomposite. Dulbecco's modified Eagle's medium (DMEM) containing 10% fetal bovine serum, 100 units/mL penicillin and 100 mg/mL streptomycin was used as culture medium. The as-built samples were sterilized, and then immersed in DMEM for three days to obtain the extracts. Then, the cells were incubated in a 96-well plate for 1 day using DMEM, subsequently substituted by the 100 centration extracts. After one, four, and seven days, Calcian-AM reagent was used to stain the cells for 15 min. Afterwards, the cells were captured using a fluorescence microscopy (BX60, Olympus, Tokyo, Japan). Furthermore, the cell counting kit-8 (CCK-8) reagent was added into the culture medium and continued to incubate for 3 h. Finally, the absorbance was detected by a microplate reader at 450 nm.

2.6. Statistical Analysis

In this work, the immersion tests, electrochemical experiments and cell experiments were performed three times. The data was expressed as means ± errors. The significant difference was investigated suing SPSS soft, in which *p* less than 0.05 was determined to be of significant difference.

3. Results

3.1. Microstructural Feature of LPBF-Processed Parts

The LPBF-processed bulk parts were shown in Figure 1a, and the corresponding XRD spectrum was depicted in Figure 1b. Results showed that strong peaks corresponding to α-Fe phase was observed for Fe and Fe/ZnS parts. Besides, some strong peaks corresponding to ZnS phase presented in Fe/ZnS composite. The microstructure was observed by SEM, as shown in Figure 1c. No obvious holes and cracks were observed in the matrix of as-built parts, indicating their good forming quality. For the LPBF of the metal parts, the relatively high porosity is easily generated due to the insufficient liquid phase or severe molten pool evaporation, thereby reducing the performance including mechanical properties and corrosion resistance [16]. However, our SEM analysis showed the high densification rate was obtained. It was reported that the stable molten pool behavior could be achieved under the optimized laser parameters, so as to promote the densification of the parts [17]. For the Fe/ZnS biocomposite, the ZnS particles (as marked by the red arrows) were uniformly distributed in the matrix.

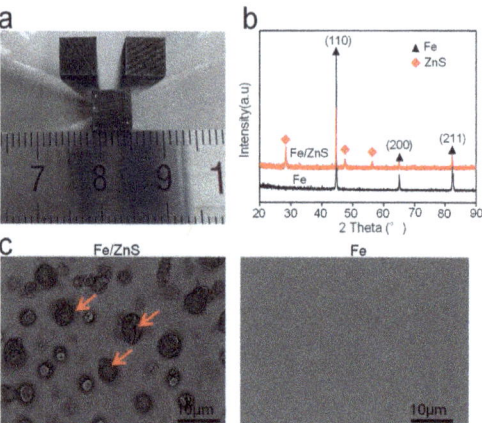

Figure 1. (**a**) LPBF processed Fe-based parts; (**b**) the XRD spectrum and (**c**) SEM for Fe and Fe/ZnS biocomposite showing the microstructure. The ZnS particles were marked by the red arrows.

3.2. Degradation Behavior

The corroded surface after immersion for 7, 14, and 28 days have bee shown in Figure 2a. Flat corrosion surface with little degradation product was observed for Fe part over the whole immersion period. As a comparison, a large amount of corrosion product was presented on the Fe/ZnS biocomposite, and a porous film with numerous corrosion pits was also observed. The corrosion pits obviously became deepened and expanded at day 28, accompanied by partial products falling off. The cross section after immersion for 28 days was examined by SEM, as exhibited in Figure 2b. Clearly, the thin corrosion film with a thickness of only ~4.8 µm was observed for Fe part. As for Fe/ZnS biocomposite, the thickness of corrosion film increased to ~23.8 µm. The element mapping analysis showed that the corrosion film mainly contained Fe and O elements, as shown in Figure 2c. Previous studies reported that the corrosion products on Fe matrix mainly contained oxides and hydroxides of Fe [18].

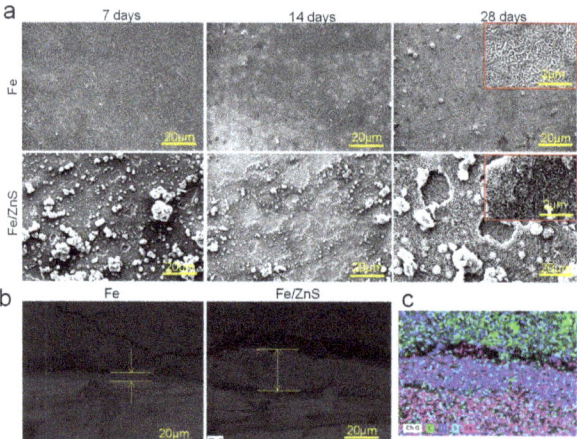

Figure 2. (**a**) SEM showing the typical corrosion surface of Fe and Fe/ZnS composite after immersion in SBF; (**b**) the cross section of corrosion production film and (**c**) the corresponding EDS mapping.

The corrosion surface after removing the corrosion product was also observed by SEM, as shown in Figure 3a. It could be seen that the corroded surface of Fe part showed small change after immersion in SBF for 7, 14, and 28 days. As a comparison, massive corrosion pits, as marked by the arrows, appeared on the matrix surface of Fe/ZnS part with the extension of immersion time. Clearly, the pits dimension gradually increased to 5–10 µm. The surface roughness after immersion for 28 days was shown in Figure 3b. For Fe part, the gradient range of surface roughness was between −1.6 and 1.0 µm. As for Fe/ZnS biocomposite, the gradient range was extended to −5.5~3.8 µm. Besides, the surface roughness profiles showed that the curve of Fe/ZnS biocomposite fluctuated sharply as compared with that of Fe, as shown in Figure 3c. It was indicated that the matrix of Fe/ZnS was severely corroded. According to the mass loss during immersion for 28 days, the degradation rates of Fe and Fe/ZnS were calculated to be ~0.05 and 0.14 mg/cm^2/year, respectively (Table 2). The significance analysis showed that the corrosion rates of Fe/ZnS was significantly higher than that of Fe ($p < 0.05$).

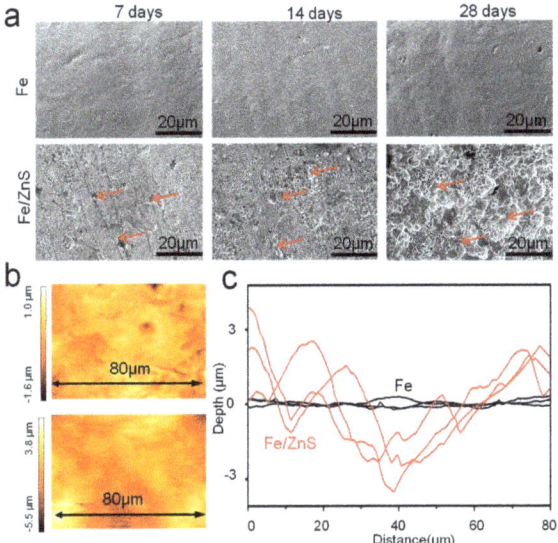

Figure 3. (a) The surface topography after removing corrosion products for Fe and Fe/ZnS composite, (b) AFM images and (c) the surface roughness profiles. The corrosion pits were marked by the red arrows.

Table 2. The corrosion rates calculated from immersion and electrochemical tests.

Samples	Cr (mg/cm^2/year)	Pi (mm/year)
Fe	0.05 ± 0.01	0.25 ± 0.02
Fe/ZnS	0.14 ± 0.03	0.72 ± 0.05

3.3. Electrochemical Behavior

The degradation mechanism of Fe/ZnS and Fe was investigated by electrochemical tests. The obtained polarization curves were shown in Figure 4a. The corrosion potential (Ecorr) and corrosion current density (Icorr) were also calculated by Tafel extrapolation method, and the result was shown in Figure 4a. The Ecorr value of Fe and Fe/ZnS composite were −0.75 V and −0.94 V, respectively. And the Icorr value of Fe/ZnS composite was significantly enhanced to 31.4 ± 0.9 µA/cm^2 as compared with that of Fe. The corrosion rate of Fe and Fe/ZnS calculated by the electrochemical parameters were ~0.25 and 0.72 mm/year, respectively, as shown in Table 2. Particularly, for Fe/ZnS composite, there was a typical pitting area in the region of anode polarization curve, as marked in

Figure 4a. It was suggested that the addition of ZnS was effectively pierced through the dense corrosion layer, and changed the corrosion type from surface corrosion to pitting corrosion. The electrochemical impedance spectra are shown in Figure 4b. The Fe/ZnS composite showed small impedance value than that of Fe, which also verified its low corrosion resistance. For the Fe part, there was only one impedance loop in the whole frequency range, indicating the formation of compact oxide film during corrosion. However, Fe/ZnS composite had a relatively small capacitive arc and impedance moduli at low frequency region, reflecting a low electron transfer resistance. Furthermore, the phase angle value and impedance value in the Bode plots were further applied to indicate the stability of corrosion films, presented in Figure 4c. For Fe/ZnS composite, the phase angle value and impedance value were smaller, which indicated the corrosion film was a loose membrane structure. It was believed that the passive film was continuously self-destroyed, thus reducing the protection efficiency for Fe/ZnS composite. The equivalent circuits of electrochemical tests for Fe and Fe/ZnS were obtained, shown in Figure 4d. There was only a semicircle in the EIS diagram, which meant only exist a corrosion layer on the Fe matrix. There were two semicircles for the Fe/ZnS composite, which indicated another transfer reaction except. Generally, it was remarked by a double-layer capacitance C_d and charge transfer resistance R_{ct}. R_{ct} and C_d were the resistance and capacitance of the passive film, respectively. As for the Fe/ZnS composite, a relatively low R_{ct} and C_d revealed its high charge transfer ability of the product layer.

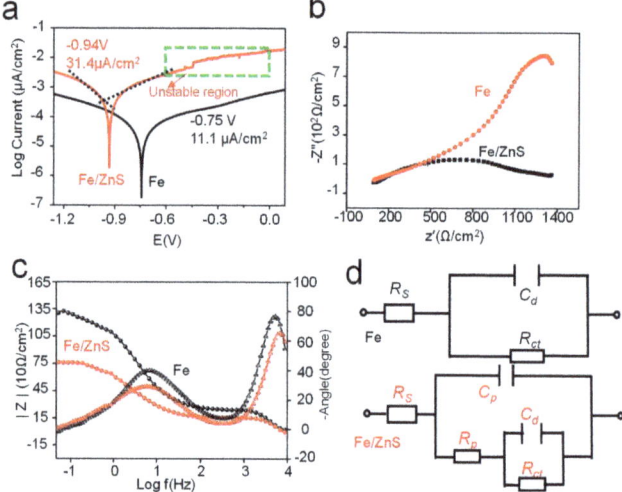

Figure 4. (a) The potentiodynamic polarization curves, (b) EIS spectra, (c) Bode plots and (d) the equivalent circuits obtained from electrochemical tests.

The cyclic voltammetry (CV) curves of Fe and Fe/ZnS samples were presented in Figure 5. The A curve was the anodic branch while C curve was the cathodic branch. Three anodic current peaks of A1, A2, and A3 were observed during the anodic scan process. The A1 peak was considered to correspond to the electro-oxidation of Fe to Fe^{2+}, which was represented the initiation of passivity formation. There was no significant difference between Fe and Fe/ZnS samples at the A1 peak. The A2 peak was characterized with the oxidation process from Fe^{2+} to Fe^{3+}, which was represented the formation process of a denser Fe_2O_3 passivation film. The A3 peak involved the transfer reaction and oxygen evolution reaction on behalf of the dissolution of Fe anode. At the A2 peak, the potential of Fe was around 0.6 V, but for Fe/ZnS samples the potential shifted to 0.7 V and caused a faster anodic reaction. With the positive shift of the potential, the anodic reaction was activated, the corrosion current increased gradually. Generally, the potential was more positive (which meant a higher anode activation energy), and thus the oxidation reaction was easier to carry out. Therefore, it was indicated that S destroyed the formation of the passivation layer and caused the continuous exposure of Fe matrix to corrosion solution, thus accelerating degradation.

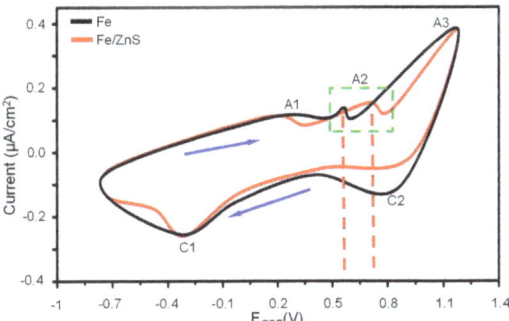

Figure 5. The cyclic voltammetry curves obtained from electrochemical tests.

3.4. In Vitro Cytotoxicity

As exhibited in Figure 6a, very few dead cells were observed during the whole incubation period. With the culture time extending, the cells were gradually increased, which proved their normal development. After seven days' culture, most the cells presented representative fusiform shape, which was known as a healthy morphology. The cell viability was quantitatively studied, as shown in Figure 6b. At day, the cell viability was relatively low, since the cells could not adapt to the new environment with high concentration of metal ions. However, with the increase of culture time, the cell activity gradually increased to 80%, which confirmed that it had acceptable biocompatibility. Moreover, there was no significant difference between the two groups.

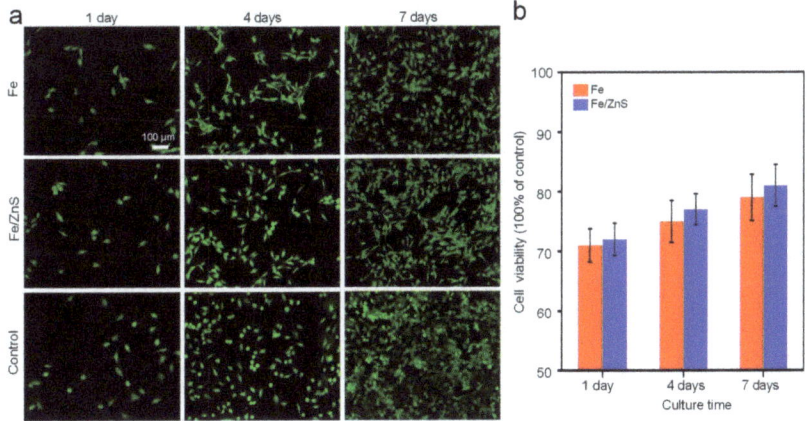

Figure 6. (**a**) The cell fluorescent images and (**b**) cell viability obtained by CCK-8 testing (n.s.).

4. Discussion

Bone implants should have a through-hole structure similar to natural bone, so as to provide a necessary microenvironment for cell growth, angiogenesis, and new bone growth after implantation [19,20]. It is generally believed that the pore size suitable for bone tissue growth ranges from 200 to 600 μm. At the same time, the porosity should reach more than 70%, which is conducive to cell adhesion, extracellular matrix deposition, oxygen, nutrition entry, and metabolite discharge [21,22]. In addition, bone implants should also have personalized and accurate shapes to improve the adaptability of implants in the process of operation and the effect of postoperative treatment [23,24].

LPBF as a typical additive manufacturing technology is especially suitable for the high-precision and high-efficiency manufacturing of personalized porous implants [25,26]. Specifically, we can use digital medical technology to conduct three-dimensional scanning of the bone defect, and then design the bone defect model through computer-aided technology. Finally, we can manufacture customized porous implants through LPBF technology, as shown in Figure 7. In light of this, the additive manufacturing technology represented by LPBF has carried out an upsurge of applied research in medical fields such as orthopedics, dentistry, and cardiovascular stents [27–29].

Figure 7. Laser additively manufactured Fe/ZnS composite scaffold with porous structure.

On the other hand, bone implants also demand suitable degradation rate to match the growth rate of new bone tissue [30]. In the present study, ZnS was introduced into the Fe matrix to accelerate degradation. Both the immersion tests and electrochemical tests proved that the Fe/ZnS showed an enhanced degradation rate, since the ZnS promoted the collapse of the passive film. In fact, the collapse of the passive film was closely related with its electronic properties. Herein, the typical Mott-Schottky plots were used to characterize the electronic property of the passive film, as shown in Figure 8a. The positive slopes of the two curves indicated that the passive film exhibited n-type semiconductor behavior. It was attributed to the present of minority carriers in the corrosion layer, and its defects were composed of oxygen vacancies. The variation range of passive film defect density with polarization potential has been shown in Figure 8b. It was obvious that the donor density in the corrosion layer increased with the increase of polarization voltage, thus resulting in the increase of vacancy accumulation between the Fe matrix and passive film interface. Therefore, the corrosion pits were formed in the Fe matrix, as verified by the corrosion surface and electrochemistry experiments. The calculated oxygen vacancy density of the passive film also demonstrated these results, as shown in Figure 8c. The electronic conductivity of the general corrosion product film was related with the defect concentration, according to other research result. The defect concentration for Fe/ZnS composite could enhance the electron transmission of the passive film, thereby resulting in the formation of corrosion pits on the surface. In this condition, the protective effect of the passive film was gradually weakened, which further accelerated the damage of passive film and the corrosion of the Fe matrix.

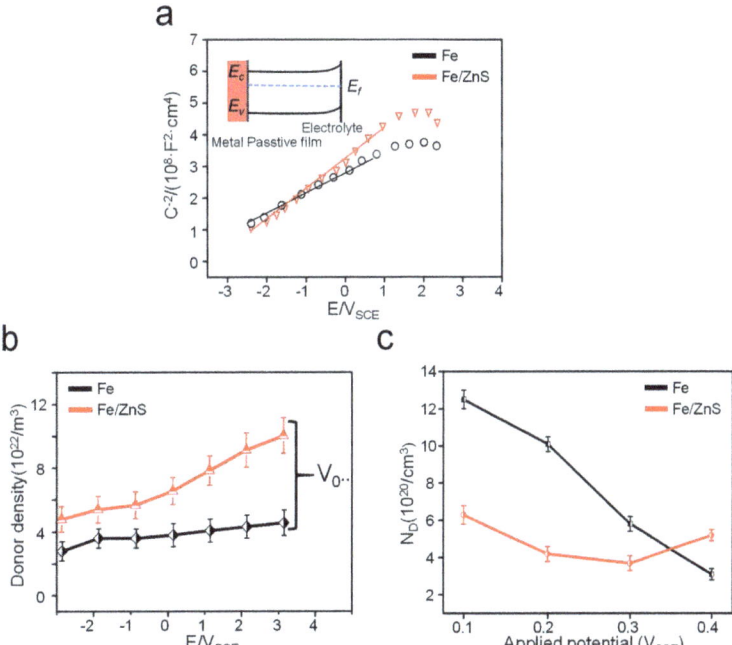

Figure 8. (a) Mott-Schottky results of Fe and Fe/ZnS composite in SBF solution, (b) the oxygen vacancy density variation of passive film, and (c) the calculated defects density of the passive film.

In the present work, the incorporation of ZnS changed the properties of the passive film. It was believed that the added ZnS would undergo a disproportionation reaction in SBF, and formed S-containing species such as S^{2-}, HS^-, and $S_2O_3^{2-}$ ions [31]. The catalysis of S^{2-} absorbed on Fe matrix through the anodic reaction led to an increase in anodic dissolution kinetics, resulting in a higher maximum current density. With a positive shift of the anode potential, S-containing species oxidized from low to high valence, resulting in the formation of adsorbed S elemental and thiosulfate ion. Previous studies demonstrated that S-containing species had a detrimental effect on Fe matrix. Marcus et al. suggest that the adsorbed S weakened the metal-metal bond, resulting in lower activation energy for the dissolution of surface metal atoms [32]. In addition, the adsorbed S might hinder or delayed passivation, as it hindered the available sites for hydroxyl ion adsorption, a precursor for passive film formation [7]. Furthermore, the formation of S-containing phases led to localized acidification, which also contributed to the degradation of the Fe matrix [33]. As our electrochemical tests proved, Fe/ZnS showed a significantly enhanced corrosion current density, which was almost five times that of the Fe part.

Generally, bone implants not only require a porous structure and appropriate degradation rate but also need good biocompatibility [34,35]. Both Fe and Zn were the trace elements of human body, and possessed good biocompatibility. As our cell testing proved, the cell viabilities of Fe/ZnS of Fe extracts were higher than 80% at day seven, which was acceptable as bone implant. It was worth noting that Fe/ZnS had a higher degradation rate, that was, a higher ion concentration for Fe/ZnS group. However, the cellular activity for Fe/ZnS group was still higher than that of the Fe group. This may be due to the fact that the released Zn ions exerted a positive role. Zn plays a significant role in the formation, development, mineralization, and maintenance of healthy bones [36,37]. Thus, it was expected that Zn ions released from the Fe/ZnS composites could promote cell growth and proliferation.

5. Conclusions

In the present work, ZnS was incorporated into Fe based implants to improve their degradation behavior. An Fe based biocomposite was fabricated by LPBF. The Mott-Schottky test analysis indicated that S destroyed the formation of the passivation layer and caused the continuous exposure of Fe matrix to corrosion solution, thus accelerating the degradation rate. After immersion in SBF for 28 days, heavy corrosion product and porous film with numerous corrosion pits presented on the Fe/ZnS composite, which revealed that it undergone severe corrosion. Besides, the incorporated ZnS had no significant effect on the biocompatibility for Fe based implants. All of the results showed that the Fe/ZnS biocomposite was a good choice for use as a bone repair material.

Author Contributions: Conceptualization, D.W. and Y.Y.; investigation, Y.Z. and L.X.; resources, J.W.; writing—original draft preparation, Y.Z.; writing—review and editing, Y.Y. and L.S. All authors have read and agreed to the published version of the manuscript.

Funding: This research was funded by: (1) Anhui Provincial Natural Science Foundation (2008085ME149); (2) Anhui Provincial Top Academic Aid Program for Discipline (Major) Talents of Higher Education Institutions (gxbjZD2020087); (3) Anhui Provincial Natural Science Research Key Program of Higher Education Institutions (KJ2021A1052, KJ2021A1062); (4) Academic Leader and Backup Candidate Research Project of Tongling University (2020tlxyxs02, 2020tlxyxs04); (5) Key Laboratory of Construction Hydraulic Robots of Anhui Higher Education Institutes, Tongling University (TLXYCHR-O-21YB01); (6) Jiangxi Key Laboratory of Forming and Joining Technology for Aerospace Components, Nanchang Hangkong University (EL202180264); (7) Jiangsu Key Laboratory of Precision and Micro-Manufacturing Technology.

Conflicts of Interest: The authors declare no conflict of interest.

References

1. Li, C.M.; Guo, C.C.; Fitzpatrick, V.; Ibrahim, A.; Zwierstra, M.J.; Hanna, P.; Lechtig, A.; Nazarian, A.; Lin, S.J.; Kaplan, D.L. Design of biodegradable, implantable devices towards clinical translation. *Nat. Rev. Mater.* **2020**, *5*, 61–81. [CrossRef]
2. Zhang, T.; Liu, C.-T. Design of titanium alloys by additive manufacturing: A critical review. *Adv. Powder Mater.* **2022**, *1*, 100014. [CrossRef]
3. Zheng, Y.F.; Gu, X.N.; Witte, F. Biodegradable metals. *Mater. Sci. Eng. R Rep.* **2014**, *77*, 1–34. [CrossRef]
4. Han, H.S.; Loffredo, S.; Jun, I.; Edwards, J.; Kim, Y.C.; Seok, H.K.; Witte, F.; Mantovani, D.; Glyn-Jones, S. Current status and outlook on the clinical translation of biodegradable metals. *Mater. Today* **2019**, *23*, 57–71. [CrossRef]
5. Cheng, J.; Liu, B.; Wu, Y.H.; Zheng, Y.F. Comparative in vitro Study on Pure Metals (Fe, Mn, Mg, Zn and W) as Biodegradable Metals. *J. Mater. Sci. Technol.* **2013**, *29*, 619–627. [CrossRef]
6. Kabir, H.; Munir, K.; Wen, C.; Li, Y. Recent research and progress of biodegradable zinc alloys and composites for biomedical applications: Biomechanical and biocorrosion perspectives. *Bioact. Mater.* **2021**, *6*, 836–879. [CrossRef]
7. Xia, D.; Song, S.; Zhu, R.; Behnamian, Y.; Shen, C.; Wang, J.; Luo, J.; Lu, Y.; Klimas, S. A mechanistic study on thiosulfate-enhanced passivity degradation of Alloy 800 in chloride solutions. *Electrochim. Acta* **2013**, *111*, 510–525. [CrossRef]
8. Gao, S.J.; Brown, B.; Young, D.; Nesic, S.; Singer, M. Formation Mechanisms of Iron Oxide and Iron Sulfide at High Temperature in Aqueous H2S Corrosion Environment. *J. Electrochem. Soc.* **2018**, *165*, C171–C179. [CrossRef]
9. Du, J.M.; Tang, Z.Y.; Li, G.; Yang, H.; Li, L. Key inhibitory mechanism of external chloride ions on concrete sulfate attack. *Constr. Build. Mater.* **2019**, *225*, 611–619. [CrossRef]
10. Yang, Y.J.; Lan, J.F.; Xu, Z.G.; Chen, T.; Zhao, T.; Cheng, T.; Shen, J.; Lv, S.; Zhang, H. Toxicity and biodistribution of aqueous synthesized ZnS and ZnO quantum dots in mice. *Nanotoxicology* **2014**, *8*, 107–116. [CrossRef]
11. Zhang, F.F.; Li, C.X.; Li, X.H.; Wang, X.; Wan, Q.; Xian, Y.; Jin, L.; Yamamoto, K. ZnS quantum dots derived a reagentless uric acid biosensor. *Talanta* **2006**, *68*, 1353–1358. [CrossRef] [PubMed]
12. Li, B.; Zhang, X.T.; Wang, T.L.; He, Z.; Lu, B.; Liang, S.; Zhou, J. Interfacial Engineering Strategy for High-Performance Zn Metal Anodes. *Nano-Micro Lett.* **2022**, *14*, 6. [CrossRef] [PubMed]
13. Zhou, L.; Yang, Q.; Zhang, G.; Fangxin, Z.; Gang, S.; Bo, Y. Additive manufacturing technologies of porous metal implants. *China Foundry* **2014**, *11*, 322–331.
14. Carluccio, D.; Bermingham, M.; Kent, D.; Demir, A.G.; Previtali, B.; Dargusch, M.S. Comparative Study of Pure Iron Manufactured by Selective Laser Melting, Laser Metal Deposition, and Casting Processes. *Adv. Eng. Mater.* **2019**, *21*, 1900049. [CrossRef]
15. Maher, S.; Wijenayaka, A.R.; Lima-Marques, L.; Yang, D.; Atkins, G.J.; Losic, D. Advancing of Additive-Manufactured Titanium Implants with Bioinspired Micro- to Nanotopographies. *Acs Biomater. Sci. Eng.* **2021**, *7*, 441–450. [CrossRef]
16. Gu, D.D.; Meiners, W.; Wissenbach, K.; Poprawe, R. Laser additive manufacturing of metallic components: Materials, processes and mechanisms. *Int. Mater. Rev.* **2012**, *57*, 133–164. [CrossRef]

17. Zhang, H.; Gu, D.; Dai, D.; Ma, C.; Li, Y.; Peng, R.; Li, S.; Liu, G.; Yang, B. Influence of scanning strategy and parameter on microstructural feature, residual stress and performance of Sc and Zr modified Al–Mg alloy produced by selective laser melting. *Mater. Sci. Eng. A* **2020**, *788*, 139593. [CrossRef]
18. Cui, L.-Y.; Gao, S.-D.; Li, P.-P.; Zeng, R.C.; Zhang, F.; Li, S.Q.; Han, E.H. Corrosion resistance of a self-healing micro-arc oxidation/polymethyltrimethoxysilane composite coating on magnesium alloy AZ31. *Corros. Sci.* **2017**, *118*, 84–95. [CrossRef]
19. Guaglione, F.; Caprio, L.; Previtali, B.; Demir, A.G. Single point exposure LPBF for the production of biodegradable Zn-alloy lattice structures. *Addit. Manuf.* **2021**, *48*, 102426. [CrossRef]
20. Wang, X.J.; Xu, S.Q.; Zhou, S.W.; Xu, W.; Leary, M.; Choong, P.; Qian, M.; Brandt, M.; Xie, Y.M. Topological design and additive manufacturing of porous metals for bone scaffolds and orthopaedic implants: A review. *Biomaterials* **2016**, *83*, 127–141. [CrossRef]
21. Wu, S.L.; Liu, X.M.; Yeung, K.W.K.; Liu, C.; Yang, X. Biomimetic porous scaffolds for bone tissue engineering. *Mater. Sci. Eng. R Rep.* **2014**, *80*, 1–36. [CrossRef]
22. Ma, K.W.; Zhao, T.Z.; Yang, L.F.; Wang, P.; Jin, J.; Teng, H.; Xia, D.; Zhu, L.; Li, L.; Jiang, Q.; et al. Application of robotic-assisted in situ 3D printing in cartilage regeneration with HAMA hydrogel: An in vivo study. *J. Adv. Res.* **2020**, *23*, 123–132. [CrossRef] [PubMed]
23. Agarwal, S.; Curtin, J.; Duffy, B.; Jaiswal, S. Biodegradable magnesium alloys for orthopaedic applications: A review on corrosion, biocompatibility and surface modifications. *Mater. Sci. Eng. C* **2016**, *68*, 948–963. [CrossRef]
24. Li, L.; Shi, J.P.; Ma, K.W.; Jin, J.; Wang, P.; Liang, H.; Cao, Y.; Wang, X.; Jiang, Q. Robotic in situ 3D bio-printing technology for repairing large segmental bone defects. *J. Adv. Res.* **2021**, *30*, 75–84. [CrossRef] [PubMed]
25. Han, C.; Babicheva, R.; Chua, J.D.Q.; Ramamurty, U.; Tor, S.B.; Sun, C.N.; Zhou, K. Microstructure and mechanical properties of (TiB + TiC)/Ti composites fabricated in situ via selective laser melting of Ti and B_4C powders. *Addit. Manuf.* **2020**, *36*, 101466. [CrossRef]
26. Wang, D.; Liu, L.; Deng, G.; Deng, C.; Bai, Y.; Yang, Y.; Wu, W.; Chen, J.; Liu, Y.; Wang, Y.; et al. Recent progress on additive manufacturing of multi-material structures with laser powder bed fusion. *Virtual Phys. Prototyp.* **2022**, *17*, 329–365. [CrossRef]
27. Li, Y.; Zhou, J.; Pavanram, P.; Leeflang, M.A.; Fockaert, L.I.; Pouran, B.; Tümer, N.; Schröder, K.U.; Mol, J.M.; Weinans, H.; et al. Additively manufactured biodegradable porous magnesium. *Acta Biomater.* **2018**, *67*, 378–392. [CrossRef]
28. Rezwan, K.; Chen, Q.Z.; Blaker, J.J.; Boccaccini, A.R. Biodegradable and bioactive porous polymer/inorganic composite scaffolds for bone tissue engineering. *Biomaterials* **2006**, *27*, 3413–3431. [CrossRef]
29. Li, H.; Li, Z.; Li, N.; Zhu, X.; Zhang, Y.F.; Sun, L.; Wang, R.; Zhang, J.; Yang, Z.; Yi, H.; et al. 3D Printed High Performance Silver Mesh for Transparent Glass Heaters through Liquid Sacrificial Substrate Electric-Field-Driven Jet. *Small* **2022**, *18*, 2107811. [CrossRef]
30. Donik, C.; Kraner, J.; Kocijan, A.; Paulin, I.; Godec, M. Evolution of the epsilon and gamma phases in biodegradable Fe-Mn alloys produced using laser powder-bed fusion. *Sci. Rep.* **2021**, *11*, 19506. [CrossRef]
31. Zhan, W.Q.; Yuan, Y.; Yang, B.Q.; Jia, F.; Song, S. Construction of MoS 2 nano-heterojunction via ZnS doping for enhancing in-situ photocatalytic reduction of gold thiosulfate complex. *Chem. Eng. J.* **2020**, *394*, 124866. [CrossRef]
32. Marcus, P. Surface science approach of corrosion phenomena. *Electrochim. Acta* **1998**, *43*, 109–118. [CrossRef]
33. Li, L.; Yan, J.; Xiao, J.; Sun, L.; Fan, H.; Wang, J. A comparative study of corrosion behavior of S-phase with AISI 304 austenitic stainless steel in H_2S/CO_2/Cl- media. *Corros. Sci.* **2021**, *187*, 109472. [CrossRef]
34. Yang, Y.; Cheng, Y.; Yang, M.; Qian, G.; Peng, S.; Qi, F.; Shuai, C. Semicoherent strengthens graphene/zinc scaffolds. *Mater. Today Nano* **2022**, *17*, 100163. [CrossRef]
35. Jiao, C.; Xie, D.; He, Z.; Liang, H.; Shen, L.; Yang, Y.; Tian, Z.; Wu, G.; Wang, C. Additive manufacturing of Bio-inspired ceramic bone Scaffolds: Structural Design, mechanical properties and biocompatibility. *Mater. Des.* **2022**, *217*, 110610. [CrossRef]
36. Yamaguchi, M. Role of nutritional zinc in the prevention of osteoporosis. *Mol. Cell. Biochem.* **2010**, *338*, 241–254. [CrossRef]
37. Yang, Y.; Wang, W.; Yang, M.; Yang, Y.; Wang, D.; Liu, Z.; Shuai, C. Laser-Sintered Mg-Zn Supersaturated Solid Solution with High Corrosion Resistance. *Micromachines* **2021**, *12*, 1368. [CrossRef]

Article

Effect of Porosity on Dynamic Response of Additive Manufacturing Ti-6Al-4V Alloys

Yihang Cui [1], Jiacheng Cai [2], Zhiguo Li [1], Zhenyu Jiao [3], Ling Hu [1] and Jianbo Hu [1,3,*]

1. Laboratory for Shock Wave and Detonation Physics, Institute of Fluid Physics, China Academy of Engineering Physics, Mianyang 621900, China; cuiyihang19@gscaep.ac.cn (Y.C.); zhiguo_li@foxmail.com (Z.L.); huling@ustc.edu (L.H.)
2. School of National Defense Science and Technology, Southwest University of Science and Technology, Mianyang 621010, China; cjc15508005281@163.com
3. State Key Laboratory for Environmentally Friendly Energy Materials, Southwest University of Science and Technology, Mianyang 621010, China; zhenyujiao1996@163.com
* Correspondence: jianbo.hu@caep.cn

Abstract: Additive manufacturing is a rapidly developing manufacturing technology of great potential for applications. One of the merits of AM is that the microstructure of manufactured materials can be actively controlled to meet engineering requirements. In this work, three types of Ti-6Al-4V (TC4) materials with different porosities are manufactured using selective laser melting using different printing parameters. Their dynamic behaviors are then studied by planar impact experiments based on the free-surface velocity measurements and shock-recovery characterizations. Experimental results indicate that the porosity significantly affects their dynamic response, including not only the yield, but also spall behaviors. With the increasing porosity, the Hugoniot elastic limit and spall strength decrease monotonically. In the case of TC4 of a large porosity, it behaves similar to energy-absorbing materials, in which the voids collapse under shock compression and then the spallation takes place.

Keywords: additive manufacturing; Ti-6Al-4V; dynamic behaviors; porosity; spall

1. Introduction

Additive manufacturing (AM) technology has a series of technical advantages such as rapid prototyping, free manufacturing, and high material utilization, thus showing great development potentials and broad application prospects in automation industry [1,2], aerospace [3,4], shipbuilding, biotechnology [5], automobile [6], parts processing [7], and other fields [8,9]. Several critical reviews regarding AM technological innovation processes have been published [10–13]. Recently, AM technology has shown one advantage which is even more attractive—that is, it can actively manipulate the microstructure of materials by adjusting printing parameters and strategies, such that some unique and excellent mechanical properties could be designed for specific applications. For example, AM components with special microstructures such as honeycomb and gradient can significantly improve heat transfer and anti-collision performance [9,14].

In many engineering cases, the mechanical properties of materials under dynamic loading are of importance and have been approved to be sensitive to the internal microstructure [15–17]. Therefore, there is an urgent demand to employ AM technology to control the microstructures, and then to manipulate the macroscopic dynamic properties of materials. By far, only a few works are available to study the dynamic response of AM materials, especially on the effect of porosity produced during printing. For instance, Valdez et al. induced various levels of porosity in Super Alloy 718 by modifying the powder bed fusion process including laser power, scan velocity, and hatch spacing, and then investigated their dynamic behaviors [18,19]. Their results show that, in the presence of high porosity, porous materials perform much like open-cell foams and are highly sensitive to densification.

Branch et al. demonstrated shock wave modulation or "spatially graded-flow" in shock wave experiments via controlling AM techniques at the micron scale by using time-resolved X-ray imaging [20]. Gangireddy et al. discovered that porous sandwich Ti-6Al-4V (TC4) AM samples exhibited greater energy absorption per unit volume than fully dense samples using split-Hopkinson Pressure Bar (SHPB) testing [21]. However, there is still a lack of systematic understanding of the effect of porosity on the dynamic mechanical properties of AM materials.

In order to better understand the effect of porosity on the dynamic response, in this work, we prepare three TC4 specimens of different porosities by using different printing parameters in the Selective Laser Melting (SLM) process and then carry out a series of planar impact experiments to investigate their dynamic behaviors. Results demonstrate that dynamic mechanical properties, including the Hugoniot elastic limit and spall strength, are significantly affected by the porosity. A small difference in porosity could remarkably reduce the dynamic properties. Depending on the porosity, the materials may present a good energy absorption capability.

2. Material Manufacturing and Characterization

TC4 specimens with different porosities are produced by the SLM process. The particle size of the used powder is 15~45 μm, which was measured by scanning electron microscope (SEM) as shown in Figure 1. The chemical composition of TC4 powder and workpiece is listed in Table 1. The porosity is controlled by adjusting the printing parameters, including scan velocity, laser power, and hatch spacing, as given in Table 2. The printing parameters are chosen based on the Refs. [22–25]. The stack of layers is along the Z-axis with the layer thickness of 30 μm and the scanning strategy of 45° rotation between layers.

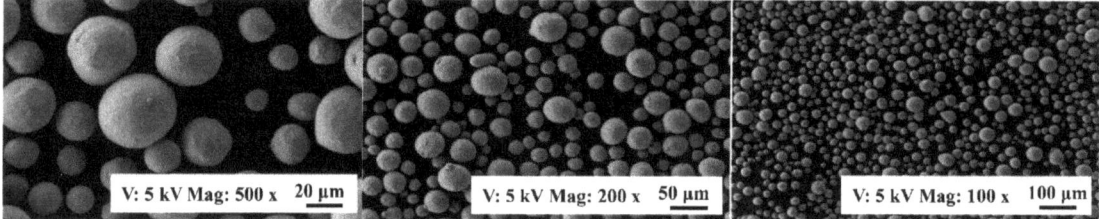

Figure 1. The SEM image of the TC4 powder.

Table 1. The chemical composition of TC4 powder and workpiece.

Condition	Element	Al (wt.%)	V (wt.%)	O (wt.%)	N (wt.%)	C (wt.%)	H (wt.%)	Fe (wt.%)	Ti (wt.%)
Powder		5.5~6.75	3.5~4.5	<0.2	<0.05	<0.08	<0.015	<0.3	Balance
Workpiece		5.98	4.11	0.12	0.022	0.013	0.010	0.029	Balance

Table 2. Summary of SLM parameters and sample's porosities and physical parameters. C_l, C_s, C_b represent the longitudinal, shear, and bulk acoustic velocities, respectively.

Sample	Power (W)	Scan Velocity (mm/s)	Track Width (mm)	C_l (km/s)	C_s (km/s)	C_b (km/s)	Density (g/cm^3)	Porosities
T1	370	1000	0.10	6.36	3.20	5.18	4.422	0.29%
T2	280	1400	0.14	6.19	3.13	5.02	4.396	0.88%
T3	200	500	0.10	5.85	3.06	4.67	4.195	5.41%

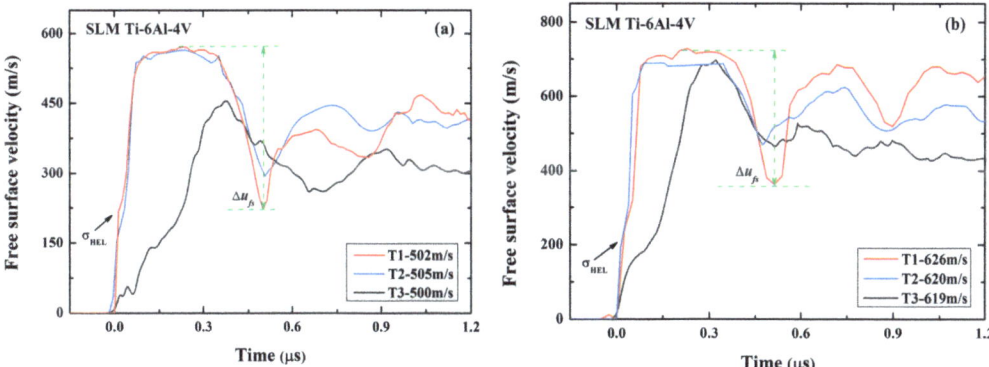

Figure 5. Free-surface particle velocity profiles of three SLM TC4 samples obtained in planar plate experiments at the impact velocity of (**a**) 500 m/s and (**b**) 620 m/s.

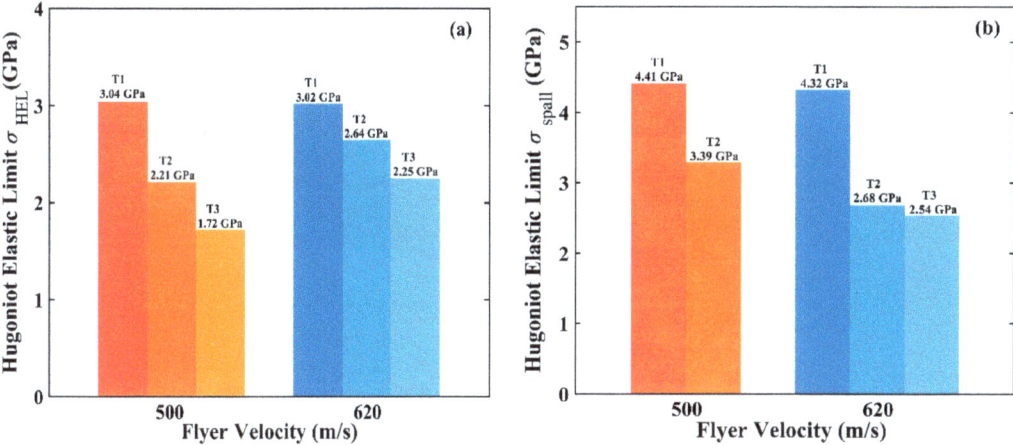

Figure 6. Hugoniot elastic limit (**a**) and spall strength (**b**) of three different TC4 materials produced by SLM. Sample number and loading velocities are indicated on each bar.

We can also calculate the spall strength (σ_{spall}) based on the observed 'pullback' signal in the free-surface velocity profiles, which is caused by the spallation damage when two rarefaction waves originated from the flyer's rear free surface and the sample's free surface, by

$$\sigma_{spall} = \rho_0 C_L \Delta u_{fs} \frac{1}{1 + \frac{C_L}{C_0}}, \quad (3)$$

where Δu_{fs} is the difference of the free surface velocity from the peak value to the first minima, as indicated in Figure 5. The calculated results are shown in Figure 6b. This demonstrates that, although the porosity difference between the two samples is only 0.59%, the difference in the spall strength is more than 25%. Therefore, the porosity produced during printing significantly affects the dynamic response, including both the yield and spall behaviors. Compared with wrought TC4 (5.28 GPa) [22], the spall strength of SLM TC4 is much smaller. The degradation in the spall strength might be due to the combined contrition of the pores and microstructure. It is well known that the microstructures of SLM and wrought materials could be remarkably different. In this work, however, it is impossible to identify which is playing the dominant role.

500 m/s and 620 m/s, respectively. A Doppler Pin System (DPS) was used to probe the free-surface particle velocity [31–33]. Figure 4b shows the configuration for shock-recovery experiments. To prevent possible secondary damage during shock recovery, a stainless-steel recovery cabin filled with low-density vacuum-sealing putty was used to capture the target samples after dynamic tensile.

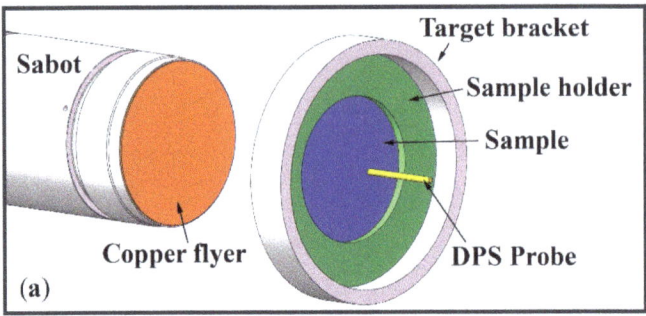

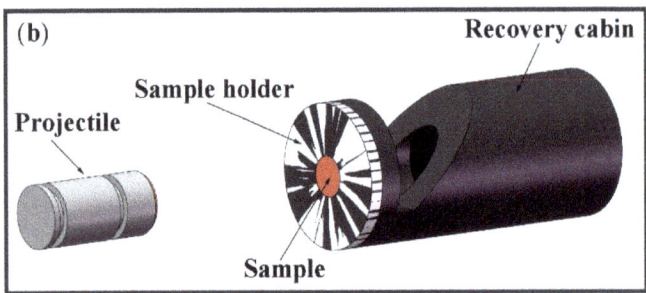

Figure 4. Experimental configurations for (**a**) free-surface particle velocity measurements and (**b**) shock-recovery characterizations. A Cu flyer is launched by a single-stage gas gun to produce spall damage in TC4 samples. DPS is used to monitor the free-surface velocity profile. A recovery cabin is used to capture the target sample for shock-recovery characterizations.

3.2. Results and Discussion

Figure 5 presents the free-surface particle velocity profiles for all the samples and at two different impact speeds. For samples T1 and T2, the velocity profiles show an obvious elastic-plastic transition during loading. The Hugoniot elastic limit (σ_{HEL}) is, thus, calculated by

$$\sigma_{HEL} = \frac{1}{2}\rho_0 C_L u_{HEL}, \tag{2}$$

where σ_{HEL} is the velocity at the transition point. It is clear in Figure 6a that σ_{HEL}, for the sample T1, is bigger than that of T2, suggesting that an increase in porosity leads to a decrease in σ_{HEL}. The difference in σ_{HEL} between the two samples is about 0.38~0.83 GPa (that is, 13%~27%), depending on the impact velocity.

The acoustic velocities (longitudinal wave velocity, C_L, and shear wave velocity, C_s) of each sample are determined by using the pulse-echo method, respectively. Then, the bulk sound velocity, C_0, is calculated by

$$C_0 = \sqrt{C_L^2 - \frac{4}{3}C_s^2} \tag{1}$$

All these results are listed in Table 2.

Figure 3 presents the sample textures. It shows that the samples represent a stronger texture in the {100} orientation than that of {110} and {111} orientations. The maximum texture intensities of T1, T2, and T3 samples observed in the {100} orientation are 42.85, 32.13, and 59.18, respectively, indicating that T3 has the most concentrated texture. This is because the change in printing parameters leads to different cooling rates and grain growth mechanisms, resulting in different grain orientations and texture intensities [30].

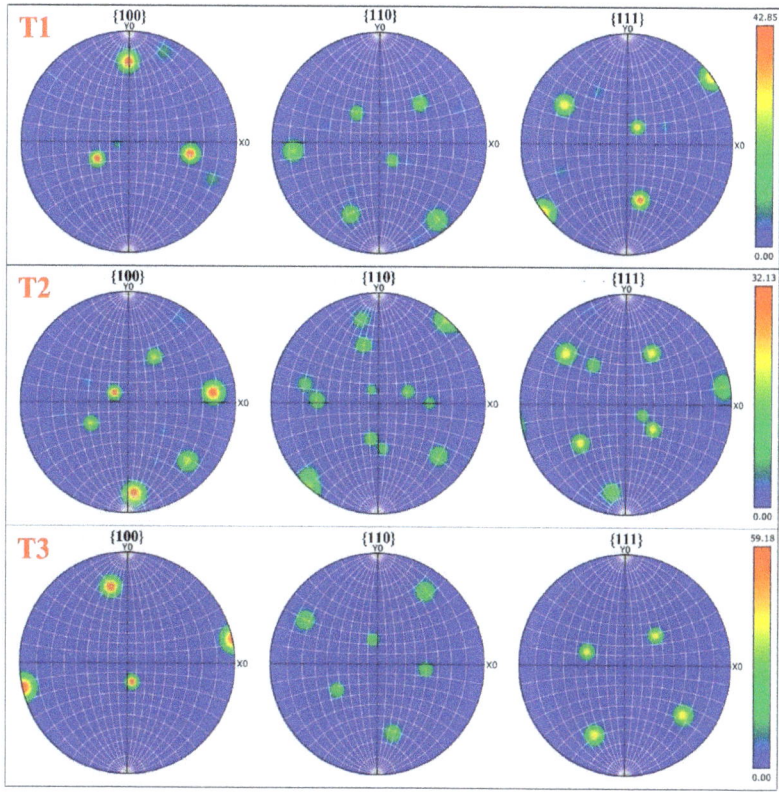

Figure 3. Pole figure maps of SLM Ti-6Al-4V materials studied.

3. Shock-Wave Experiments and Results

3.1. Planar Impact Experiments

Planar impact experiments were performed on a single-stage gas gun with a caliber of 14 mm. Figure 4a schematically shows the experiment configuration for measuring the free-surface particle velocity history. SLM TC4 samples were wire-cut to 1.8 mm in thickness and 12 mm in diameter for experiments to make sure that the shock compression was along the z-axis. The oxygen-free copper flyer plate was 0.9 mm in thickness to drive tensile damage in the center of the TC4 sample. The impact speeds of the flyer impact are

To quantify the porosity of the samples, the micro-morphology of three samples is analyzed by optical microscopy (OM), as shown in the left column of Figure 2. It is obvious that the porosity of the three samples increases from T1 to T3. The average pore sizes of T1, T2, and T3 samples are 1.8 µm, 3.2 µm, and 81.5 µm, respectively. Using the drainage method, we determine the density (ρ_0) of each sample and then calculate the porosity according to (1-ρ_0/ρ_d), where ρ_d is the density of dense wrought TC4. As given in Table 2, the samples T1-T3 have the porosity of 0.29%, 0.88%, and 5.41%, respectively.

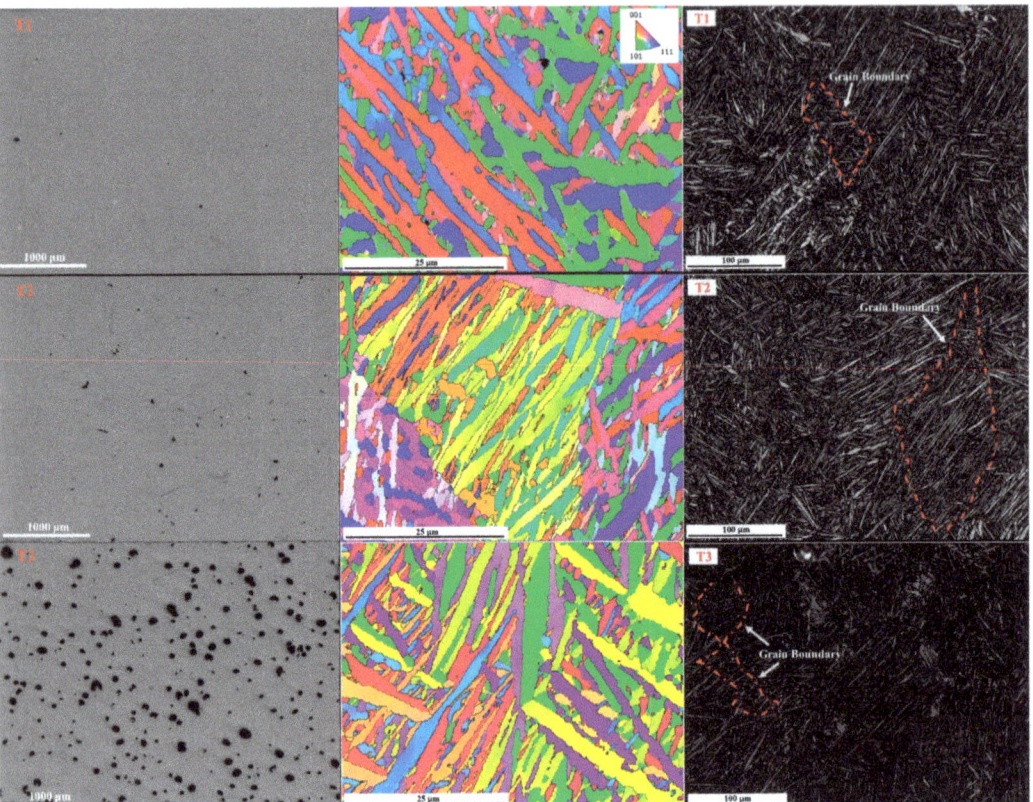

Figure 2. Optical microscopy (**left**), EBSD (**middle**), and band-contrast images (**right**) of the SLM Ti-6Al-4V materials studied. Top view: sample T1 of porosity 0.29%; middle view: sample T2 of porosity 0.88%; bottom view: sample T3 of porosity 5.41%.

The inverse pole-figure obtained via electron backscatter detection (EBSD), shown in the middle column of Figure 2, presents needle-shaped textures, indicating the existence of acicular α martensite [26,27]. The size of acicular α martensite for all three samples is almost the same, with an average grain width of 1.9–2.0 µm and a length of 10–100 µm. Further, acicular α martensite forms complex β columnar grains which appear along the building direction with the size of 100s µm. Continuous grain boundaries are observed in three band-contrast images, as marked by the red dotted line shown in the right column of Figure 2 [28,29].

For sample T3, the free-surface velocity profile is extremely different from that of samples T1 and T2. The free-surface velocity increases gradually and no shock wave is formed. This is due to the compaction of numerous voids under dynamic compression. Note that, at the impact velocity of 500 m/s, the peak free-surface velocity is remarkably lower than that of the samples T1 and T2, while at 620 m/s the peak velocity is almost the same as the ones in samples T1 and T2. It thus suggests that the voids in the sample T3 are completely compacted and the sample becomes dense at higher shock pressure.

The Hugoniot elastic limit and spall strength of the sample T3, as shown in Figure 6, are calculated by using Equations (2) and (3), respectively. Results demonstrate that, as expected, both σ_{HEL} and σ_{spall} of the sample T3 are further lowered due to the large porosity. Based on the above comparison, it is concluded that the increase in porosity markedly reduces the Hugoniot elastic limit and spall strength of SLM TC4 materials, thus degrading their mechanical performance.

Jones et al. have also investigated the spall strength of SLM TC4 [22]. In the case of the loading direction parallel to the printing one which is the same as us, SLM TC4 samples had the spall strengths of 3.03 GPa and 3.34 GPa at the impact velocities of 310 m/s and 415 m/s, respectively. In comparison with our result, we speculate that SLM TC4 materials used in Ref. [22] may have a porosity of more than 0.5%.

To further understand the dynamic behaviors of shocked materials, shock-recovered samples were characterized by using OM, as shown in Figure 7. At the impact velocity of 500 m/s, no cracks are visible in sample T1, while in the samples T2 and T3 there exist obvious isolated cracks. Some cracks have rounded tips, suggesting the coalescence of voids [34,35]. Such an observation is somehow inconsistent with the observed velocity profiles in which the spallation occurs in the samples T1 and T2. We attribute this inconsistency to the incomplete spallation in sample T1 because it has a higher resistance to dynamic tension than the samples T2 and T3 as indexed by the spall strength. In the sample T3 recovered, the collapse of voids is clearly observed, providing direct evidence for the shock-driven compaction. At the impact velocity of 620 m/s, complete spallation takes place in all three samples, thus obvious spallation damages are visible. In the sample T1, a large number of small cracks with a length of less than 50 μm locate at the zone with a width of 270 μm. While in the samples T2 and T3, the damage region expands significantly, suggesting that cracks originate from widely distributed voids.

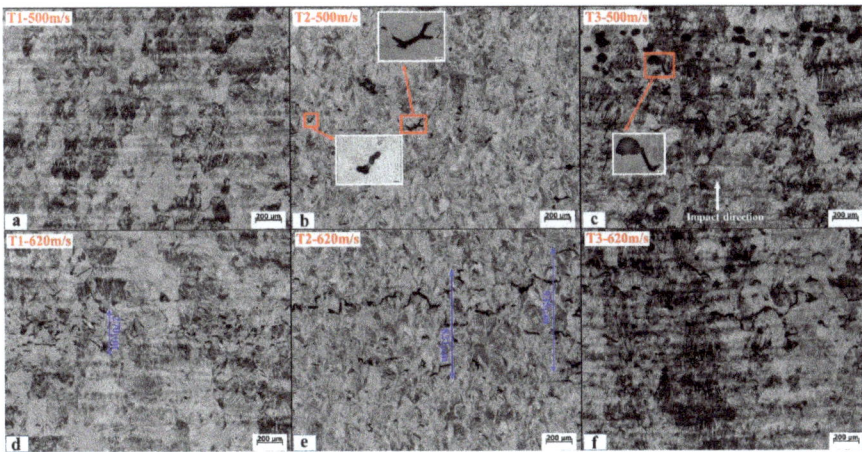

Figure 7. Optical metallography of three SLM TC4 samples shock-recovered at the impact velocity of (**a**–**c**) 500 m/s and (**d**–**f**) 620 m/s.

4. Conclusions

We have printed TC4 materials with different porosities by designing the printer parameters and experimentally investigated the effect of porosity on the dynamic response of SLM TC4 materials. Experimental results indicate that:

- SLM TC4 with different porosities show observably different dynamic characteristics in the free-surface particle velocity profiles. The increase in porosity significantly degrades the dynamic mechanical properties, including the Hugoniot elastic limit and the spall strength.
- Dense samples show better tensile-resistant properties than porous samples, while porous samples show a good energy absorption capability than dense samples.
- Shock-recovery characterizations indicate that, in porous samples, void collapse and energy absorption occur in the impact stage and cracks mainly originate from widely distributed voids.

Therefore, materials with various dynamic properties can be produced by selecting certain printing parameters. This work may guide AM technology to design materials for certain applications under extreme conditions.

Author Contributions: Conceptualization, J.H.; methodology, Z.L.; validation, Y.C., Z.J. and J.C.; formal analysis, Z.L., L.H. and Y.C.; investigation, Y.C.; resources, J.H.; writing—original draft preparation, Y.C.; writing—review and editing, Z.L. and J.H.; project administration, J.H. and Z.L.; funding acquisition, J.H. and Z.L. All authors have read and agreed to the published version of the manuscript.

Funding: This research was funded by the National Natural Science Foundation of China grant number No. 12072331 And the Science Challenge Project grant number No. TZ2018001. Z.L. thanks the support of the project of Key Laboratory of Impact and Safety Engineering (Ningbo University), Ministry of Education (CJ202007).

Acknowledgments: We thank Yang Liu and Jinhui Meng of Ningbo University for their support in characterization analysis.

Conflicts of Interest: The authors declare no conflict of interest.

References

1. Yap, C.; Chua, C.; Dong, Z.; Liu, Z.; Zhang, D.; Loh, L.; Sing, S. Review of selective laser melting: Materials and applications. *Appl. Phys. Rev.* **2015**, *2*, 041101. [CrossRef]
2. Aboutaleb, A.M.; Bian, L.; Shamsaei, N.; Thompson, S.M. Systematic optimization of Laser-based Additive Manufacturing for multiple mechanical properties. In Proceedings of the 2016 IEEE International Conference on Automation Science and Engineering (CASE), Fort Worth, TX, USA, 21 August 2016; pp. 780–785.
3. Antonysamy, A.A. Microstructure, Texture and Mechanical Property Evolution during Additive Manufacturing of Ti6Al4V Alloy for Aerospace Applications. Master's Thesis, University of Manchester, Manchester, UK, 2012.
4. Yakout, M.; Cadamuro, A.; Elbestawi, M.A.; Veldhuis, S.C. The selection of process parameters in additive manufacturing for aerospace alloys. *Int. J. Adv. Manuf. Technol.* **2017**, *92*, 2081–2098. [CrossRef]
5. Yadroitsev, I.; Krakhmalev, P.; Yadroitsava, I. Selective laser melting of Ti6Al4V alloy for biomedical applications: Temperature monitoring and microstructural evolution. *J. Alloy Compd.* **2014**, *583*, 404–409. [CrossRef]
6. Maamoun, A.; Xue, Y.; Elbestawi, M.; Veldhuis, S. Effect of selective laser melting process parameters on the quality of al alloy parts: Powder characterization, density, surface roughness, and dimensional accuracy. *Materials* **2018**, *11*, 2343. [CrossRef] [PubMed]
7. Maamoun, A.; Xue, Y.; Elbestawi, M.; Veldhuis, S. The effect of selective laser melting process parameters on the microstructure and mechanical properties of Al6061 and AlSi10Mg alloys. *Materials* **2018**, *12*, 12. [CrossRef] [PubMed]
8. Yang, K.V.; Rometsch, P.; Jarvis, T.; Rao, J.; Cao, S.; Davies, C.; Wu, X. Porosity formation mechanisms and fatigue response in Al-Si-Mg alloys made by selective laser melting. *Mater. Sci. Eng. A* **2018**, *712*, 166–174. [CrossRef]
9. Baufeld, B. Wire based additive layer manufacturing: Comparison of microstructure and mechanical properties of Ti-6Al-4V components fabricated by laser-beam deposition and shaped metal deposition. *J. Mater. Process. Technol.* **2011**, *211*, 1146–1158. [CrossRef]
10. Gardan, J. Additive manufacturing technologies: State of the art and trends. *Int. J. Prod. Res.* **2016**, *54*, 3118–3132. [CrossRef]
11. Frazier, W.E. Metal additive manufacturing: A review. *J. Mater. Eng. Perform.* **2014**, *23*, 1917–1928. [CrossRef]

12. Vaezi, M.; Seitz, H.; Yang, S. A review on 3D MIcro-additive manufacturing technologies. *Int. J. Adv. Manuf. Technol.* **2013**, *67*, 1721–1754. [CrossRef]
13. Seifi, M.; Salem, A.; Beuth, J.; Harrysson, O.; Lewandowski, J.J. Overview of materials qualification needs of metal additive manufacturing. *JOM* **2016**, *68*, 747–764. [CrossRef]
14. Sames, W.J.; List, F.A.; Pannala, S.; DeHoff, R.R.; Babu, S.S. The metallurgy and processing science of metal additive manufacturing. *Int. Mater. Rev.* **2016**, *61*, 315–360. [CrossRef]
15. Hudson, J.A.; Liu, E.; Crampin, S. The mechanical properties of materials with interconnected cracks and pores. *Geophys. J. Int.* **1996**, *124*, 105–112. [CrossRef]
16. Lu, G.; Xiao, G. Mechanical Properties of Porous Materials. *J. Porous Mater.* **1999**, *6*, 359–368. [CrossRef]
17. Zhao, B.; Gain, A.K.; Ding, W. A review on metallic porous materials: Pore formation, mechanical properties, and their applications. *Int. J. Adv. Manuf. Technol.* **2018**, *95*, 2641–2659. [CrossRef]
18. Valdez, M. Induced porosity in super alloy 718 through the laser additive manufacturing process: Microstructure and mechanical properties. *J. Alloy Compd.* **2017**, *725*, 757–764. [CrossRef]
19. Martin, A.A.; Calta, N.P.; Khairallah, S.A. Dynamics of pore formation during laser powder bed fusion additive manufacturing. *Nat. Commun.* **2019**, *10*, 1987. [CrossRef]
20. Branch, B.; Lonita, A.; Clements, B.E. Controlling Shockwave dynamics using architecture in periodic porous materials. *J. Appl. Phys.* **2017**, *121*, 135102. [CrossRef]
21. Gangireddy, S.; Faierson, E.J.; Mishra, R.S. Influences of Post-processing, Location, Orientation, and Induced Porosity on the Dynamic Compression Behavior of Ti–6Al–4V Alloy Built Through Additive Manufacturing. *J. Dynam. Behav. Mater.* **2018**, *4*, 441–451. [CrossRef]
22. Jones, D.R.; Fensin, S.J.; Dippo, O.; Beal, R.A.; Iii, G. Spall fracture in additive manufactured Ti-6Al-4V. *J. Appl. Phys.* **2016**, *120*, 1–8. [CrossRef]
23. Thijs, L.; Verhaeghe, F.; Craeghs, T.; Humbeeck, J.V.; Kruth, J.P. A study of the microstructural evolution during selective laser melting of Ti–6Al–4V. *Acta Mater.* **2018**, *58*, 3303–3312. [CrossRef]
24. Simonelli, M.; Tse, Y.Y.; Tuck, C. On the texture formation of selective laser melted Ti-6Al-4V. *Metall. Mater. Trans. A* **2014**, *45*, 2863–2872. [CrossRef]
25. Kelly, S.M.; Kampe, S.L. Microstructural evolution in laser-deposited multilayer Ti-6Al-4V builds: Part i. microstructural characterization. *Metall. Mater. Trans. A* **2004**, *35*, 1861–1867. [CrossRef]
26. Liu, Y.; Xu, H.; Zhu, L.; Wang, X.; Wang, D. Investigation into the microstructure and dynamic compressive properties of selective laser melted Ti–6Al–4V alloy with different heating treatments. *Mater. Sci. Eng. A* **2020**, *805*, 140561. [CrossRef]
27. Kobryn, P.A.; Semiatin, S.L. The laser additive manufacture of Ti-6Al-4V. *JOM* **2001**, *53*, 40–42. [CrossRef]
28. Keist, J.S.; Palmer, T.A. Role of geometry on properties of additively manufactured Ti-6Al-4V structures fabricated using laser based directed energy deposition. *Mater. Des.* **2016**, *106*, 482–494. [CrossRef]
29. Kanel, G.I. Distortion of the wave profiles in an elastoplastic body upon spalling. *J. Appl. Mech. Tech. Phys.* **2001**, *42*, 358–362. [CrossRef]
30. P'erez-Ruiz, J.D.; Lacalle, L.N.L.; Urbikain, G.; Pereira, O.; Martínez, S.; Bris, J. On the relationship between cutting forces and anisotropy features in the milling of LPBF Inconel 718 for near net shape parts. *Int. J. Mach. Tool Manuf.* **2021**, *170*, 103801. [CrossRef]
31. Weng, J.; Wang, X.; Ma, Y.; Tan, H.; Cai, L.; Li, J.; Liu, C. A compact all-fiber displacement interferometer for measuring the foil velocity driven by laser. *Rev. Sci. Instrum.* **2008**, *79*, 113101. [CrossRef]
32. Weng, J.; Tan, H.; Wang, X.; Ma, Y. Optical-fiber interferometer for velocity measurements with picosecond resolution. *Appl. Phys. Lett.* **2006**, *89*, 4669. [CrossRef]
33. Xiao, D.; Fan, Q.; Xu, C.; Zhang, X. Measurement methods of ultrasonic transducer sensitivity. *Ultrasonics* **2016**, *68*, 150–154. [CrossRef] [PubMed]
34. Ren, Y.; Wang, F.; Tan, C.; Wang, S.; Yu, X.; Jiang, J. Shock-induced mechanical response and spall fracture behavior of an extra-low interstitial grade Ti–6Al–4V alloy. *Mater. Sci. Eng. A* **2013**, *578*, 247–255. [CrossRef]
35. Gray, G.T., III; Livescu, V.; Rigg, P.A.; Trujillo, C.P.; Fensin, S.J. Structure/property (constitutive and spallation response) of additively manufactured 316l stainless steel. *Acta Mater.* **2017**, *138*, 140–149. [CrossRef]

Article

Microstructure and Wear of W-Particle-Reinforced Al Alloys Prepared by Laser Melt Injection

Zhidong Xu, Dengzhi Wang *, Wenji Song, Congwen Tang, Pengfei Sun, Jiaxing Yang, Qianwu Hu and Xiaoyan Zeng

Wuhan National Laboratory for Optoelectronics, Huazhong University of Science and Technology, Wuhan 430074, China; m202073063@hust.edu.cn (Z.X.); wjsong@hust.edu.cn (W.S.); m201972902@hust.edu.cn (C.T.); d202181055@hust.edu.cn (P.S.); m202073085@hust.edu.cn (J.Y.); huqw@hust.edu.cn (Q.H.); xyzeng@hust.edu.cn (X.Z.)
* Correspondence: dzwang@hust.edu.cn; Tel.: +86-189-711-93217

Abstract: W-particle-reinforced Al alloys were prepared on a 7075 aluminum alloy surface via laser melt injection to improve their wear resistance, and the microstructure, microhardness, and wear resistance of the W/Al layers were studied. Scanning electron microscopy (SEM) results confirmed that a W/Al laser melting layer of about 1.5 mm thickness contained W particles, and Al_4W was formed on the surface of the Al alloys. Due to the reinforcement of the W particles and good bonding of the W and Al matrix, the melting layer showed excellent wear resistance compared to that of Al alloys.

Keywords: laser melting injection; W particles; Al alloys; wear

1. Introduction

Aluminum alloys are widely used in automobiles, ships, aerospace, and other fields because of their low density, high specific strength, and good corrosion resistance. However, low hardness and poor wear resistance restrict their application in various fields [1,2]. Preparing particle-reinforced metal matrix composite coatings on aluminum alloys via surface engineering techniques is an effective way to improve their surface properties [3]. Laser technology has been a research hotspot in recent years [4,5]. Ayers et al., first proposed laser melt injection (LMI), which injects the additive particles into the laser melt pool directly, and then particle-reinforced metal matrix composite coatings can be formed on various metal substrate surfaces [6]. Compared with laser cladding, LMI has the advantages of low particle solubility, high surface performance, and low cracking tendency [7]. Ayers et al., prepared TiC- and WC-reinforced metal matrix composite layers on aluminum alloy substrates, and the wear resistance of the aluminum alloys was improved [8–12]. Vreeling et al., prepared SiC/Al composite layers on Al substrates via LMI, and found that preheating the Al substrate is an effective means of injection of SiC into the Al melt [13]. Wang et al., modified Al substrate surfaces via LMI using CeO_2 particles, and the microstructure of the surface was suitably modified in terms of corrosion resistance [14].

In the existing literature, the most commonly used injection particles are ceramics, such as WC, SiC, TiC, etc. Ceramics are well known for their high hardness and good wear resistance, but low room-temperature toughness. Compared with ceramics such as WC, SiC, TiC, etc., W has better room-temperature toughness, and W/Al have better interface compatibility and smaller thermal and physical differences, making W an ideal reinforcing particle for aluminum alloys. Over the past few years, high-performance W/Al composite layers have been prepared via stirring friction, laser metal deposition, and laser alloying [15–17]. However, no studies on the laser melt injection of W-particle-reinforced metal matrix composite layers have been reported to date.

In this study, a W-particle-reinforced aluminum matrix composite layer was prepared via LMI, and the microstructure and wear behavior of the composite layer were studied.

2. Materials and Methods

Tungsten (W) particles with diameter of 5–25 μm were selected as the injection particles (Figure 1), and a 7075 aluminum alloy block with dimensions of 200 mm × 150 mm × 50 mm was used as the substrate, the chemical composition of which is shown in Table 1.

Figure 1. SEM of the W particles.

Table 1. Chemical composition of the 7075 Al alloy (wt.%).

Elements	Si	Cu	Mg	Zn	Mn	Ti	Cr	Fe	Al
Wt.%	0.40	1.2–2.0	2.1–2.9	5.1–6.1	0.30	0.20	0.18–0.28	0.50	Bal.

As shown in Figure 2, the LMI apparatus included a 6 kW continuous-wave fiber laser (IPG, YLR-6000, IPG Photonics, Oxford, MA, USA) with a laser wavelength of 1.06 μm, a homemade laser head, a 6-axis robot, and a powder feeder (HUST-III, Huazhong University of Science and Technology, Wuhan, China).

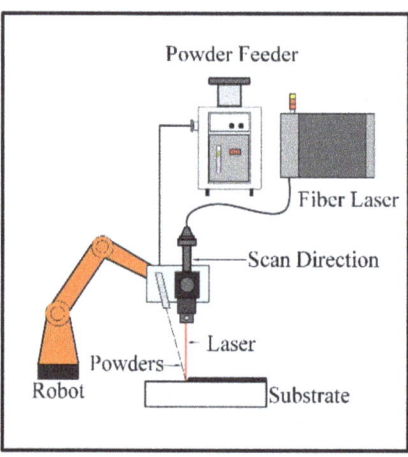

Figure 2. Equipment for the laser melt injection.

www.ingramcontent.com/pod-product-compliance
Lightning Source LLC
LaVergne TN
LVHW072339090526
838202LV00019B/2442